ASTRONAUTICAL RESEARCH 1972

ASTRONAUTICAL RESEARCH
1972

PROCEEDINGS OF THE 23RD CONGRESS OF THE
INTERNATIONAL ASTRONAUTICAL FEDERATION
VIENNA, 8–15 OCTOBER 1972

Editor-in-chief

L. G. NAPOLITANO

Editors

P. CONTENSOU and W. F. HILTON

SPRINGER-SCIENCE+BUSINESS MEDIA, B.V.

First printing: December 1973

Library of Congress Catalog Card Number 72–97959

ISBN 978-94-010-2578-2   ISBN 978-94-010-2576-8 (eBook)
DOI 10.1007/978-94-010-2576-8

TABLE OF CONTENTS

PART III / UTILIZATION AND APPLICATIONS
OF SPACE TECHNOLOGY

PART IV / SECOND I.A.F. STUDENT CONFERENCE

PREFACE

Space scientists and engineers belonging to the professional societies associated with the International Astronautical Federation gathered together in Vienna to hold the Federation's 23rd Congress.

A selected number of papers and critical surveys that were presented and debated at this Congress and which span the widely diversified field of astonautics are collected in the present Proceedings, together with a number of summaries of Round Table Discussions and/or Forum Sessions.

As its predecessors in the series, *Astronautical Research 1972* constitutes an indispensable reference for several groups of people: those who are actively engaged in astronautics; those who are interested in following and assessing, year by year, the developments in astronautics, its progress, its new directions in research; and those who are concerned with its many applications.

Space science and technology are bound to play an increasing role in the immediate future, now that greater effort is being devoted to the exploitation of their relevance to other fields of human activity. Problems posed by the scarcity of earth resources and by their inadequate management, pollution problems, problems created by man's indiscriminate and often irresponsible action in vital sectors of the biosphere can be tackled and successfully alleviated, if not solved, by means of the soft and hard advanced technology developed for space systems. Space research is witnessing a widening of its scope and a diversified proliferation of its objectives as more emphasis is being placed on its application to natural and human sciences and to the optimization and control of ecological systems.

Astronautical Research 1972 reflects and emphasizes these new trends by presenting, as a prelude to the Proceedings, the fourth International Astronautical Federation Invited Lecture, 'Impact of Space on World Development' delivered by Dr Guyford Stever, Director of the National Science Foundation of the United States. Dr Stever's concluding remarks set the stage: "Space applications can be used to advance knowledge generally, to support local and regional applications, to stimulate the economy, to improve the environment and to provide further means of international cooperation". These Proceedings provide ample substantiation for such statements and should be perused by those who are either curious or sceptical – or both – about them.

After the summary of the Forum Session, devoted to the theme of the Congress: 'Space for World Development', there follows the main body of the book which consists of four parts.

The first part deals with basic problems and contains papers in the field of astrodynamics and bioastronautics.

The second part deals with the engineering and management aspects of space technology. It contains papers on such classical subjects as structure, materials, and propulsion, as well as papers discussing the technological problems of and the mission objectives for scientifiec spacecraft.

The third part is concerned with the utilization and application of space science and technology. It contains papers dealing with remote sensing and its application to hydrology, with Earth and ocean physics, with navigation and weather satellites, and with materials processing in zero gravity.

The novelty and timeliness of the topics discussed during the I.A.F. Congresses are particularly evident in this section of the Proceedings. Two instances, out of the many available, seem worth mentioning explicitly. The first one is the Round Table Discussion on the results from the mission of the Earth Resource Technology Satellite (ERTS-1) which constituted a 'world première' on an international stage. The second is the paper on the 'Skylab Student Project' describing and commenting the experiments which were devised by students and will be performed during the Skylab mission.

The fourth and last part contains papers from the Third I.A.F. Student Conference. They present, in a significant and interesting perspective, a cross section of the scientific activities of the younger generation and of its assessment of the problems of world development.

In closing, I wish to thank all those who have helped me in editing these Proceedings: the session chairmen for the timely and valuable help in the selection of papers, the language editors Dr P. Contensou and Dr W. F. Hilton for their valid work; the Secretariat in Paris for untiring cooperation and much needed coordinating action; the managing editor of D. Reidel Publishing Company for coping efficiently with the many problems arising from the publication of Proceedings of International Congresses and, last but not least, the authors who have graciously complied with the stringent requirements I have been compelled to impose on the length of their manuscripts.

LUIGI G. NAPOLITANO
Editor-in-chief

IMPACT OF SPACE ON WORLD DEVELOPMENT

H. GUYFORD STEVER

Director, National Science Foundation

Today, we meet to discuss the impact of space on world development. As a scientist; as the director of an agency of the United States Government, the National Science Foundation; and as a participant in this congress who is concerned about peace on Earth and, therefore, world development, I endorse the theme which brings us together and extend my compliments of those of you who have made this assembly possible. I am grateful that you invited me to join in this worthy task.

We are meeting at a time which is quite significant in man's exploration of outer space. This is the year of ever-increasing cooperative interests in space ventures, including the joint, manned space effort by the Soviet Union and the United States. Thanks to the genius of our early pioneers, many of whom are in this great hall, the results of an era of experimentation now show us clearly the promise which the orderly and determined use of space and space vehicles has to offer. Space science and space technology are already going to work.

The perspective of space gives us new insights and fresh outlooks. In these remarks, I will speak about some of these as a prelude to a discussion of what I refer to as a space ethic. Since ethics have meaning only in the experience of life, I will then examine some past space experiences and prospective space uses in the light of that ethic. World development includes all countries, but our regard for developing countries prompts me to emphasize their special needs.

1. Perspectives of Space

To a scientist, man's leap into space means new data and a way to test theory against reality. To the world-minded, it is access to events everywhere and instantly. But, does space have a meaning which is common to all? I believe that it does. Oddly enough, it is not space itself but the view of earth from space that shows us that Earth is a spaceship. Seeing the earth as a whole for the first time, with its unique colors and complement of resources, and comparing its protective environment with the stark landscape of the Moon was an awesome experience from which man emerged never to be the same again.

Looking at Earth from space, there are no developed or developing countries – just land, sea, and atmosphere. But, looking at space from Earth, it would be unrealistic to overlook or deny the fact that there is a spectrum of space capability and a diversity of national resources for space operations which correlates quite closely with the spectrum of national development.

L. G. Napolitano et al. (eds.), Astronautical Research 1972, 1–11. *All Rights Reserved*

As long as the primary emphasis in space was on manned or unmanned exploration or scientific experiment, only those few technologically advanced countries capable of supporting space flight were involved. Now, as the emphasis shifts from exploration to application, and as man is beginning to use space platforms for Earth-oriented applications, the focus of space is broadening. All nations can be involved and can profit – developed as well as developing, those with strong space capabilities as well as countries with none at all. As a result, there is emerging a tension between the concept of earth as a unitary system as seen from space and the earthly realities of men and countries with individual and sometimes conflicting interests. In these circumstances, space becomes an exploitable resource, not just a medium, and the old problems of allocation of a new but scarce commodity face the involved parties. Is space to become tomorrow's tragic commons?

Truly, it may, While there are exceptions, in space and on Earth, man has not usually had a good record when it comes to the allocation of resources. In part, this poor record stems from the concept of allocation – a word I use deliberately because it is so familiar and yet so wrong. It is wrong because allocation divides resources and is contrary to the unitary concept of a a spaceship earth in which the emphasis is on global resources and system management.

2. A Space Ethic

When we shift concepts in this way, the need for an entirely new basis for action arises. To divide a pie, simple rules are enough. Priorities of time, possession, or strength usually govern – whether they are right or wrong. To manage a system, simple rules are not enough. We must know the purposes of the system and, therefore, what we want. If we are to do this, and do it right, we must have an ethic which is grounded in the new perspectives of space – in other words, a space ethic to guide our actions and decisions.

From his earliest beginnings, man has moved from simple to complex modes of cooperation. Such cooperative modes, or symbioses, first involved individuals only, then individuals and society. The next great ethical step forward is the symbiosis of space, Earth, and man. Before we can take such a step, we need to know more about the nature of a space ethic. and what are its implications to man.

It appears to me that a space ethic has some of the following features:

(1) It is holistic in that it emphasizes the organic relation and continuity between the Earth's parts and wholes. Its concern is with the total planetary system, not piecemeal aggregates of unrelated fragments and processes.

(2) It expands ethical concerns which typically have only involved man to man, or man to society relationships. Copernicus taught us that the Earth was not the center or our solar system. Our voyage in space tells us that man is not the center of our ethical system. Indeed, in a holistic system there is no center, and man shares the system with other environmental creatures and objects which warrant and have become the subject of serious ethical concern. To illustrate the change in our thinking,

in a recent case before the United States Supreme Court, Justice Douglas advocated that "valleys, estuaries, beaches, ridges, groves of trees, swampland, or even air" be permitted to be parties in litigation. He justified this on the ground that, and I quote:

The river, for example, is the living symbol of all the life it sustains or nourishes – fish, aquatic insects, water ouzels, otter, deer, elk, bear, and all other animals, including man, who are dependent on it or enjoy it for its sight, its sound, or its life. The river, as plaintiff, speaks for the ecological unit of life that is part of it.

(3) A space ethic deals with the processes of natural systems. The space ethic stresses action which preserves natural regulatory systems. It avoids action which taxes the capacity of any kind of natural system, especially when there is a danger of irreversible damage or waste of nonrenewable resources. Fortunately, from our new positions in space, we are able to engage in environmental monitoring on an adequate scale for the first time. The Earthwatch Program, approved in June 1972 at the U. N. Conference on the Human Environment in Stockholm, is a welcome step in the direction of monitoring hazardous conditions in the environment. I am sure that satellites will play an ever-increasing role in the effort to maintain the natural systems on which we all depend.

It is not enough to describe an ethic, we must also think about its viability in the emerging space world. For those of us who have 'lived' space for so many years, its implications for world development seem as endless as space itself. Let us, however, bring space down to Earth by noting that satellites are likely to change our values and social priorities in the future as much as the automobile did in the past. When autos were first put into use, we had no idea that they would affect where we live; that they would change the air we breathe, the work we do, and, in many cases, what we want. Space may not change all that – in fact, I hope it doesn't – but man's values and desires will surely be transformed.

It may be too early to tell how far and fast our values will evolve, but we ought to think of space applications more as social evolution than as technical innovation. Not only can we expect changes in social institutions and the kinds of decisions we will make, but perhaps our personality may change as it did when we got behind the wheel of a fast, powerful auto.

Space applications also enlarge and, at the same time, contract the perceived boundaries of the human community. When we see ourselves as part of a single, planetary system, the notions of city, state or national boundaries seem at best only conveniences and at worst anachronistic. Regardless of man made boundaries, exposure to other cultures and personalities – mainly through the medium of electronic satellite communication – is tending to move us toward becoming a 'global village'.

People now take rapid communication for granted, forgetting that television is only about a third of a century old, radio use on a large scale is barely a half century old, and telephones not yet a full century. In these comparatively few years, we have witnessed a homogenization of world culture as well as the weakening of local institutions. Even the way we use our senses and react to things begins to be more and more of a kind. We seem to be in a process of 'implosion' as we recreate on a global scale, through space technology, the ancient and feudal way of communicating by one person telling another. At the same time, the explosion of social forces is no longer dissipated and absorbed into distant territories, but reverberates from all part of the globe.

Obviously, with major changes in values and social institutions, together with the way people decide matters and relate to one another, the future will be much different than the past. The need for a space ethic that can shape and remold the future is heightened. As different decisions are made, new uses for old data will be conceived and new data developed. New options will open up, and old ones will no longer be feasible. To structure such a dynamic world is the crucial problem of our era, and its timely solution is by no means assured.

I have been speaking of extensive changes that might take place as a result of our space activities. But the need for and the needs of international cooperation tend to continue in the same direction and can be increased by space activities. Consider, for example, the International Telecommunication Union. The ITU, formed in 1865, has often been criticized for limiting its activities to regulating policy among national telecommunications administrations. And yet, it has lasted longer than most other international organizations. The reasons which explain this past durability for ITU, and which will be operative in space activities of the future, include:

– Acceptance of the need to insulate some scientific and technological activities from the unpredictable turns of international politics;

– The desire to maintain benefits derived from cooperation; and

– The expansion of the range of available national opportunities.

These continuing imperatives of protecting the search for and use of knowledge, maximizing their benefits and the number of participants, will, however, take place in a new context of world development. At first, we viewed development of emerging countries as a problem of capital investment. When it became clear that effective capital investment depended on managerial skill, we turned to technicians and experts of various kinds to administer capital productively. Then, it seemed that they could not function unless they were part of an institutional framework manned by a well-trained local citizenry. This too had its problems, and for the past decade there has been gradual but steady decline in the percent of gross national product committed by donor countries to less developed countries. In part, this decline may also be due to the increasing awareness on the part of so called developed countries that they too are underdeveloped in many respects. This sensitivity is especially true for those aspects of society included in the phrase 'quality of life', such as the physical envorinment, culture, art, and the like.

Whatever the reason may be for current developmental problems, it seems clear that we are standing at a new threshold, in which the *aims* of world development are more important than how we do it. Later in these remarks, I will try to show how the space ethic can help us identify the proper aims of development and cope with change caused by space activities. In so doing, it can provide a basis for assessing the negative and positive aspects of production and access to goods. It can provide a rationale for solidarity among men within regions and throughout the world. Finally, it bounds the proper use of developmental technology by protecting natural regulatory systems.

3. Space Experience and Its Lessons

All of this sounds rather abstract and difficult to apply. What are the pertinent lessons of our experience in space? What are the policy guidelines which will help us in our increasing effort to use space for world development? In this context, the significant space activities relate to scientific research, meteorological services, communication, navigation, and Earth resources.

I admit to a certain personal bias in placing scientific research at the head of my list of space activities which are significant for world development. But this attitude is based on positive evidence that scientific research is indispensable for long-term progress in world development and other activities as well. Research provides a liveliness and a flow of ideas far out of proportion to its cost. It is a highly reactive ingredient, without which developmental progress would begin to lose its momentum. While our research potential usually outruns our pocketbook, I think it fair to say that the funding levels and the criteria of choice in selecting experiments in space do show a continuing and encouraging support for scientific research. What is equally important is the maintenance of the ambiance of space research. By this I mean open programs, sharing of information, and access to experiments.

Substantial progress has been and is being made in this direction. The broader question is: can the openness, sharing, and broad participation so characteristic of the scientific enterprise be extended into operational areas, as a model for the space ethic? To a considerable extent, we have done so in regard to meteorological services. When experimental evidence made it clear that meteorology was at a turning point because satellites could measure atmospheric conditions over the entire surface of the Earth at frequent intervals, a single coherent world service was established – the World Weather Watch. Members of the World Meteorological Organization receive or can expect to receive, through the World Weather Watch, such benefits as extended weather forecasts, storm warnings, and environmental information related to pollution. We now enjoy worldwide availability of automatic picture transmissions which may be received by comparatively simple equipment in any country over which the satellite is passing. With such equipment, it is possible to secure photocopies of the surrounding areas within a distance of about 2000 kilometers, and I understand that about 100 countries are now using this transmission. One can hardly find a better illustration of open sharing with wide operational participation.

It is important to note that a widespread network to collect meteorological data existed before the advent of the satellite. More than 8000 land stations were in operation, and more than 5000 merchant ships reported weather observations when at sea. Using a global telecommunications system, these data were exchanged rapidly and continuously.

I cite the preexistence of an operational meteorological network because of its effect on how the new space satellite capability was put into use. Initially, meteorological satellites were an experimental problem in space science and technology. As the satellite systems became operational, the problem became one of data collection and handling. Fortunately, a framework for coping with the problems already existed and could be extended readily through customary arrangements.

Even with this advantage, it is clear that data management is becoming a critical issue. In the words of D. A. Davies, secretary-general of the World Meteorological Organization:

The price for using (satellite meteorological) data to maximum advantage will inevitably increase; even the most highly developed countries will be forced (as they are at present) to establish priorities for data acquisition and use. For countries and organizations which remain completely in the data-user class the cost of receiving and processing the data to meet their particular needs will rise sharply. A possible result is that most users will be forced by economic necessity to accept the fruits of increased space activity in the form that the provider is able to supply them with little or no modifications or adaptation.

Two lessons emerge from this experience. First, we must distinguish between the problems which are typical of the scientific as compared to the operational phases of space activities. Problems in the scientific phase may tax our intellectual capacities but they do not normally involve the scale or level of activity dealt with operationally. The sharing of scientific knowledge is not usually severely limited by funding, but the cost of data sharing may limit what is shared, how often, and by whom. Even with the best of intentions, we have to accommodate to the real world where sharing and participation take on different meanings as circumstances change.

The second lesson relates to the virtues of precedent. Existing institutions and customary services which preceded satellites have generally offered the best means to take advantage of the new capabilities they provide. Expertise, contacts, treaty and other legal agreements are powerful incentives to work within traditional arrangements. This is particularly true for meteorological observations, where the satellites improved the quantity and quality of data available rather than creating entirely new possibilities, as in the case of remote sensing of Earth resources.

It is clear that putting space to work in world development is far more complicated than only using it for scientific study. That is not to say that openness, sharing, and wide participation are not desirable. As scientists, we believe they are essential, but there is a lot more to the problem.

The development of the INTELSAT global telecommunications system provides important lessons. Under INTELSAT, international satellite communications, after passing from the experimental into the commercial phases, followed commercial arrangements that are very little different from those used for international radio and cable communications. What made INTELSAT different was the possibility of any country having direct communications access to any other country through a satellite. This was in sharp contrast to presatellite international commercial communications, in which the bulk of the countries in the world could only communicate with each other by going through carriers controlled by just a few countries.

INTELSAT is also different because international communications are an entirely quantitative thing. Rates can be estimated, charged, and adjusted; capital investment can be computed, and return on profit estimated. Consequently, it becomes possible to have a commercial enterprise devoted to optimizing communications services at the lowest cost. Of course, there are a lot of national and political interests involved, but month by month, year by year, the operation of INTELSAT is a business enterprise.

Given the important new precept of direct, country to country communication on a global basis, the impact on existing methods of communications, and the commercial 'utility' character of the enterprise, it was clear that a new type of organization or consortium was needed to capitalize and regulate this new international capability. This was no easy task. Even negotiation methods needed to be changed. In the past, telecommunications entities had been able to negotiate directly with their counterparts in international communications. Now in the space age, governments have national interests in outer space, and political factors have to be taken into account. As a result, negotiations have to be conducted on a multi-national basis, and at a governmental level. Despite these difficulties, interim arrangements were established in 1964 and new, definitive arrangements agreed to in May 1971. It is not my purpose to analyze this agreement except to note that through a process of compromise the essential business character of INTELSAT has been maintained while creating a broader international quality to the management of the system.

Global communications satellites have had more impact on world development than any other space activity. Are there any lessons for us out of this experience to date that will help us in the broader application of the space ethic? As in the case of meteorological satellites, a significant distinction exists between scientific and technological experimentation and the administration and management of satellite operations. Operationally, we see repeated the general practice of retaining customary arrangements.

On the other hand, important new ideas were introduced regarding organization and economics. Given the capabilities existing in the world in the early 1960's, the United States could have created its own global communications system. The United States could have just built it, flown it, and sold the services at rates it determined. It would have been a U.S. system, and such profits as it made would have all come to the United States. As you know, it did not do so. Nor did it seek to establish some international agency to control operations. Having voted for the United Nations resolution of December 1961 that communications should be available to the nations of the world as soon as practicable on a global and nondiscriminatory basis, the United States felt that the most effective mechanism would be some form of commercial, international, public utility. Notwithstanding the many difficulties of reconciling national interests and adjusting to different viewpoints, the wisdom of choosing a multi-national, quasi-governmental organizational alternative is demonstrated by the later course of events which are familiar to this audience.

Experience in testing or operating new satellite systems is not the only way we learn how to apply a space ethic to world development. Policy commitments hammered out through intensive negotiation at the United Nations, and formally accepted by the governments of the world, also represent important guidelines. Let me cite a few:

- space should be used for the benefit of all peoples, irrespective of their degree of economic or scientific development
- space is not subject to national appropriation
- space shall be free for exploration, use, and scientific investigation
- space activities shall be conducted in accordance with international law

4. Space Ethics and Remote Sensing of Earth Resources

I have, to this point in my remarks, described a space ethic, factors affecting its viability, and some experiential and policy guidelines which might be useful in its application. Now we must see how all this might work in practice. As a test case, I have selected remote sensing of the Earth's resources by satellite because it is most relevant to the theme of this Congress.

Earth-orbiting satellites give us the potential ability to secure repetitive, synoptic coverage of energy emanating, at various wavelengths, from the Earth's surface. Collection and study of these energy emanations will allow us to identify objects and secure a variety of information which has economic, social, cultural, and ecological value.

Obviously, if successful, such a satellite system would be able to perform a wide variety of tasks. Not all of these tasks, however, are equally well understood. For example, the problems of environmental monitoring are not as familiar to us as those of mapping or cultural and resource data collection. Also, we don't know the optimum configuration for an Earth resources survey, the best sensors to use, the most efficient way to process data, and the economics of such surveys.

Clearly, we are in an experimental phase and, as you know, the Earth Resources Technology Satellite called ERTS-A has recently been launched. As was the case in other satellite programs, a substantial fundamental research effort is part of the total package, and many countries in different stages of development are participating. Representatives from 40 countries and 16 international organizations attended a U.S.-sponsored International Workshop in the Earth Resources Survey System. Canada is active with ERTS data acquisition and data processing facilities, and in preparation for a remote sensing system, Brazil and Mexico have entered into cooperative agreements with the United States. Under these agreements, remote sensing aircraft programs are helping to pave the way for use of spacecraft data by development of necessary skills and ground truth data. India, too, has undertaken a similar domestic aircraft program.

I am mentioning these details in order to highlight the fact that the program has proceeded on an open, unrestricted basis with full sharing of available information and wide participation. At the same time, it is proceeding deliberately, and we hope prudently, so that developing countries can make their own decision, in the light of their special circumstances, as to the proper degree and timing of any commitments of money and manpower.

Anticipating that earth resource satellites should be capable of yielding valuable data, President Nixon has promised to make the data of its future programs available to all "as this program proceeds and fulfills its promise." Assuming a validation of the experimental phase, what are some of the attributes of an earth resources satellite system that are relevant to an ultimate operational system?

Of first importance, we must recognize that an Earth resources satellite system represents a new capability for providing different and additional kinds of information not hitherto available. But it will not be 'selfsufficient'. For maximum effectiveness, it must be supplemented by aircraft and some ground observation. In such circumstances, a new ingredient is added – national sovereignty. No government has yet contested the proposition that satellites can overfly countries without interference. But nobody denies that the required supplemental aircraft ground observations will be subject to national control.

A desire to maintain national control of earth resource data is not an archaic attitude which has no place in the space age. The provision of meteorological data or communication links are services which can't be prejudicial to a country, because they don't have to accept them. Countries can decide for themselves if the benefits of a service are worth the cost. In the case of remote sensing of Earth resources, countries may see the capability as being prejudicial but don't have the same opportunity to prevent its application in their territory. Accordingly, there will be a strong interest in having quite explicit, binding international agreements which govern the use of this capability.

Another significant consideration that bears on operational characteristics of remote sensing of earth resources is the variety of taks to be performed. Unlike the relatively simple objectives of meteorological or communications satellites, an Earth resource satellite will have many purposes to be served, representing a broad range of interests in both kind and intensity. For example, there will be resource surveying tasks to do over areas of the globe that are not subject to national sovereignty – such as the oceans and polar regions. There will be resource surveying to be done which by its content has an international, scientific, or economic interest over and above the interest of individual countries. Then, of course, there will be tasks which arise directly from the national interests of individual countries. Certainly, individual and industrial purposes are also to be served. Given the range and complexity of tasks, it is evident that no simple operational scheme is going to suffice.

Taking these considerations into account, as well as the fact that there is still much to do in the experimental phase, it takes a brave if not a foolhardy individual to try to be precise about the nature of operational Earth resource satellite systems. Being one but not the other, I emphasize that I speak only as a private individual with a long-term interest in space affairs. With that qualification, I will now offer some thoughts on how the space ethic can influence the application of Earth resources satellite systems to world development.

Let me recall for you the holistic space ethic which emphasizes the organic relation and continuity between the Earth's parts and wholes. The proper policy guidelines for remote earth sensing already exist in the concept that it should be used for the benefit of all peoples, whatever the degree of their economic or scientific development. How shall we do so?

It seems clear that the new capability should and will be applied and organized on a global, not purely national, basis. Many alternative multinational organizational models exist, ranging from governmental preemption, as is the case of the meteorological satellites; to quasigovernmental, multinational entities, such as INTELSAT; to U.N. involvement, possibly in the form of a new agency following the model of the International Atomic Energy Agency. I suspect that the ultimate arrangement, reflecting the absence of direct antecedents and the variety of tasks, will be something entirely new, and yet within the limits of the existing organization spectrum. Whatever structure is worked out must provide safeguards against fragmentation of the capabilities of the Earth resources satellite, and it should foster direct relationships between countries.

While data collection is inherently global in character, data handling is not necessarily so. Data handling is expected to be so much more expensive and complicated than collection that smaller countries or those who wish to participate on a limited basis probably will want to band together in regional groupings. As we expand our ethical and practical concerns to a host of environmental objects, we can anticipate requirements from regional groups of countries for diverse agricultural, cultural, and general mapping purposes. From the satellite's single image, the data required can then be identified, extracted, and made available. If it proves to be true that most countries are going to have to associate themselves with others to overcome data handling problems, it means they will have to change their thinking about exclusive, national proprietary interests in the data. Some have already done so. For others, benefits to be received from the program, and development of some necessary protections, may be persuasive.

This is not to say that national interests have no place in such a plan. On the contrary, if we are to follow the space ethic of preserving our natural regulatory systems, it will be done primarily at the national level. Similarly, if we are to achieve our goals of world development, it will be through the stimulus of nationally based efforts. It would be naive to expect that the same degree of openness and sharing that are characteristic of the experimental and even collection phase would extend to the final steps of data processing and collection. At this payoff point, flexible national arrangements are needed for a multiplicity of tasks, including proprietary industrial applications.

As a somewhat simplistic way of summing up some of the nesting-like attributes of a possible Earth resources satellite operational system, one might say that global concepts are to data collection as regional concepts are to data handling as national concepts are to use or exploitation.

Whatever forms, devices, and mechanisms we may contrive in the future for satellite Earth sensing systems, when such systems join the family of space applications for world development, we can expect the following:

- Developing countries, which may be cautious about investing substantial scarce resources in space activities, can in fact acquire the product or benefits from space activities at a very nominal cost.
- The nature and character of space satellites are such that very few of the data handling services can be efficiently provided on a purely national basis.
- Space capabilities, therefore, whether they relate to weather, communications, or resources will have to be applied and organized on an international cooperative basis. It would be reasonable to expect that similar arrangements will evolve for prospective applications in navigation, traffic control, and other uses of space satellites.

In this connection, I am happy to note that a special Committee of the American Institute of Aeronautics and Astronautics is using the occasion of this congress to meet informally with heads of societies of a number of nations to explore possible avenues for closer cooperation.

About a year ago, I presented a keynote address to a Panel on Remote Sensing of Earth Resources before a Committee of the House of Representatives of the United States Congress. Although I was speaking to a more limited subject at that time, I believe now, as I did then, that space applications can be used to:

- advance knowledge generally,
- support local and regional applications,
- stimulate the economy,
- improve the environment, and
- provide further means of international cooperation.

That, in short, is what space and world development is all about.

We are in continuing dramatic period of space experimentation. Space science, like all the sciences, is being asked to contribute evermore to the urgent needs of mankind. Space can indeed offer much to people everywhere. We now have a new perspective of ourselves and of our planet. If we use our new knowledge wisely, the human family will gain and we in science will see that the best is yet to come. Let us work together in this important task. We can continue our science and exploration and, at the same time, we can contribute applications to society.

SPACE FOR WORLD DEVELOPMENT
(Summary of the Forum Session)

F. DE MENDONCA (CHAIRMAN)

and

RICHARD W. PORTER (CO-CHAIRMAN)

What role will the newly won space capabilities, such as communication, Earth observation and data gathering, play in world development? This was the question around which the Theme Forum Session of the XXIIIrd Astronautical Congress was structured. It was answered in turn by eminent specialists in agriculture and forestry, fish and wildlife, water and mineral resources, industry and commerce, health care and population control, education and culture (including sports), and demography. Then representatives of various national and international space administrations, including those of France, Italy, U.S.A., U.S.S.R., ESRO and the United Nations, were asked to comment on the statements by these experts. It was an interesting plan; unfortunately, however, it was only partially fulfilled because of the unavoidable but unexpected last-minute absence of several key participants.

The first statement, by Dr Roger M. Hoffer, Department of Forestry and Conservation, Purdue University, U.S.A., made the point that effective development, conservation and utilization of agricultural and forest resources of any country requires relatively complete and reliable sources of information about these resources. In reviewing the information needs of various user groups, Dr Hoffer concluded that two primary requirements are cited most frequently: (a) the need for more accurate, timely, and economical information on the extent and location of the resource base, and (b) the need to be able to determine the current condition of the material, as for example, the density, vigor, and health of agricultural and forest resources.

Data to meet these needs can be obtained by the use of earthward looking sensors on aircraft or on Earth satellites, such as ERTS-1, with automatic data processing techniques. For these techniques to be effective, the cover types or situations of interest must be spectrally different from other cover types or situations, and the spectral categories that can be distinguished must have information value to the user. (There is no purpose in defining and mapping twenty spectral classes of rangeland if there is no economic or utilitarian significance to each of the twenty classifications.) Spectral data alone may not be able to achieve some of the information requirements shown in Figure 1, but can be used in conjunction with temporal and spatial data inputs to provide more complete information than any of these inputs separately.

The potential for achieving even the most detailed level of information were demonstrated by a large scale research project conducted in the United States during 1971, which involved the monitoring of Southern Corn Leaf Blight over a

L. G. Napolitano et al. (eds.), Astronautical Research 1972, 13–17. All Rights Reserved
Copyright © 1973 by D. Reidel Publishing Company, Dordrecht-Holland

seven state area every two weeks by means of airborne sensors. Researchers using automatic data processing were able to identify and map corn (maize) and other crop species, delineate acreage infected with moderate or severe levels of blight during July and August, and to track the spread of this disease with relatively high confidence.

Dr Hoffer also described results of the analysis of the first ERTS-1 data taken over the central United States. Forested areas sprayed with herbicide as part of a rangeland improvement project were clearly defined by their low infrared reflectance, even though some of them could not be visually distinguished from surrounding areas. Within twenty-four hours after ERTS data were collected over an area in Southern Indiana, a reasonably accurate mapping of cover types (row crops, forest and woodlot areas, diverted acres, pasture areas, nonproductive grasslands, and water) was produced. He concluded with a very optimistic prognosis for better management of natural and cultural resources by virtue of information obtained using these techniques.

In the regretable absence of Adm. Moreira da Silva, Ministerio da Marinha, Brasil, the Co-Chairman, Dr Porter, made a brief presentation on the role of space technology in development and conservation of fish and wildlife resources. He pointed out that edible fish, which represent a substantial part of the protein diet for many people of the world, are not now cultivated, but rather hunted, and that the process is relatively inefficient and costly. Fish, being highly mobile, go where the conditions are attractive to them. If the attractive conditions are known, regions where they exist can be located by remote sensing from spacecraft, thereby improving the efficiency of fishing operations. However as fishing becomes more efficient, there is growing need to avoid overfishing and to 'manage' this resource, as in forestry, by intelligent limitation of catch, operation of hatcheries, and providing an improved environment for desirable species, including control of their diseases and natural enemies. Space capabilities can be of great assistance in many of these management functions.

An example given was the recent 'red tide' episode off the New England Coast in the United States, a population explosion of *Gymnodinium*, *Gonyaulax*, and other marine dinoflagellate protozoans, which killed vast numbers of edible fish and crustaceans and rendered those which were not killed poisonous to man. Cost of this episode to the fisheries of the area amounted to many millions of dollars. The growth and extent of this 'red tide' can be exactly delineated from ERTS data; hopefully, such information will be useful in minimizing the effects of this menace and eventually eliminating it.

Although wild birds and animals are no longer a major factor in food production or industry, Dr Porter pointed out that there are moral, scientific, and esthetic reasons for maintaining the wild species which now inhabit the Earth. The use of space technology to track the migrations of animals and birds can be helpful in understanding the migratory behavior and thus the whole life cycle of many wild species, thereby making possible more intelligent efforts to preserve them.

The Chairman, Prof. F. de Mendonca, read a brief summary of the remarks which would have been made by Dr O. Dominguez, *Centro de Investigaciones de Recursos Naturales, Secretaria de Agricultura y Ganaderia*, Argentina, who was also, unfortuately prevented at the last minute from coming to Vienna. Although, as is well known, remote sensing data from satellites can be very helpful in understanding the gross geology of inaccessible regions and thereby in locating desirable areas in which to prospect for minerals or to drill for oil, such data may eventually be of even greater value in conserving and utilizing the fresh-water resources of marginally arid regions. In the long run, world development may indeed be limited by availability of this resource. Remote sensing and data collection by satellites can provide information about snow cover, rainfall, soil moisture, content of natural and artificial storage areas, stream flow, evaporation, and contamination.

Mr D. E. Findley of the U.S.A. Dept. of Transporation, substituting for Dr R. H. Cannon, Assistant Secretary for Systems Development and Technology, made a statement on the role of space technology in development of world transportation. This statement dealt primarily with the use of space communication and navigation aids to international air and marine traffic, but also mentioned the use of satellite geodesy and photography for surveying and mapping purposes, a prerequisite for the engineering and construction of highways and railroads in developing areas of the world.

The statement on industry and commerce was presented by Dr Ir Th. P. Tromp, former Chairman of the Philips Company, Netherlands. After a wryly humorous reference to unemployment problems and financial losses to industry in the space field, he went on to mention the so called 'spin-offs' of the space program, and what he chose to call 'spin-ups', or instances where technological developments, such as integrated circuits, had been greatly accelerated by the requirements and resources of the space program and were now of crucial importance for industry.

Industry, he stated, is a major user as well as a maker of communication systems. Philips spends roughly 2% of its sales on travel and communications. The cost of one television channel on the Intelsat IV satellite is about the same as the cost of an executive aircraft. The implication is clearly that the availability of this new inexpensive, reliable form of communication will have a profound effect on the way large international corporations conduct their business. Another point is the need of industry and commerce for skilled and educated people. The use of satellites for large-scale educational and instructional programs may be important to the commerical and industrial development of the presently underdeveloped countries.

Finally the point was made that the future will necessarily involve industrial collaboration on a global scale and that this collaboration requires giant systems for management. Space projects have provided and will continue to provide the experience in setting up and handling giant systems.

Dr Charles Berry, the U.S. astronauts doctor, spoke about the embryonic role of communication satellites in providing extended medical care. Experiments with paramedical personnel using audio and visual communications with hospitals where

competent physicians are continuously on call, are being tried in remote areas of
the Southwestern United States and Alaska, and show great promise. Mass communication by direct television broadcasting to village receivers can also provide
improvements in public health, sanitation, and epidemic control. However, as death
rates are decreased by modern medicine, it becomes even more important to limit
the birth rates to levels that will produce a relatively stable population. Indoctrina-
tion and education by direct broadcasting now appears to be the only socially
acceptable and economically feasible means to achieve this goal in certain areas
of the world.

Dr C. Terzani of *Radio Televisione Italiana*, Rome, Italy, delivered a statement
of general interest, in which he made the point that communication satellites give
news, including sporting events the immediacy that is the greatest factor of their
interest, and have filled the gaps between terrestrial television links (oceans, deserts,
and great uninhabited areas), thus extending the transmission of live picture information all over the world. Live coverage of the exploration of the Moon was cited as a
dramatic example of how modern television coverage surmounts the boundaries
of nations, continents, and even of the Earth itself. Another example was the live
international coverage of recent Olympic Games in Japan and Munich. Although
Mr Terzani did not mention it, one could also note the strong political and social
impact of television broadcasts (by satellite) of the anguish and devastation of the
fighting in Viet Nam. Perhaps this will be a new force for peace in the world.

Also mentioned in this statement were the various efforts now underway in the
Soviet Union, Canada, and elsewhere to provide economic television program
distribution service by means of satellites. It concludes that, at least in Western
Europe, the technical advantages of the satellite distribution systems seem not to
compensate for the additional cost over that of existing terrestrial systems.

In the area of satellite direct broadcasting service, although no systems are now
in operation, a mixed distribution-community reception experiment is envisaged
in India using the ATS-F satellite of the U.S.A., and an extended educational program
is foreseen in Brazil, involving 500 schools initially, and later 12 Earth stations and
152000 community receivers all over the Country. Satellite systems have the special
advantage that they can assure practically total coverage of the population from the
beginning of the service, especially in areas where lack of conventional communications makes installation of terrestrial systems very difficult. It is concluded that a
satellite system is financially advantageous in countries of large extent not having
a widespread terrestrial television system already in existence.

Demography deals primarily with the study of land use by man for cities, towns,
factories, and transportation networks, and of the migrations of people from farm
to city, one city or region to another, and so on. Dr John Fry, O.S.T., Agency for
International Development, U.S.A., described in his statement how remote sensing
data from satellites could be used to measure the changing patterns of land use,
provide a rough measure of energy consumption, detect air and water pollution,
and to estimate gross changes in population. Inasmuch as these are the basic data

needed for rational planning on an urban, regional, or national level, their value to the intelligent shaping of world development is clearly evident.

Discussion of these statements was presented by Dr W. von Braun of the U.S.A., Prof. A. Hocker of ESRO, and Mr Marvin Robinson of the United National Secretariat. There was general agreement on the principle that space technology can and should contribute to world development in many ways, some of which have probably not yet been envisaged. Dr von Braun listed as needs the following items: (1) A complete worldwide inventory of fish. Beginnings are already being made. It is possible to sense a thin natural oil slick on the ocean surface and identify the species of fish which produce it. Ripple patterns also aid the identification and quantification. (2) A study of the life cycle of salmon, to the point where it would be possible to estimate whether 1, 2, or 3% will return to the spawning stream as mature fish. (3) A study of the pollution of continental shelves, as an aid to rational regulation of their use as dumping grounds. (4) Location of fresh water aquifers terminating in coastal sea bottom areas, for use as additional sources of usable water, and for the information they can yield about the general network of subsurface water flow in the adjoining land mass.

The overall message of this Theme Session seemed to be that space activity has finally 'come of age', and that it is now expected to contribute in its own special way to the development of a better world for people to inhabit.

PART I

BASIC PROBLEMS

SOME PROBLEMS OF THE THEORY OF A LATITUDINAL SPACE VEHICLE MOTION IN A NONCENTRAL GRAVITATIONAL FIELD

V. K. KAISIN

Academy of Sciences of the U.S.S.R., Moscow, U.S.S.R.

In [1–4] the problems of a space vehicle with continuous reactive thrust in a central gravitational field are considered. The lifetime of such space vehicles can vary from some days to a year. In this connection the problem arises of the influence on vehicle motion of an attracting body having a non-central gravitational field.

In [5] a vehicle motion is studied in a normal gravitational field of a stationary Earth under the influence of additional forces, a part of which should be considered reactive. Let us consider a more common problem in which we assume that the motion takes place in a gravitational field of the rotating Earth (or of another celestial body). System of differential equations of such vehicle motion take the form:

$$\frac{d^2x}{dt^2} - 2\omega\frac{dy}{dt} - \omega^2 x = \frac{\partial}{\partial x}(U + \bar{U}),$$

$$\frac{d^2y}{dt^2} + 2\omega\frac{dx}{dt} - \omega^2 y = \frac{\partial}{\partial y}(U + \bar{U}), \tag{1}$$

$$\frac{d^2z}{dt^2} = \frac{\partial}{\partial z}(U + \bar{U}),$$

where x, y, z are rectangular Cartesian coordinates of a vehicle in the system $Oxyz$ rigidly connected with the central body rotating uniformly with the angular velocity ω relative to the Oz axis. The equatorial plane coincides with the xOy plane, the origin of coordinates O coincides with the body mass center.

U is the body potential approximated by the potential of a generalized problem of the two fixed centers [6, 7].

$\bar{U}$ is the potential of additional forces [5]:

$$U = f\left(\frac{m_1}{r_1} + \frac{m_2}{r_2}\right), \qquad \bar{U} = \frac{\delta}{\bar{\varrho}^2} + \frac{\lambda}{z^2} - \mathit{b}U + \frac{U(l)}{\bar{\varrho}^2}, \tag{2}$$

where f is the gravitation constant; m_1, m_2 are the masses of the centers placed on the Oz axis at the distances $\bar{c}_1 = -2m_2\bar{c}/(m_1 + m_2)$, $\bar{c}_2 = 2m_1\bar{c}/(m_1 + m_2)$; $r_1 = \sqrt{x^2 + y^2 + (z - \bar{c}_1)^2}$, $r_2 = \sqrt{x^2 + y^2 + (z - \bar{c}_2)^2}$, $\bar{\varrho} = \sqrt{x^2 + y^2}$; δ, λ, b are constants; $U(l)$ is the function approximating some longitudinal terms in the body potential series. Parameters $\bar{c}$, m_1, m_2 for the Earth have the values: $\bar{c} = -ic$, $m_1 = m_2 = M/2$ ($c = 210$ km, M is the Earth mass) [6, 7], l is the vehicle longitude.

Let us consider first two variants of reactive acceleration $\mathbf{g}_r$, assuming

$$\mathbf{g}_r = \mathbf{f}_\lambda + \mathbf{f}_\delta, \tag{1}$$

L. G. Napolitano et al. (eds.), Astronautical Research 1972, 21–29. All Rights Reserved
Copyright © 1973 by D. Reidel Publishing Company, Dordrecht-Holland

$$\mathbf{g}_r = \mathbf{f}_\delta + \mathbf{f}_b; \tag{2}$$

$$\mathbf{f}_\lambda = \{0, 0, -2\bar{\lambda}/z^3\}, \quad \mathbf{f}_\delta = \{-2\delta x/\bar{\varrho}^4, -2\delta y/\bar{\varrho}^4, 0\}, \quad \mathbf{f}_b = \{-\bar{b}U'_x,$$
$$-\bar{b}U'_y, -\bar{b}U'_z\} \tag{3}$$

and study the vehicle motion relative to the Earth. In Equations (1) we shall proceed to the variables u, v, w, θ according to the formulas:

$$x = cu, \qquad y = cv, \qquad z = cw, \qquad t = \frac{c\theta}{V_0},$$

where V_0 is the vehicle initial velocity.

In the equation system obtained we shall turn to the spheroidal flattened coordinates ξ, η, l:

$$u = \sqrt{(\xi^2+1)(1-\eta^2)}\cos l, \qquad v = \sqrt{(\xi^2+1)(1-\eta^2)}\sin l, \qquad w = \xi\eta$$

$$\left(l = \operatorname{arctg}\frac{v}{u}, \qquad 0 \leqslant \xi < \infty, \qquad -1 \leqslant \eta \leqslant 1\right).$$

Using Jacoby's method we find the general integral of the system (1):

$$\int \frac{\xi^3\,d\xi}{\sqrt{R_1(\xi)}} + \int \frac{\eta^3\,d\eta}{\sqrt{R_2(\eta)}} = \theta + \beta,$$

$$l + \alpha_1\left[\int \frac{\xi\,d\xi}{(\xi^2+1)\sqrt{R_1(\xi)}} - \int \frac{\eta\,d\eta}{(1-\eta^2)\sqrt{R_2(\eta)}}\right] +$$

$$+ \mu_1\left[\int \frac{\xi^3\,d\xi}{\sqrt{R_1(\xi)}} + \int \frac{\eta^3\,d\eta}{\sqrt{R_2(\eta)}}\right] = \beta_1,$$

$$\int \frac{\xi\,d\xi}{\sqrt{R_1(\xi)}} - \int \frac{\eta\,d\eta}{\sqrt{R_2(\eta)}} = 2\beta_2; \tag{4}$$

$$\xi^2(\xi^2+\eta^2)^2\left(\frac{d\xi}{d\theta}\right)^2 = R_1(\xi), \qquad \eta^2(\xi^2+\eta^2)^2\left(\frac{d\eta}{d\theta}\right)^2 = R_2(\eta),$$

$$(1+\xi^2)(1-\eta^2)\left(\frac{dl}{d\theta} + \mu_1\right) = \alpha_1, \tag{5}$$

where $R_1(\xi)$ and $R_2(\eta)$ have a form of

$$R_1(\xi) = [2\bar{h}\xi^4 + 4(\mu_2 - \mu_3)\xi^3 + \alpha_2\xi^2 + 2\lambda](\xi^2+1) + (\alpha_1^2 - 2\delta)\xi^2, \tag{6}$$

$$R_2(\eta) = (2\bar{h}\eta^4 - \alpha_2\eta^2 + 2\lambda)(1-\eta^2) - (\alpha_1^2 - 2\delta)\eta^2, \tag{7}$$

and h, β, α_1, α_2, β_1, β_2 ($\bar{h} = h + \mu_1\alpha_1$) are the integration constants. The λ, δ, μ_1, μ_2, μ_3 parameters in (4)–(7) are defined by the formulas:

$$\lambda = \frac{\bar{\lambda}}{V_0^2 c^2}, \qquad \delta = \frac{\bar{\delta}}{V_0^2 c^2}, \qquad \mu_1 = \frac{\omega c}{V_0}, \qquad \mu_2 = \frac{fM}{2V_0^2 c}, \qquad \mu_3 = \frac{\bar{b}}{V_0^2 c^2}.$$

Proceeding in (1) to cylindrical coordinates ϱ, l, w by the formulas

$$u = \varrho\cos l, \qquad v = \varrho\sin l, \qquad w = w\,(\varrho = \bar{\varrho}/c)$$

we shall obtain for the first variant in (3) $(\mu_3 = 0)$:

$$\frac{d^2\varrho}{d\theta^2} - \frac{4\sigma^2}{\varrho^3} = \frac{\partial}{\partial\varrho}(W + \bar{W}),$$

$$\frac{d\sigma}{d\theta} = 0, \qquad \frac{d^2w}{d\theta^2} = \frac{\partial}{\partial w}(W + \bar{W}), \tag{8}$$

where $\sigma = \frac{1}{2}\varrho^2(l' + \mu_1)$ is the sector velocity,

$$W = U(\varrho, l, w, \mu_2), \qquad \bar{W} = \frac{\delta}{\varrho^2} + \frac{\lambda}{w^2}.$$

The system (8) has a particular solution

$$\varrho = \varrho_0, \qquad \sigma = \sigma_0, \qquad w = w_0, \qquad \varrho' = \varrho'_0 = w' = w'_0 = 0 \tag{9}$$

(index 0 denotes the initial values of the variables).

The vehicle circular motion at the distance $w = w_0$ from the equatorial plane corresponds to the solution of (9), when

$$\xi = \xi_0 = \xi_1 = \xi_2, \qquad \eta = \eta_0 = \eta_1 = \eta_2,$$

where ξ_1, ξ_2 and η_1, η_2 are real roots of $R_1(\xi)$ and $R_2(\eta)$.

As

$$\eta^2 = \left(1 + \frac{1-\eta^2}{\xi^2}\right)\sin^2\varphi, \tag{10}$$

where φ is a vehicle latitude and $\xi \geqslant 30$ for all real motions in the vicinity of the Earth, then from (10) it follows that: $\eta \sim \sin\varphi$. Thus the solution (9) corresponds to the vehicle motion at a constant latitude.

Let us divide the vehicle trajectories into two classes: (1) the elliptic type of motion V_0 does not exceed a satellite initial velocity on the Keplerian elliptic orbit; (2) the hyperbolic type of motion V_0 exceeds a satellite velocity in the perigee of the Keplerian parabolic orbit or is equal to it.

The trajectories of the first class comprise synchronous orbits $(l'_0 = 0)$. Asynchronous orbits $(l'_0 \neq 0)$ are in both classes.

From (8) the conditions of a geosynchronous vehicle must obey the equations

$$g_{r\varrho} = \frac{\mu_2}{\varrho_0^2}(k_1 \cos^2\varphi_0 - n_0^3), \qquad g_{rw} = \frac{\mu_2 \sin\varphi_0 \cos^3\varphi_0}{\varrho_0^2} k_2,$$

where $g_{r\varrho}, g_{rw}$ are the components of dimensionless acceleration, $n_0 = \varrho_0/\varrho_s$ (ϱ_s is a long orbital semiaxis of a geosynchronous satellite), k_1 and k_2 are the Earth compression parameters of a form:

$$k_1 = 1 - 7.5 \sin^2\varphi_0/\varrho_s^2 + 1.5/\varrho_s^2, \qquad k_2 = 1 - 7.5 \sin^2\varphi_0/\varrho_s^2 + 4.5/\varrho_s^2.$$

Routh's conditions [8] allow us to define the region of possible motions while keeping the relation $\alpha_1^2 = 2\delta$:

$$0 < \varrho_0 \leqslant \sqrt{\tfrac{6}{5}}, \qquad \arcsin \sqrt{\tfrac{3}{5}} \leqslant |\varphi_0| < \frac{\pi}{2}.$$

For the second variant g_r in (3) analogous results are obtained.

For the first variant $\mathbf{g}_r$ using the Jacobian integral we obtain

$$\Phi(\varrho, \lambda, \delta) = \mu_1^2 \varrho^2 + \frac{2\delta}{\varrho^2} + \frac{2\lambda}{\xi^2 \eta^2} + \frac{4\mu_2 \xi}{\xi^2 + \eta^2} \geqslant -2k. \tag{11}$$

At $\delta > 0$, $h < 0$, $\lambda < 0$ from (11) it follows that for $\varrho = (2\delta/\mu_1^2)^{1/2}$ the function $\Phi(\varrho, \lambda, \delta)$ has its minimum, that is why the vehicle limited motions correspond to these values of ϱ. As V_0 can have any value, both classes of trajectories appear possible in the annular region:

$$\xi_1 \leqslant \xi_0 \leqslant \xi \leqslant \xi_2, \qquad \eta_1 \leqslant \eta_0 \leqslant \eta \leqslant \eta_2.$$

The rest of sign combinations δ, λ and h leads in general to the same results. In the second variant of $\mathbf{g}_r$ the case when $\delta = 0$ is of interest; here the vehicle trajectories are in the plane crossing the equatorial plane.

In some applications of a latitudinal vehicle, for example for meteorological observations, it should be useful to study motions at small values of latitude and a constant value of coordinate ξ.

At the same time there appears a problem of defining reactive acceleration parameters. In the general case from (6) and (7) equations for defining the values of h, α_2, λ, δ follows:

$$\begin{aligned}
A_1 h + B_1 \alpha_2 + C_1 \lambda + D_1 \delta &= E_1, \\
A_2 h + B_2 \alpha_2 + C_2 \lambda + D_2 \delta &= E_2, \\
A_3 h + B_3 \alpha_2 + C_3 \lambda + D_3 \delta &= E_3, \\
A_4 h + B_4 \alpha_2 + C_4 \lambda + D_4 \delta &= E_4.
\end{aligned} \tag{12}$$

The coefficients in system (12) are the functions of ξ_0, η_1, η_2, μ_1, μ_2, α_1 ($\xi = \xi_1 = \xi_2 = \xi_0$) values. This system determinant nowhere, but in the points $\xi_1 = \xi_2$, $\eta_1 = \eta_2$, turns into zero; that is why there exists a non-zero solution depending on α_1.

Consider now the second equation in (5). Let us substitute the variable η by

$$\eta^2 = \bar{q} - \xi_0^2.$$

The equation for η is reduced to

$$\tfrac{1}{2}\bar{q}\, \frac{d\bar{q}}{d\theta} = \sqrt{R_2(\bar{q})},$$

where

$$\begin{aligned}
R_2(\bar{q}) &= a_1 \bar{q}^3 + a_2 \bar{q}^2 + a_3 \bar{q} + a_4, \\
a_1 &= -2\bar{h}, \qquad a_2 = 6\bar{h}\xi_0^2 + 2\bar{h} + \alpha_2, \\
a_3 &= -6\bar{h}\xi_0^4 - (4\bar{h} + \alpha_2)\,\xi_0^2 - (\alpha_2 + \alpha_1^2 - 2\delta + 2\lambda), \\
a_4 &= 2\bar{h}\xi_0^6 + (2\bar{h} + \alpha_2)\,\xi_0^4 + (\alpha_2 + \alpha_1^2 - 2\delta + 2\lambda)\,\xi_0^2 + 2\lambda.
\end{aligned}$$

Introducing an independent variable τ by the relation

$$d\theta = \tfrac{1}{2}\bar{q}\, d\tau$$

we transform the initial system (5):

$$\frac{d\bar{q}}{d\tau} = \sqrt{R_2(\bar{q})}, \qquad \theta = \theta_0 + \tfrac{1}{2}\int_0^\tau \bar{q}\, d\tau,$$

$$l = l_0 + \mu_1(\theta - \theta_0) + \frac{\alpha_1}{2(1+\xi_0^2)}\int_0^\tau \frac{\bar{q}\, d\tau}{1+\xi_0^2 - \bar{q}}. \tag{13}$$

For a sufficiently general case of various real roots $q_1 < q_2 < q_3$ of the polynomial $R_2\ldots(\bar{q})$ we have

$$\bar{q} = q_1 + (q_2 - q_1)\, \mathrm{sn}^2\, \bar{\tau},$$

where

$$\bar{\tau} = \varkappa\tau + \varkappa_0, \qquad \varkappa_0 = \int_0^{\gamma_0} \frac{d\gamma}{\sqrt{1 - k^2 \sin^2 \gamma}}, \qquad \varkappa = \sqrt{2|\bar{h}|\,(q_3 - q_1)},$$

$$\gamma_0 = \arcsin\sqrt{\frac{q_0 - q_1}{q_2 - q_1}}, \qquad k^2 = \frac{q_2 - q_1}{q_3 - q_1},$$

$$\gamma = \mathrm{am}\,\bar{\tau};$$

$$l = l_0 + \mu_1(\theta - \theta_0) + \frac{\alpha_1}{2(1+\xi_0^2)}\left\{ \frac{1+\xi_0^2}{\varkappa(1+\xi_0^2 - q_1)} \times \right.$$
$$\left. \times [\Pi(\gamma, n, k^2) - \Pi(\gamma_0, n, k^2)] - \tau \right\},$$

$\Pi(\gamma, n, k^2)$ being an elliptical integral of the third type with the parameter $n : n = = (q_2 - q_1)/(1 + \xi_0^2 - q_1)$.

To define the coordinates as functions of θ we use the second equation in (13). Using small parameters k^2 and $\varepsilon = (q_2 - q_1)/q_1$ we obtain:

$$2\gamma - e_1(\sin 2\gamma - e^* \sin 4\gamma) = g, \tag{14}$$

where the values e_1, g and e^* take the form:

$$e_1 = k^2\, \frac{(64 + 48k^2 + 32\varepsilon/k^2 + 4\varepsilon)}{4(64 + 16k^2 + 9k^4 + 32\varepsilon + 48\varepsilon k^2)},$$

$$g = 2e_2\psi(\theta) + 2\gamma_0 - e_1 \sin 2\gamma_0 + e_1 e^* \sin 4\gamma_0 + \cdots$$

$$e^* = \frac{192 + \varepsilon}{256(4 + 3k^2 + 8\varepsilon/k^2 + 4\varepsilon)};$$

$$e_2 = \frac{64\varkappa}{64 + 16k^2 + 9k^4 + 32\varepsilon + 12\varepsilon k^2},$$

$$\psi(\theta) = \frac{2(\theta - \theta_0)}{q_1}.$$

Defining γ from (14) we obtain

$$\gamma = \tfrac{1}{2}\{g + (\sin g - e^* \sin 2g)\, e_1 + \\ + \tfrac{1}{2}[\sin 2g + e^*(3 \sin 3g - \sin g) + 2e^{*2} \sin 4g]\, e_1^2 + \cdots\}.$$

According to [5] the vehicle mass variation law m is defined by the formula:

$$m = m_0 \exp\left[-\frac{2V_0}{|\mathbf{u}_r|} \int_{\theta_0}^{\theta} \left(\frac{\delta^2}{\varrho^6} + \frac{\lambda^2}{w^6}\right)^{1/2} d\theta\right],$$

where $|\mathbf{u}_r| = \text{const}$ is the modulus of the jet exhaust velocity. Using a technique analogous to that mentioned above, we find that

$$\int_{\theta_0}^{\theta} \sqrt{\frac{\delta^2}{\varrho^6} + \frac{\lambda^2}{w^6}}\, d\theta = \tfrac{1}{2} \cdot \frac{q_1}{\varkappa} \left\{ (T_0 + T_1 \varepsilon + \cdots)(\gamma - \gamma_0) + T_0\left(\frac{k^2}{2} + \varepsilon\right) \times \right.$$

$$\left. \times \left[\tfrac{1}{2}(\gamma - \gamma_0) - \tfrac{1}{4}(\sin 2\gamma - \sin 2\gamma_0)\right] + \cdots \right\},$$

$$T_0 = \left(\frac{\delta^2}{\varrho_0^6} + \frac{\lambda^2}{w_0^6}\right)^{1/2}, \qquad T_1 = \tfrac{3}{2}\left\{ \frac{q_1(\xi_0^2 + 1)^{-3/2}}{(\xi_0^2 + 1 - q_1)^{5/2}\, \xi_0^3 (q_1 - \xi_0^2)^{3/2}} \times \right.$$

$$\times \left[\xi_0^6 (q_1 - \xi_0^2)^3\, \delta^2 + \lambda^2(\xi_0^2 + 1)^3 \times \right.$$

$$\times (\xi_0^2 + 1 - q_1)^3]^{1/2} - \frac{q_1^2 \lambda(\xi_0^2 + 1 - q_1)}{[(\xi_0^2 + 1)(q_1 - \xi_0^2)]^{1/2}} \times$$

$$\times [\xi_0^6 (q_1 - \xi_0^2)^3\, \delta^2 + (\xi_0^2 + 1)^3 \times$$

$$\left. \times (\xi_0^2 + 1 - q_1)^3\, \lambda^2]^{-1/2} \right\}.$$

The components of jet acceleration are determined by the formulas:

$$g_{ru} = \psi(\gamma)\cos l, \quad g_{rv} = \psi(\gamma)\sin l, \quad g_{rw} = -2\lambda\xi_0^{-3}[\xi_0^2 + 1 - q_1(1 + \varepsilon \sin^2 \gamma)]^{-3/2},$$

where

$$\psi(\gamma) = -2\delta\{(\xi_0^2 + 1)[\xi_0^2 + 1 - q_1(1 + \varepsilon \sin^2 \gamma)]\}^{-3/2}.$$

The problem of stability of Equations (1) stationary solution is of great interest. Let us consider the stability of such a solution for the case of a lunar latitudinal vehicle.

For lunar artificial satellites an approximation can be made for the longitudinal terms in the Moon potential series by a function of the form $U_1(l)/\bar{\varrho}^2$ for small eccentricities and orbital inclination [7].

Introduce now into (1) the reactive force potential with the function $U_2(l)$ when $U_2(l) \equiv U_1(l)$, i.e. $U(l) = U_2(l) - U_1(l) = 0$. This allows us to investigate the latitudinal vehicle motions, taking into account the Moon triaxiality and asymmetry. In Equation (1) values m_1, m_2, $\bar{c}_1$, $\bar{c}_2$ accept the meaning:

$$m_1 = \frac{M}{2}(1 - \bar{l}), \qquad m_2 = \frac{M}{2}(1 + \bar{l}),$$

$$\bar{c}_1 = c(\bar{l} + 1), \qquad \bar{c}_2 = c(\bar{l} - 1),$$

where

$$c = \frac{R_0}{2}\sqrt{\Delta}, \qquad \bar{l} = \frac{J_{30}}{J_{20}\sqrt{\Delta}}, \qquad \Delta = \frac{J_{30}^2}{J_{20}^2} + 4J_{20} > 0,$$

R_0 the Moon equatorial radius, J_{20}, J_{30} are the coefficients of zonal harmonics of the second and third orders in the Moon potential series, M is the Moon mass. By calculation one obtains $\bar{l} = 1.01354$, $c = 15.08$ km.

To transform Equation (1) prolonged spheroidal coordinates ξ, η, l should be used:

$$u = \sqrt{(\xi^2 - 1)(1 - \eta^2)}\cos l, \qquad v = \sqrt{(\xi^2 - 1)(1 - \eta^2)}\sin l, \qquad w = \xi\eta + \bar{l}$$

$$(1 \leqslant \xi < \infty, \qquad -1 \leqslant \eta \leqslant 1).$$

The general solution would be of a form

$$\int \frac{\xi^2\,d\xi}{\sqrt{R_1(\xi)}} - \int \frac{\eta^2\,d\eta}{\sqrt{R_2(\eta)}} = \theta + \beta_1,$$

$$\alpha_1\left[\int \frac{d\xi}{(\xi^2 - 1)\sqrt{R_1(\xi)}} + \int \frac{d\eta}{(1 - \eta^2)\sqrt{R_2(\eta)}}\right] - $$

$$-l - \mu_1\left[\int \frac{\xi^2\,d\xi}{\sqrt{R_1(\xi)}} - \int \frac{\eta^2\,d\eta}{\sqrt{R_2(\eta)}}\right] = \beta_2,$$

$$\int \frac{d\xi}{\sqrt{R_1(\xi)}} - \int \frac{d\eta}{\sqrt{R_2(\eta)}} = \beta_3;$$

(15)

$$(\xi^2 - \eta^2)^2\left(\frac{d\xi}{d\theta}\right)^2 = R_1(\xi), \qquad (\xi^2 - \eta^2)^2\left(\frac{d\eta}{d\theta}\right)^2 = R_2(\eta),$$

$$(\xi^2 - 1)(1 - \eta^2)\left(\frac{dl}{d\theta} + \mu_1\right) = \alpha_1.$$

(16)

Polynomials $R_1(\xi)$ and $R_2(\eta)$ are defined in such a way that

$$R_1(\xi) = (\xi^2 - 1)(2\bar{h}\xi^2 + 4\mu_2\xi + \alpha_2) - \alpha_1^2,$$
$$R_2(\eta) = (\eta^2 - 1)(2\bar{h}\eta^2 + 4\mu_2\bar{l}\eta + \alpha_2) - \alpha_1^2.$$

To study stationary motion stability (9) we use system (15). Let us introduce in-

V. K. KAISIN

dependent variable τ by the relation

$$\tau = \int\limits_{\eta 0}^{\eta} \frac{d\eta}{\sqrt{R_2(\eta)}}$$

and define perturbations x_1, y_1 by the formulas

$$x_1 = \xi - \xi_0, \qquad y_1 = \eta - \eta_0.$$

Considering h, α_1 and α_2 in $R_1(\xi)$ and $R_2(\eta)$ independent of perturbations x_1, y_1 [9, 10] one obtains that

$$\frac{dx_1}{d\tau} = x_2, \qquad \frac{dx_2}{d\tau} = \tfrac{1}{2}R_1'(\xi_0 + x_1), \qquad \frac{dy_1}{d\tau} = y_2, \qquad \frac{dy_2}{d\tau} = \tfrac{1}{2}R_2'(\eta_0 + y_1). \quad (17)$$

From the system (17) we derive

$$h_1 = x_2^2 - \left(\frac{\partial R_1}{\partial x_1}\right)_0 x_1 - \tfrac{1}{2}\left(\frac{\partial^2 R_1}{\partial x_1^2}\right)_0 x_1^2 - \cdots \equiv \text{const},$$

$$h_2 = y_2^2 - \left(\frac{\partial R_2}{\partial y_1}\right)_0 y_1 - \tfrac{1}{2}\left(\frac{\partial^2 R_2}{\partial y_1^2}\right)_0 y_1^2 - \cdots \equiv \text{const}.$$

For the trajectories satisfying the condition $\hbar < 0$ in points $\xi = \xi_0$, $\eta = \eta_0$ there exist the correlations:

$$\left(\frac{\partial R_1}{\partial x_1}\right)_0 = \left(\frac{\partial R_2}{\partial y_1}\right)_0 = 0, \qquad \left(\frac{\partial^2 R_1}{\partial x_1^2}\right)_0 < 0, \qquad \left(\frac{\partial^2 R_2}{\partial y_1^2}\right)_0 < 0.$$

Thus the form

$$V(x_1, x_2, y_1, y_2) = x_2^2 + y_2^2 - \tfrac{1}{2}\left(\frac{\partial^2 R_1}{\partial x_1^2}\right)_0 x_1^2 - \tfrac{1}{2}\left(\frac{\partial^2 R_2}{\partial y_1^2}\right)_0 y_1^2 - \cdots \qquad (18)$$

is a positive definite value. In accordance with Equation (17) from (18) we conclude that $dV/d\tau \equiv 0$. In view of Lyapunov's stability theorem a stationary solution $\xi = \xi_0$, $\eta = \eta_0$ is stable for all small initial perturbations.

An analogous investigation of a near-Earth vehicle stationary motion stability showed that for the first variant g_r there exists a locus of stable artificial points of a libration in the near space, i.e. a region limited by axis Oz and the Earth surface. In case of the second variant g_r stable stationary motions are absent.

A near-Moon latitudinal vehicle can be used for meteorological observations of the Earth's surface as well as for radio communication between different points of lunar surface with the Earth. This last mentioned fact should be borne in mind when expeditions and automatic vehicles are sent to the invisible side of the Moon. As one application of a near-Earth latitudinal vehicle one should point to a possibility of constant observation with the help of such a vehicle of hurricanes or cyclones.

References

[1] Nita, M. M.: 1961, 'Über die Bewegung der Rakete in einen Zentralkraftfeld', *Rev. Méch. Appl.* **6**, No. 2.

[2] Beletzkii, V. V.: 1964, 'On Trajectories of Space Flights with Reactive Acceleration. Constant Vector', *Space Res.* **2**, issue.

[3] Djomin, V. G.: 1966, 'A Method of a Space Vehicle Motion Investigation in a Planet's Activity Sphere', *Proc. of P. Lumumba*, University of Peoples' Friendship, **17**, issue 4.

[4] Kunitzyn, A. L.: 1969, 'On a Vehicle Motion in a Central Force. Field with Reactive Acceleration Constant Vector', *Space Res.* **7**, issue 5.

[5] Kaisin, V. K.: 1969, 'Space Vehicle Motion in a Normal Gravitational Field of the Earth Under the Influence of Additional Forces', *Space Res.* **7**, issue 5.

[6] Aksjonov, E. P., Grebenikov, E. A., and Djomin, V. G.: 1961, 'A General Solution of an Artificial Satellite Motion, Problem in a Normal Gravitational Field of the Earth', *Artificial Satellites of the Earth*, issue 8.

[7] Djomin, V. G.: 1968, *Artificial Satellite Motion in a Noncentral Gravitation Field*, Publishing House 'Nauka', M.

[8] Routh, E. J.: 1884, *The Advanced Part of a Treatise on the Dynamics of a System of Rigid Bodies*, London.

[9] Lyapunov, A. M.: 1954, 'On Constant Helical Motion of a Rigid Body in a Fluid', *Collection of Works* **1**, M.

[10] Pozharutzkii, G. K.: 1958, 'On Lyapunov's Function Construction on the Base of Disturbed Motion Integrals', *Appl. Math. Mech.* **22**, issue 2.

THE ESTIMATION OF ACCURACY OF SHORT-TERM
ATMOSPHERE DENSITY PREDICTION

P. E. ELYASBERG, B. V. KUGAENKO, V. M. SYNITSYN, and M. I. VOISKOVSKY

Academy of Sciences of the U.S.S.R., Institute for Space Research, Moscow, U.S.S.R.

Abstract. Attainable accuracy of short-term prediction of atmosphere density for operational determination and prediction of close Earth satellite orbits is estimated in this paper. Using available atmospheric models short-term variations of atmospheric parameters are represented generally as a function of the 10.7 cm solar flux $F_{10.7}$ and the planetary geomagnetic index a_p (or A_p). Using such models the accuracy of description of atmosphere density changes within the interval of AES orbit determination and prediction depends considerably on the accuracy of predicted quantities $F_{10.7}$, a_p or A_p. Various methods of prediction are considered here and their accuracies are compared. Errors of atmosphere density prediction are recalculated from errors of predicted quantities $F_{10.7}$, a_p and A_p. In this case empirical relationships between variations of the quantities $F_{10.7}$, a_p, A_p and corresponding atmospheric density changes calculated from AES drag data are used. The prediction intervals considered are from some hours up to three days. Some tables and diagrams given allow estimation of the accuracy obtainable of atmosphere density calculations for the altitudes of 150–350 km.

The accuracy of operational prediction of the flight of a close artificial Earth satellite (AES) is mainly defined by the capabilities of mathematical description and prediction of short-term atmospheric density changes. In the available models of the atmosphere the short-term changes of parameters are usually presented as a function of both solar radioradiation $F_{10.7}$ intensity and values of geomagnetic index a_p (or A_p). So, the accuracy for describing the short-term changes of the atmospheric density ϱ within the interval of AES flight prediction depends primarily on the accuracy in prediction of the values $F_{10.7}$ and a_p (or A_p) and on the fit of analytic relations (as functions of the given indexes) selected in the models, to the real variations of the atmospheric density. This paper estimates the influence of the first source of errors, i.e. the influence of errors in predicting indexes $F_{10.7}$, a_p and A_p.

Let us analyse the following versions of defining the values of these indexes within the interval of prediction:

I version: extrapolation of indexes by the method based on the theory of random functions [1, 2];

II version: using the last measured value of each index under assumption on the constancy of the index values within the prediction interval;

III version: using the predicted values of indexes $F_{10.7}$ and A_k (instead of A_p) received by the Institute of Earth magnetism, ionosphere and propagation of radiowaves (IZMIRAN), ASUSSR.

While calculating we will use the model of the atmospheric density given in [3]. The duration of interest for the prediction intervals is not more than three days.

It has been shown in [1] that for stationary normal sequences the best extrapolation equation is the following:

$$\bar{\xi}(t_1+\delta)=\sum_{k=1}^{n} a_k \xi(t_k), \tag{1}$$

L. G. Napolitano et al. (eds.), Astronautical Research 1972, 43–53. All Rights Reserved

where t_1 is a moment of the last measured value of index ξ; $\bar{\xi}(t_1+\delta)$ is a predicted value of index ξ for the moment $t_1+\delta$; δ is a duration of prediction period; $\xi(t_k)$ are the known, from the measurements, values of index at moments t_k. The coefficients a_k are determined by solving the set of n algebraic equations:

$$\sum_{k=1}^{n} a_k R(t_k-t_j)=R[(t_1-t_j)+\delta], \qquad (2)$$

where $j=1, 2, ..., n$; $R(t_k-t_j)$, $R[(t_1-t_j)+\delta]$ are the values of autocorrelation function of index ξ that correspond to the time displacements equal to t_k-t_j and $(t_1-t_j)+\delta$.

To show the possibility of using the given algorithm of prediction of indexes $F_{10.7}$ and a_p (or A_p) the verification of the constancy of sequences of these indexes was carried out, as well as the fit of a law of their distribution to the normal one.

The calculations for the verification of the fit of the normal distribution law were performed for the period from 1964 to 1971, taking each year taken separately. The calculated probability of fitting the distribution to the normal one (according to χ^2-criterion) lies in the range 0.75–0.90 for all three indexes.

The constancy of the sequences $F_{10.7}$, a_p and A_p was analyzed by means of comparing the autocorrelation functions of appropriate sequences calculated for several years (given on the plots) and for some displacements of the reading beginning. In this case the following equation is applied

$$\hat{R}(\tau)=[1/(N-M)]\sum_{i=1}^{N-M}[\xi(t_i)-\bar{\xi}][\xi(t_i+\tau)-\bar{\xi}], \qquad (3)$$

where N is the total number of the values of the sequence within the interval of one year; $M=\tau/\Delta$; Δ is a sequence step; $\bar{\xi}$ is an arithmetical mean value of ξ index within the considered interval.

The resulting normalized autocorrelation functions $K(\tau)$ (without displacement) are shown in Figures 1–3. The shaded sections of these figures correspond to the region of $K(\tau)$ changes for the time periods given on the plots. The calculations carried out under conditions with displacements of the first reading have shown that in the case of small displacements (up to 10–20 days) the changes of the appropriate normalized autocorrelation functions are insignificant.

Figure 4 shows the effect of changing the average-annual values of indexes $F_{10.7}$, a_p and A_p and the root-mean-square (rms) deviations of indexes $F_{10.7}$ and a_p. This plot can describe the variability of the mean and the variance for each sequence in time

Analysis of the curves shown in Figures 1–4 shows that, in the general case none of the analyzed sequences meets the requirements of constancy within long intervals of time. Taking into account, however, that this paper deals with the short-term prediction of indexes (up to several days) we can consider the suggestion on constancy of these sequences to be true for these small intervals of time, and can apply the above method of prediction.

To choose the number of addenda n in Equation (1) the analysis of the dependence

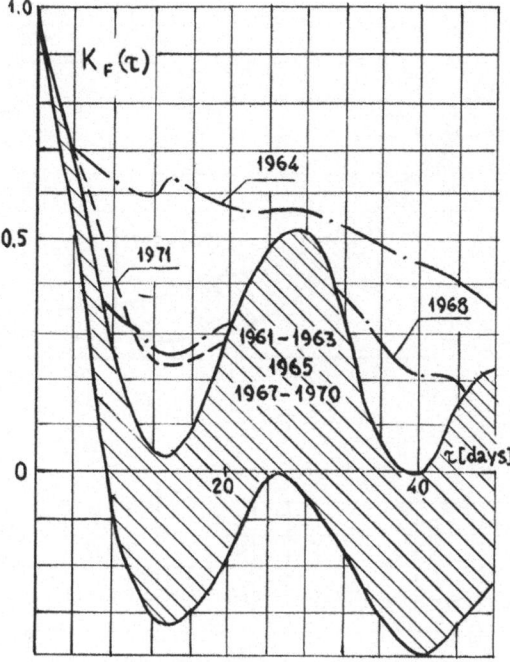

Fig. 1.

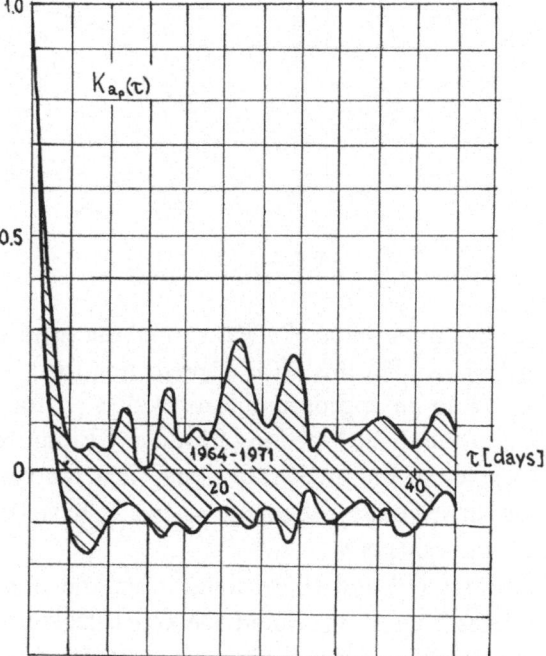

Fig. 2.

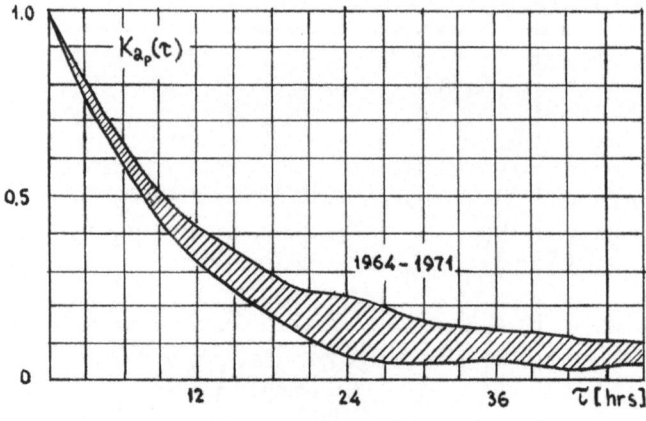

Fig. 3.

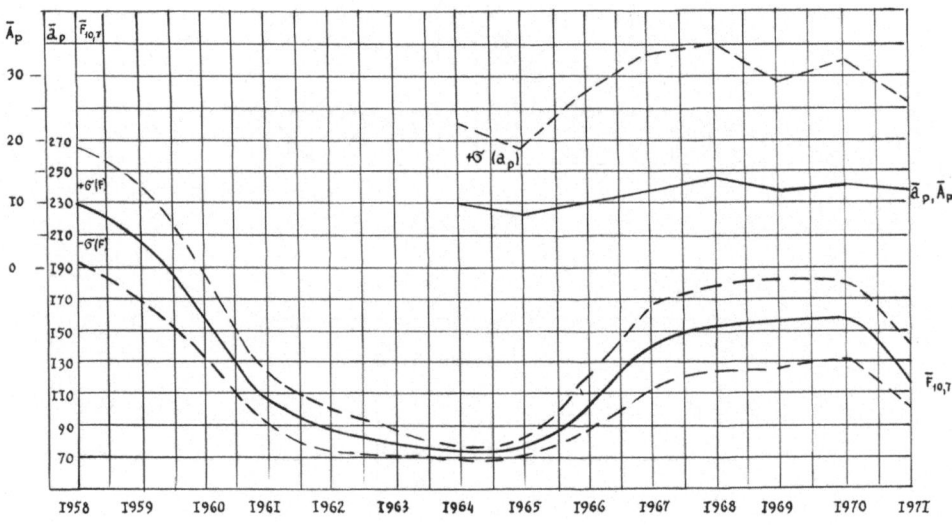

Fig. 4.

of the prediction accuracy on the value of n was carried out (with $1 \leqslant n \leqslant 8$). For this purpose the ratio is calculated of a rms error of prediction for 1 day of the value of index $F_{10.7}$ with $1 \leqslant n \leqslant 8$ to an appropriate error with $n = 1$ for several years. All calculated values fell into the shaded section of Figure 5. From this figure it is seen that, provided the accuracy of the prediction is sufficient, we may take $n = 4$. The analogous calculations carried out for sequence a_p and A_p have shown that in these cases we may use the same value of n.

To compare the accuracy of I and III versions, mean and rms deviations of the predicted values of indexes from the measured one were calculated. In the case of III version, the predicted values $F_{10.7}$ and A_k obtained in IZMIRAN for the interval of time from 1 January, 1969 to 31 December, 1971 were used.

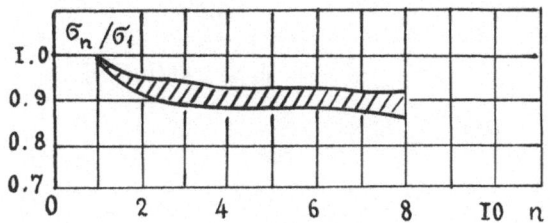

Fig. 5.

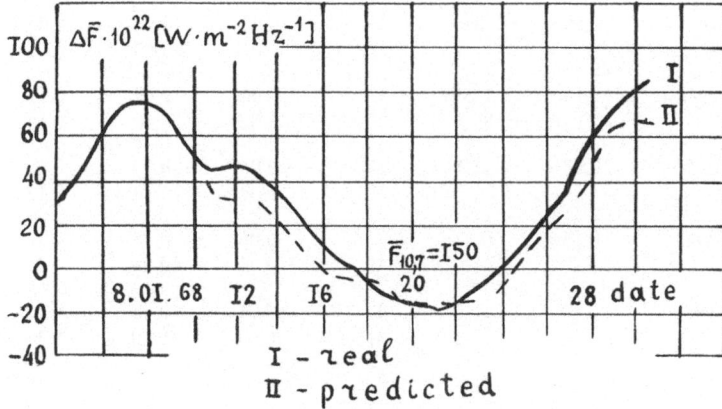

Fig. 6.

The estimation of the II version accuracy of defining the values of indexes within the interval of the prediction was performed by means of calculating a variance for increments of the indexes by the equation:

$$D_\xi(\delta) = M|\xi(t_1+\delta) - \xi(t_1)|^2. \tag{4}$$

Table I gives the results of the accuracy estimation of the first two versions for periods from 1964 year to 1971 year while predicting index $F_{10.7}$ for 1; 2 and 3 days and index A_p for 1 day.

2 and 3 columns of Table I give the ratios of the rms deviation for sequence to its arithmetical mean value for each year (1 column). 5, 6, 7 and 8 columns present the appropriate ratios of the prediction errors for both versions.

It follows from Table I that I version of the prediction is somewhat more accurate than II version.

The relative rms error of index $F_{10.7}$ prediction is less than the relative deviation of sequence $F_{10.7}$ within the whole period of the predictions, while for A_p these values become approximately equal while predicting for one day that evidences in the impossibility of A_p prediction by the considered methods (the predicted value of index A_p slightly differs from the mean of sequence A_p).

P. E. ELYASBERG ET AL.

TABLE I

Year	$\sigma(F)/\bar{F}$	$\sigma(A_p)/\bar{A}_p$	δ version	$\sigma[\xi(t_1+\delta)]/\xi$ $F_{10.7}$			A_p
				1	2	3	1
1	2	3	4	5	6	7	8
1964	0.05	0.84	I	0.02	0.02	0.02	0.80
			II	0.02	0.02	0.04	0.95
1965	0.06	1.01	I	0.02	0.03	0.05	1.02
			II	0.03	0.03	0.05	1.20
1966	0.18	1.24	I	0.06	0.09	0.13	1.20
			II	0.06	0.10	0.14	1.39
1967	0.18	1.25	I	0.06	0.10	0.15	1.19
			II	0.07	0.10	0.15	1.28
1968	0.15	0.97	I	0.05	0.08	0.12	0.92
			II	0.05	0.08	0.13	0.92
1969	0.18	1.13	I	0.04	0.08	0.11	1.05
			II	0.05	0.10	0.13	1.15
1970	0.14	1.23	I	0.04	0.07	0.10	1.21
			II	0.05	0.08	0.11	1.41
1971	0.16	0.91	I	0.04	0.06	0.08	0.88
			II	0.04	0.07	0.09	0.98

TABLE II

δ version	$\sigma[\xi(t_1+\delta)]/\xi$ $F_{10.7}$			A_p			a_p		
	1	2	3	1	2	3	3^h	6^h	9^h
I	0.04	0.07	0.10	1.06	1.21	1.24	1.02	1.15	1.37
II	0.05	0.08	0.12	1.20	1.46	1.54	1.08	1.38	1.55
III	0.05	0.06	0.09	1.02	1.10	1.13	–	–	–

The plot of Figure 6 gives an example of prediction by I version for one day of index $F_{10.7}$ for January, 1968. One can see that the predicted values approximately track the course of the measured values of index trying to smooth its fluctuations.

Table II presents the relative rms errors in predicting indexes $F_{10.7}$, A_p and a_p by three versions for period from 1 January 1969 to 31 December, 1971.

As seen from Table II, all versions give approximately the same accuracy. The third version is somewhat more accurate, the second one is the least accurate.

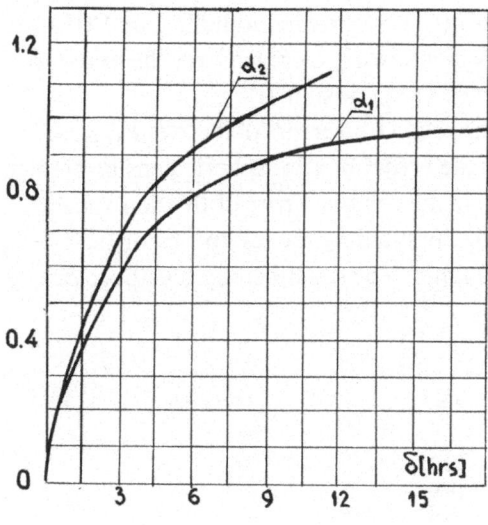

Fig. 7.

The possible accuracy of index a_p prediction for several hours was also estimated as follows. The initial portion of the mean normalized autocorrelation function $K_{a_p}(\tau)$ corresponding to Figure 3 was approximated by the following expression:

$$k_{a_p}(\tau) = \exp(-\alpha\tau). \tag{5}$$

The value obtained $\alpha(1/\alpha \approx 12^h)$ was substituted in the expressions for estimations of the prediction errors by I and II versions which are given in [1]:

$$\sigma[\bar{a}_p(t_1 + \delta)] = \sigma(a_p)[1 - \exp(-2\alpha\delta)]^{1/2} \quad \text{for I version} \tag{6}$$

$$\sigma[\bar{a}_p(t_1 + \delta)] = \sigma(a_p)\{2[1 - \exp(-\alpha\delta)]\}^{1/2}, \quad \text{for II version} \tag{7}$$

where $\sigma(a_p)$ is a rms deviation of a_p sequence.

The equations

$$\alpha_1 = [1 - \exp(-2\alpha\delta)]^{1/2} \tag{8}$$

and

$$\alpha_2 = \{2[1 - \exp(-\alpha\delta)]\}^{1/2} \tag{9}$$

are presented in Figure 7. As seen from Figure 7 the prediction of a_p index is essentially permissible for the interval of about nine hours using the first version and for about six hours for the second version. For the computation of absolute errors of the index a_p prediction Table III gives the values of $\sigma(a_p)$ for several years corresponding to low, middle and high solar activities (during current cycle of solar activity).

From the relation (7) one can find the maximum length of the prediction interval δ for which the rms error of prediction for indexes ξ by the second version does not exceed rms deviation of indexes. Determining the values α from the plots for the appropriate normalized autocorrelation functions, we find that for index $F_{10.7} \delta \approx 4$

days, for index A_p $\bar{\delta} < 1$ day and for index a_p $\bar{\delta} \approx 8$ h. This suggests that in the absence of the predicted values $\bar{\xi}(t_1 + \delta)$ corresponding to I and III versions it is expedient to use for $\delta > \bar{\delta}$ the values of indexes ξ averaged over the long-term intervals, rather than the last measured values $\xi(t_1)$ (version II).

The estimation of the prediction errors of indexes $F_{10.7}$, a_p and A_p permits determination of the appropriate errors of atmospheric density prediction. Using the model density [3] we find that the relative error of the density calculation $(\varDelta\varrho/\bar{\varrho})_\xi$ ($\bar{\varrho}$ is an average density value) stipulated by the relative error $\varDelta\xi/\xi$ of index ξ within the altitude range of 160–600 km can be described by the approximate relations of the form

TABLE III

Year	$\sigma(a_p)$
1964	12.3
1965	10.0
1966	16.6
1967	20.9
1968	20.3
1969	17.1
1970	19.7
1971	13.7

(h is taken in km):

$$(\varDelta\varrho/\bar{\varrho})_\xi = (a + bh)(\varDelta\xi/\bar{\xi}). \qquad (10)$$

The values of coefficients a and b for indexes $F_{10.7}$ and a_p at different levels of solar activity F_0 are given in Table IV.

TABLE IV

ξ	F_0	75	100	125	150
$F_{10.7}$	a	-0.630	-0.750	-0.710	-0.765
	b	0.00506	0.00560	0.00562	0.00571
a_p	a	-0.109	-0.109	-1.112	-0.103
	b	0.00089	0.00087	0.00083	0.00079

Table V presents the relative rms errors of density prediction (%) for all three versions at altitudes of 200 km and 300 km. In the same table the rms deviations of the density values are given for comparison which result from the short-term changes of indexes ξ that conforms to the rms error of ϱ prediction taking in account only the long-term (eleven years) changes of indexes. The errors given in Table V are computed with the use of Table II and correspond to the period of time from 1 January, 1969 to 31 December, 1971.

It follows from Table V that, at the altitudes considered, all three methods of density prediction are essentially equal in accuracy: the rms errors of prediction do not differ more than 3–5% within the whole prediction interval of interest. From this table it also follows that using the above methods of prediction of index $F_{10.7}$, we can reduce the error of density prediction at the whole interval; the use of prediction of the geomagnetic disturbance is justified at intervals up to six hours; prediction of an average

TABLE V

δ h[km]		$F_{10.7}$			$\left(\dfrac{\sigma\varrho}{\bar{\varrho}}\right)_F$	a_p			$\left(\dfrac{\sigma\varrho}{\bar{\varrho}}\right)_{a_p}$	A_p			$\left(\dfrac{\sigma\varrho}{\bar{\varrho}}\right)_{A_p}$
		1	2	3		3^h	6^h	9^h		1	2	3	
200	I	1.5	2.6	3.8		5.6	6.3	7.5		5.8	6.7	6.8	
	II	1.9	3.0	4.5	7.5	5.9	7.6	8.5	8.1	6.6	8.0	8.5	6.0
	III	1.9	2.3	3.4						5.6	6.0	6.2	
300	I	3.8	6.6	9.5		13.7	15.4	18.4		14.2	16.2	16.6	
	II	4.7	7.6	11.4	18.9	14.5	18.5	20.8	19.8	16.1	19.6	20.6	14.6
	III	4.7	5.7	8.5						13.7	14.7	15.1	

daily change of the geomagnetic disturbance does not lead to reductions of the appropriate error in the density prediction. When solving the problem on the expediency of using the predicted values of indexes $F_{10.7}$ and a_p while determining and predicting AES orbit it is also necessary to take into account the lag in time of the atmospheric density changes corresponding to the variations of indexes $F_{10.7}$ (of an order of one day) and a_p (about six hours).

Figures 8 and 9 present the variation of the relative rms errors for density prediction with altitude, in predicting index $F_{10.7}$ for 1, 2 and 3 days and index a_p for 3, 6 and 9 h

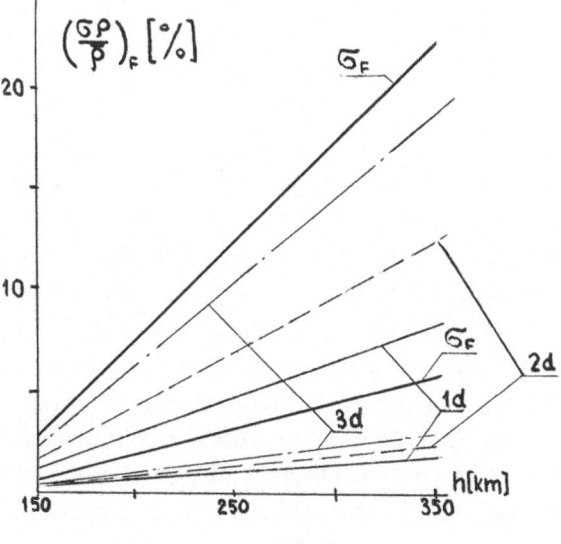

Fig. 8.

(I version of prediction). The same plots show the rms deviations of density stipulated by short-term changes of indexes $F_{10.7}$ and a_p (equivalent to using only the mean, long-term, values of indexes for density prediction). The lower curves conform to minimum, and the upper ones to maximum errors within the interval from 1966 to 1971. The graphs permit estimation of the appropriate errors for the computed density values in the altitude range from 150 km up to 350 km.

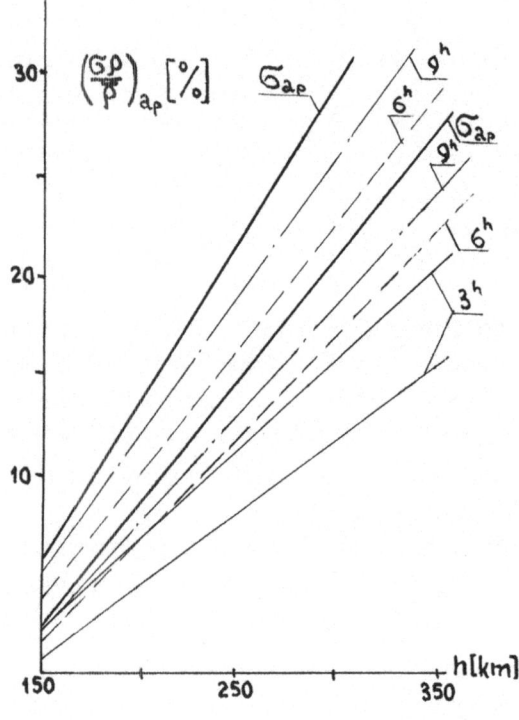

Fig. 9.

Figure 10 presents a graph of the summary rms deviations of the diurnal density values (straight line β) plotted by Table V data for the same altitude range, as well as graphs of the rms density changes for 1, 2 and 3 days (corresponding to errors of density ϱ in the orbit prediction for the first, second and third days and under condition of availability of the measured AES average drag parameters between the two previous orbit determinations separated by one day). The graphs are plotted on the assumption of independence of the atmospheric density changes corresponding to indexes $F_{10.7}$ and A_p. The small values of the cross-correlation normalized coefficients K_{A_pF} computed for several years at different displacements τ can serve as a substantiation for this assumption. The maximum values of coefficients K_{A_pF} are given in Table VI.

TABLE VI

Year	$K_{A_p F}$
1964	0.26
1965	0.08
1966	0.07
1967	0.15
1968	0.05
1969	0.32
1970	0.12
1971	0.16

The values of deviations ϱ presented in Figure 10 depend slightly on the level of solar activity due to the relatively high influence of the component $(\sigma\varrho/\bar{\varrho})_{A_p}$ which does not change significantly during the solar activity cycle.

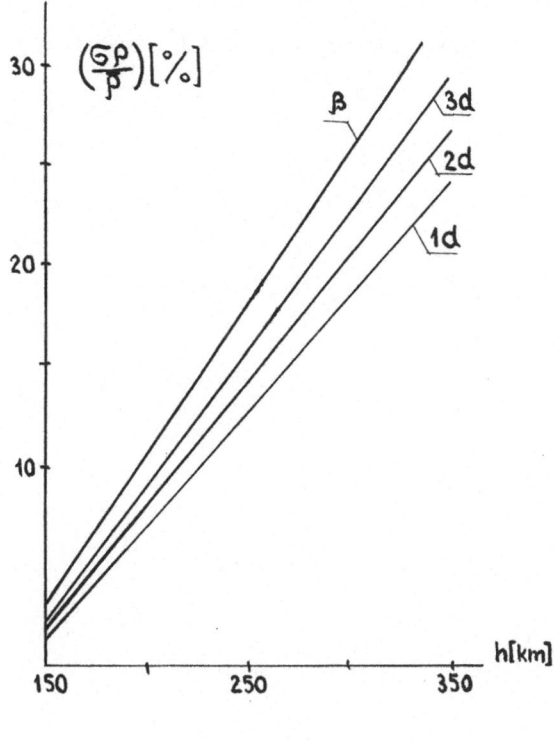

Fig. 10.

As seen from Table V and Figures 8 and 9, prediction of index a_p for three hours reduces the possible errors by 30–45%, for 6 h – approximately by 20% and for 9 h by 5–10%.

From Figures 8 and 9 it also follows that the density variations proportional to flux $F_{10.7}$ change within wide limits with variation of the solar activity level. Thus at altitude of 200 km $(\sigma\varrho/\bar{\varrho})_F$ is changed from 2% up to 7–8% as well as at 300 km – from 4% up to 7–18%. The values of density variations corresponding to index of geomagnetic disturbancy a_p are somewhat higher, but their change within the eleven year cycle is relatively less. At an altitude of 200 km $(\sigma\varrho/\bar{\varrho})_{a_p}$ is changed from 8–9% up to 13–14% and at 300 km – from 19–22% up to 30%.

During the quiet Sun years the use of the predicted values of index $F_{10.7}$ permits reduction of the appropriate error of density prediction by 2–3 times over the whole period of prediction.

During the active Sun years this advantage is limited to 1–2 days. However, from Table III it follows that the corresponding part of the summary errors is relatively small because of the considerable average daily density changes correlating with index A_p.

References

[1] А. М. Яглом, 'Успехи математических наук', *выпуск* **5**, 1962 г.
[2] В. М. Вяткина, *Радиоизлучение Солнца*, Сборник I. Изд-во Ленинградского Университета, 1969 г.
[3] P. E. Elyasberg, B. V. Kugaenko, V. M. Synitsyn, and M. I. Voiskovsky, *Space Res.* **XII**, 727 (1972).

OPTIMAL THREE DIMENSIONAL MANEUVERING
OF A ROCKET POWERED HYPERVELOCITY VEHICLE

JOSEPH W. GRIFFIN, JR.

The Bendix Corporation, Aerospace Systems Division, Ann Arbor, Mich. 48107, U.S.A.

and

NGUYEN X. VINH

Dep. of Aerospace Engineering, The University of Michigan Ann Arbor, Mich. 48104, U.S.A.

Abstract. Optimal lift, bank angle and thrust magnitude control laws are derived for a hypervelocity vehicle along a flight trajectory such that the thrust direction is aligned with the velocity vector. The optimal control laws are derived in a general form and a complete analytical solution is obtained for the free range problem. An optimal skip maneuver is used to demonstrate their use in a specific problem.

1. Introduction

Rocket-powered shuttle type vehicles capable of producing aerodynamic lift have stimulated new interest in the study of hypervelocity flight. Due to the highly non-linear nature of the optimization problem, analytical solutions have in the past been obtained only for specific problems restricted to either the vertical or horizontal plane. Furthermore, most of the available solutions associated with optimal reentry rely upon numerical methods to solve the resulting general two point boundary value problem.

In light of the classical results of Leitmann (1956), Miele (1958), Contensou (1965) and the recent contributions by Busemann *et al.* (1969), Fave (1968a, b), the works have been extended by Griffin and Vinh (1971) and Speyer and Womble (1971) to the optimal three dimensional flight of a non-thrusting vehicle.

In the present paper we investigate the optimal dimensional maneuvering of a rocket-powered hypervelocity vehicle which possesses a throttable rocket engine to provide a means of control in addition to the variable lift and bank angle control parameters. To ease the analysis we assume that the thrust direction is aligned parallel to the velocity vector during the thrusting phase, thus only the thrust magnitude control need be considered.

The problem consists of finding the modulation of the thrust magnitude, the angle-of-attack and the bank angle in some optimal manner to transfer the vehicle from prescribed initial conditions, to partially prescribed final conditions such that a certain final flight element is maximized.

2. Problem Formulation

The problem is formulated as an optimal control problem. At time t, the state of the

L. G. Napolitano et al. (eds.), Astronautical Research 1972, 43–53. All Rights Reserved

vehicle is defined by

$$\bar{r}(t) = \text{position vector}$$
$$\bar{V}(t) = \text{velocity vector} \tag{1}$$
$$m(t) = \text{instantaneous mass}.$$

The motion of the vehicle along its flight path is controlled by the aerodynamic force and the thrust magnitude. We have the equations of motion (see nomenclature of Figure 1).

$$R \, d\delta/dt = V \cos\gamma \cos\psi$$
$$R \, d\varepsilon/dt = V \cos\gamma \sin\psi$$
$$dz/dt = V \sin\gamma$$
$$dV/dt = (T - D)/m - g \sin\gamma \tag{2}$$
$$d\gamma/dt = (L/mV) \cos\phi + (1/V)(V^2/R - g) \cos\gamma$$
$$d\psi/dt = L \sin\phi/mV \cos\gamma$$
$$dm/dt = -T/c.$$

In these equations, although the altitude is varying, we have assumed that the distance r from the vehicle to the center of attraction is practically equal to the radius R of the Earth. Furthermore, we assume that the acceleration due to gravity, g, is constant and the assumption of Fave (1968a) concerning short lateral range has been

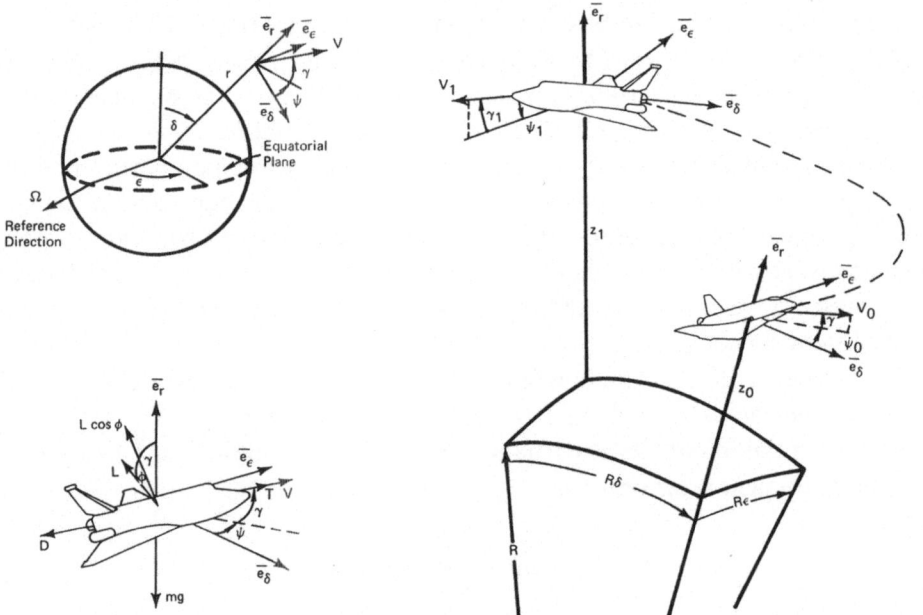

Fig. 1.   Nomenclature.

used. We shall make the usual assumption for the lift and drag that

$$L = \varrho C_L S V^2/2$$
$$D = \varrho C_D S V^2/2.$$

(3)

For flight in the hypervelocity regime, the lift and drag coefficients are assumed independent of the Mach number and the Reynolds number. They are related by the equation for a parabolic drag polar,

$$C_D = C_{D_0} + k C_L^2.$$

(4)

The atmospheric mass density varies as a function of the altitude according to the exponential law,

$$\varrho(z) = \varrho_0 e^{-z/\beta}.$$

(5)

By relations (3) and (4) the control parameters in the state equations are now the thrust magnitude, T, the lift coefficient C_L, and the bank angle ϕ. We shall use a lift control parameter λ defined by

$$\lambda = C_L/C_L^*,$$

where C_L^* is the lift coefficient corresponding to the maximum lift to drag ratio. From (4) we have for the drag coefficient

$$C_D = (C_D^*/2)(1 + \lambda^2),$$

(7)

where C_D^* is the drag coefficient corresponding to the maximum lift-to-drag ratio.

Then

$$C_L^* = (C_{D_0}/k)^{1/2}, \quad C_D^* = 2C_{D_0},$$

and

$$E = C_L^*/C_D^* = \tfrac{1}{2}(k C_{D_0})^{1/2}.$$

8)

In general, the controls are bounded with

$$0 \leqslant T \leqslant T_{\max}$$
$$0 \leqslant \lambda \leqslant \lambda_{\max}$$
$$0 \leqslant |\phi| \leqslant \phi_{\max}.$$

(9)

We should note that $T_{\max}$ and $\lambda_{\max}$ are constants while $\phi_{\max}$ is usually a function of the state variables. Here we shall assume that $\phi_{\max}$ is also constant.

We define the nondimensional variables,

$$\xi = R\delta/\beta, \quad \eta = R\varepsilon/\beta, \quad \omega = \varrho C_L^* S \beta/2m_0$$
$$u = V/c, \quad \mu = m/m_0, \quad \tau = \beta T/m_0 c^2.$$

Using the flight path angle γ as the new independent variable, we rewrite the state equations with nondimensionalized variables

$$d\xi/d\gamma = (\cos\gamma \cos\psi)/\omega\Lambda$$
$$d\eta/d\gamma = (\cos\gamma \sin\psi)/\omega\Lambda$$
$$d\omega/d\gamma = -(\sin\gamma)/\Lambda$$
$$du/d\gamma = [(\tau/\omega u\mu) - u(1+\lambda^2)/2E\mu]/\Lambda \tag{11}$$
$$d\psi/d\gamma = (\lambda \sin\phi)/\mu\Lambda \cos\gamma$$
$$d\mu/d\gamma = -\tau/\omega u\Lambda.$$

The small term $g \sin\gamma$ has been neglected in the equation for u. The parameter Λ is defined as

$$\Lambda = (\lambda/\mu) \cos\phi + G, \tag{12}$$

where

$$G = \frac{2m_0}{C_L^* \varrho S} [1/R - g/V^2] \cos\gamma. \tag{13}$$

The quantity G is varying but it has been observed by Loh (1968) that this term is insensitive to integration with respect to either ϱ or γ. Speyer and Womble (1971) have applied Loh's assumption to the optimization of a glide vehicle and have demonstrated using extensive numerical integration that even for the case of lift and bank modulation, Loh's second order theory is very accurate. This represents a significant improvement over the usual first order assumption used by Allen and Eggers (1958). Hence in this paper we shall treat G as a constant.

To write the variational equation for optimal trajectories we introduce the adjoint variables $p_i (i = 1, ..., 6)$, to form the Hamiltonian,

$$H = \frac{\cos\gamma}{\omega\Lambda} (p_1 \cos\psi + p_2 \sin\psi) - \frac{p_3 \sin\gamma}{\Lambda}$$
$$- \frac{p_4 u(1+\lambda^2)}{2\mu\Lambda E} + \frac{p_5 \lambda \sin\phi}{\mu\Lambda \cos\gamma} + \frac{\tau(p_4 - \mu p_6)}{\omega\mu u\Lambda}. \tag{14}$$

These adjoint variables are defined by

$$dp_1/d\gamma = 0$$
$$dp_2/d\gamma = 0$$
$$dp_3/d\gamma = \cos\gamma(p_1 \cos\psi + p_2 \sin\psi)/\omega^2\Lambda + \tau(p_4 - \mu p_6)/\omega^2 u\mu\Lambda$$
$$dp_4/d\gamma = p_4(1+\lambda^2)/2E\mu\Lambda + \tau(p_4 - \mu p_6)/\omega u^2 \mu\Lambda \tag{15}$$
$$dp_5/d\gamma = \cos\gamma(p_1 \sin\psi - p_2 \cos\psi)/\omega\Lambda$$
$$dp_6/d\gamma = -\lambda \cos\phi \cos\gamma(p_1 \cos\psi + p_2 \sin\psi)/\omega\mu^2\Lambda^2 +$$
$$+ p_3\lambda \cos\phi \sin\gamma/\mu^2\Lambda^2 - Gp_4 u(1+\lambda^2)/2E\mu^2\Lambda^2 +$$
$$+ Gp_5\lambda \sin\phi/\mu^2\Lambda^2 \cos\gamma + \tau(p_4 G + p_6\lambda \cos\phi)/\omega u\mu^2\Lambda^2.$$

The solution to the minimizing problem is obtained by integrating the system of state Equations (11) and adjoint Equation (15) with the appropriate end conditions, while selecting the controls τ, λ and ϕ subjected to the constraints (9), in such a manner that, at each instant, the Hamiltonian given by (14) is an absolute maximum.

3. Optimal Controls

The problem has three first integrals. From (14) we have

$$p_1 = a_1, \qquad p_2 = a_2, \tag{16}$$

where a_1 and a_2 are constants. They are identically zero if the longitudinal range and the lateral range are not specified. Next we write

$$dp_5/d\gamma = d/d\gamma(a_1\eta - a_2\xi).$$

Upon integration

$$p_5 = a_1\eta - a_2\xi + a_5, \tag{17}$$

where a_5 is another constant of integration.

To determine the optimal thrust magnitude control, we define the switching function

$$K = (p_4 - \mu p_6)/\Lambda. \tag{18}$$

To maximize H, we have the following optimal thrusting law.
If

    $K < 0$, select $\tau = 0$ (coasting phase)

    $K > 0$, select $\tau = \tau_{max}$ (boosting phase) $\tag{19}$

    $K = 0$ for a finite interval of time, select $\tau = $ variable thrust (sustaining phase).

Hence, from the point of view of thrusting, optimal trajectories consist of three types of arcs: the boosting arc, the coasting arc and the sustaining arc.

For aerodynamic force control, H is maximized at ϕ_{max} or at an interior point and at λ_{max} or an interior point. We shall use the terms normal bank and normal lift to designate the optimal controls at an interior point. These controls are obtained using the necessary conditions $\partial H/\partial \phi = 0$ and $\partial H/\partial \lambda = 0$. Explicitly, we have after lengthly algebraic manipulations the optimal control laws,

$$\sin\phi = Ep_5/p_4 u\lambda \cos\gamma \tag{20}$$

and

$$\lambda^2 = \left\{ 2G^2\mu^2 + \frac{2E\mu}{p_4 u}[p_3 \sin\gamma - (p_1 \cos\psi + p_2 \sin\psi)/\omega] + \right.$$

$$\left. + 1 - \tau/(p_4 - \mu p_6)/\omega u \right\} \pm$$

$$\pm 2G\mu \left\{ G^2\mu^2 + \frac{2E\mu}{p_4 u}[p_3 \sin\gamma - (p_1 \cos\psi + p_2 \sin\psi)/\omega] + 1 - \right.$$
$$\left. -\tau(p_4 - \mu p_6)/\omega u - E^2 p_5^2/p_4^2 u^2 \cos^2\gamma \right\}^{1/2}. \tag{21}$$

4. Solutions for the Free-Range Problems

In many key maneuvers where the terminal conditions in altitude and velocity as well as the heading and flight path angle of the vehicle are of prime importance, one does not place restrictions on the longitudinal and lateral ranges of the flight trajectory. For this case a more complete solution can be obtained. First we have

$$p_1 = a_1 = 0, \qquad p_2 = a_2 = 0, \qquad p_5 = a_5. \tag{22}$$

For a high thrust rocket engine operating over a short time interval, we can assume that the thrusting phase is conducted impulsively. In this case the variables ξ, η, ω, ψ and γ remain essentially unchanged along the short boasting arc. Replacing μ as the independent variable in the state equation for the velocity u,

$$du/d\mu = -1/\mu + O(1/\tau_{max}).$$

Using the assumption the $\tau_{max} \to \infty$, upon integration we obtain a relation for the impulsive boosting arc,

$$\mu e^u = C, \tag{23}$$

where C is a constant of integration. It remains to consider the coasting and sustaining arcs.

4.1. Coasting arc

Along a coasting arc $\tau = 0$, and $\mu = \mu_c = $ constant. Using (22) in the adjoint equation for p_3 we see that

$$p_3 = a_3 = \text{constant}. \tag{24}$$

Furthermore, one easily verifies that

$$d/d\gamma(p_4 u) = \tau(2p_4 - \mu p_6)/\omega u \mu \Lambda. \tag{25}$$

Hence, along a coasting arc

$$p_4 u = a_4 = \text{constant}. \tag{26}$$

For free range, coasting flight, the set of adjoint equations are integrated completely. The optimal control laws for the bank and lift control becomes

$$\sin\phi = b_5/\lambda \cos\gamma \tag{27}$$

$$\lambda^2 = [2G^2\mu_c^2 + 2\mu_c b_3 \sin\gamma + 1] \pm 2G\mu_c[G^2\mu_c^2 + 2\mu_c b_3 \sin\gamma + \\ + 1 - b_5^2/\cos\gamma]^{1/2}, \tag{28}$$

where

$$b_3 = Ea_3/a_4 \quad \text{and} \quad b_5 = Ea_5/a_4 \tag{29}$$

are two constants of integration to be evaluated using the given initial and terminal conditions. For a pure aerodynamic coasting arc, $\mu_c = 1$, and we have the results of Speyer and Womble (1971). Using Egger's assumption for the skip trajectory, we set $G = 0$ and obtain the classical result of Contensou (1965) which we see now is also valid in the extension to three dimensional flight. It was shown by Griffin and Vinh (1971) that when the terminal conditions on the state variables are free the constants are easily determined and when the terminal conditions are specified, the determination of the constants b_3 and b_5 involves the computation of elliptic integrals. For the case of small flight path angles the problem is reduced to solving a system of two algebraic equations.

4.2. SUSTAINING ARC

This type of arc is characterized by $K \equiv 0$ during a finite time interval. From the adjoint equation we see that p_3 is also a constant along this arc. Consider the equation $K = 0$ written as

$$p_4 = \mu p_6. \tag{30}$$

Since this relation is valid along the entire length of the sustaining arc, we take its derivative to have,

$$dp_4/d\gamma = \mu dp_6/d\gamma + p_6 d\mu/d\gamma.$$

Replacing the derivative by the appropriate expressions from the state and adjoint equations and rearranging, we have

$$p_4 u (1 + \lambda^2)(G + \Lambda/u)/2E = a_3 \lambda \cos\phi \sin\gamma + a_5 G\lambda \sin\phi/\cos\gamma. \tag{31}$$

Using this additional relation, we can eliminate $(p_4 u)$ in the optimal control Equations (20) and (21) with the result

$$\begin{aligned} [\mu\lambda \sin\phi \sin 2\gamma] \, a_3 + [(1 - \lambda^2) - 2G\mu\lambda \cos\phi] \, a_5 &= 0 \\ [\lambda^2 \sin\phi \cos\phi \sin 2\gamma] \, a_3 + [2G\lambda^2 \sin^2\phi - (1 + \lambda^2)(G + \Lambda/u)] \, a_5 &= 0. \end{aligned} \tag{32}$$

For nontrivial solutions in a_3 and a_5, the determinate of the coefficients of this homogeneous system must vanish. This yields the relation for sustaining flight

$$\mu\Lambda [(\lambda^2 + 1)/u - (\lambda^2 - 1)] = 0. \tag{33}$$

Since $\mu\Lambda \neq 0$, we obtain the normal lift control law along a sustaining arc in terms of the flight velocity

$$\lambda^2 = (u + 1)/(u - 1). \tag{34}$$

With this condition, system (32) is reduced identically to a single equation. Let

$$\sigma = a_3/a_5 \tag{35}$$

and rewrite the first Equation (32)

$$(\sigma\mu\lambda \sin 2\gamma) \sin \phi - (2G\mu\lambda) \cos \phi + (1 - \lambda^2) = 0. \tag{36}$$

Solving for the optimal bank angle,

$$\tan(\phi/2) = \frac{-\sigma\mu \sin 2\gamma \pm [\mu^2(\sigma^2 \sin^2 2\gamma + 4G^2) - 4/(u^2 - 1)]^{1/2}}{2[\mu G - 1/\sqrt{u^2 - 1}]}. \tag{37}$$

The thrust magnitude control along the sustaining arc is obtained by taking the derivative of (31). After some algebraic manipulation we have

$$\tau = \omega u^4/E(u - 1)(1 + u - u^2) - \\ - \omega u^2\mu(u^2 - 1) \sin \phi (\sigma \cos \phi \cos^2\gamma + G \sin \phi \tan \gamma)/(1 + u - u^2). \tag{38}$$

Using (37) for ϕ we see that the thrust magnitude control is also expressed in terms of the state variables and the constant parameter σ. We also notice that the adjoint variables p_4 and p_6 are obtained from (20) and (30), respectively, without further integration. Hence, the optimal control problem for the free-range case is completely solved.

5. Example of an Optimal Skip Maneuver

To illustrate the use of the general optimal control laws, consider a free range ($a_1 = = a_2 = 0$) optimal skip maneuver in which it is desired to minimize fuel consumption and satisfy the following set of initial and terminal conditions.
   At the initial time,

$$\gamma = \gamma_0, \qquad \omega = \omega_0, \qquad u = u_0, \qquad \mu = 1. \tag{39}$$

At the final time,

$$\gamma = \gamma_1, \qquad \omega = \omega_1, \qquad u = u_1. \tag{40}$$

To simplify the equations we apply Egger's assumption for the skip trajectory by taking $G = 0$. Since the final heading is unspecified we obtain from a transversality condition $a_5 = 0$ and it is clear that the entire flight is effected in a vertical plane containing the initial velocity.
   With $\phi = 0$, the three types of optimal subarcs for the normal lift problem are,
   (1) Impulsive boosting subarc,

$$\mu e^u = C. \tag{41}$$

   (2) Coasting subarc,

$$\lambda^2 = 1 + 2\mu_c b_3 \sin \gamma \tag{42}$$

$$\tau = 0. \tag{43}$$

(3) Sustaining subarc

$$\lambda^2 = (u+1)/(u-1), \qquad u > 1 \tag{44}$$

$$\tau = \omega u^4 / E(u-1)(1+u-u^2), \tag{45}$$

where it is necessary that $1 < u < (1+\sqrt{5})/2$ for a sustaining arc to exist.

Integrals are easily obtained during each phase of flight. During impulsive boost $\mu e^u = C$ and all other state variables remain unchanged. For the coast subarc we have $\eta = 0$, $\psi = 0$, $\mu = \mu_c$, $p_1 = a_1 = 0$, $p_2 = a_2 = 0$, $p_3 = a_3$, $p_4 u = a_4$, $p_5 = a_5 = 0$ and $p_6 = -a_3\omega/\mu_c + c_6$ where the ratio a_3/a_4 is determined from either of the two transcendental equations,

$$\omega_1 - \omega_0 = \mu_c \int_{\gamma_0}^{\gamma_1} \frac{\sin\gamma \, d\gamma}{\sqrt{1 + 2\mu_c E (a_3/a_4)\sin\gamma}} \tag{46}$$

$$\ln(u_1/u_0) = \frac{-1}{E} \int_{\gamma_0}^{\gamma_0} \frac{1 + \mu_c E(a_3/a_4)\sin\gamma \, d\gamma}{\sqrt{1 + 2\mu_c E(a_3/a_4)\sin\gamma}}. \tag{47}$$

Along a singular thrust arc, we have $p_1 = a_1 = 0$, $p_2 = a_2 = 0$, $p_3 = a_3$, $p_5 = 0$ as before with $p_4 = \bar{a}_4 - a_3\omega$ and $p_6 = p_4/\mu$ where we have noted that $dp_4/d\gamma = -a_3 \, d\omega/d\gamma$ while using the singular conditon, $dK/d\gamma = 0$.

Integrals of the state equations during sustaining flight are obtained using the relations for the optimal thrust, lift and thrust magnitude controls. We obtain for the altitude variation

$$\omega + E\mu \frac{(u-1)}{u} \sin\gamma = \bar{a}_4/a_3 \tag{48}$$

and flight path angle,

$$\frac{1}{E}(\gamma + c_1) = \sqrt{u^2 - 1}/u + \operatorname{arcsec} u - \log(u + \sqrt{u^2 - 1}) \tag{49}$$

using the state equation for the velocity. Variation of the mass ratio with velocity is

$$\mu = c_2/\sqrt{u^2 - 1}. \tag{50}$$

Due to the limited realm of velocities in which (44) and (45) are valid we shall rule out the existence of the singular arc so that the total trajectory is composed of only coasting and impulsive boosting subarcs. Furthermore, for the purpose of discussion we shall assume that K has at most one zero crossing so that multiple switching between boost and coast subarcs do not occur. Hence either the sequence (i) boost-coast or (ii) coast-boost is optimal. A zero crossing does not occur if the specified value of u_1 is less than u_1^* achieved by using only pure aerodynamic lift. In this case is possible to satisfy the specified terminal conditions with no fuel consumption. However, if

$u_1 > u_1^*$ then an impulsive thrust is applied to adjust the velocity to satisfy the terminal condition on u_1.

Finally, we can derive conditions for the optimal sequence of arcs. For example, let us find the condition such that we have the coast-boost sequence. Along the coasting phase $K = p_4 - \mu p_6 < 0$. Let us calculate K along the coast arc for $u \in [u_0, u^*]$. Using state and adjoint equations for ω and p_6 we have

$$p_6(\gamma) = -\frac{a_3}{\mu_c} [\omega(\gamma) - \omega(\gamma_0)] + p_6(\gamma_0) \tag{51}$$

with $p_4(\gamma) = a_4/u(\gamma)$, then

$$K = a_4/u + a_3 [\omega(\gamma) - \omega(\gamma_0)] + \mu_c p_6(\gamma_0). \tag{52}$$

At the final altitude, with $K = 0$, thrust is applied at $u = u_1^*$. Hence

$$\mu_c p_6(\gamma_0) = a_4/u_1^* + a_3(\omega_1 - \omega_0) \tag{53}$$

so that

$$K = a_4(u_1^* - u)/uu_1^* + a_3 [(\omega - \omega_0) - (\omega_1 - \omega_0)]. \tag{54}$$

Since coasting starts at the initial point, a condition is that $K < 0$ at $\gamma = \gamma_0$. Hence,

$$a_4 \left(\frac{u_1^* - u_0}{u_0 u_1^*} \right) < a_3 \Delta\omega. \tag{55}$$

The constant a_4 can be shown positive using the Legendre condition. To maximize H,

$$\sum_{ij} \frac{\partial^2 H}{\partial v_i v_j} \pi_i \pi_j \leqslant 0 \tag{56}$$

must be satisfied for all variations π_i and π_j where $v = [\lambda, \phi]$. Hence, the condition for coast boost becomes

$$\frac{a_3}{a_4} \Delta\omega > (u_1^* - u_0)/u_0 u_1^*, \tag{57}$$

where $\Delta\omega = \omega(\gamma_1) - \omega(\gamma_0)$ is prescribed. The ratio (a_3/a_4) is determined through computation of the elliptic integral (46) with $\mu_c = 1$. It is then used in (47) to evaluate u_1^*. This value of u_1^* must satisfy (57) for the coast boost trajectory.

As a particular case when the prescribed final altitude equals the initial altitude, $\Delta\omega = 0$ and the condition for coast-boost is $u_1^* < u_0$.

References

Allen, H. and Eggers, A., Jr.: 1958, 'A Study of the Motion and Aerodynamic Heating of Missles Entering the Earth's Atmosphere at High Supersonic Speeds', NACA Report No. 1381.

Busemann, A., Vinh, N. X. and Kelly, G. F.: 1969, 'Optimum Maneuver of a Skip Vehicle with Bounded Lift Constraints', *Journal of Optimization Theory and Applications* 3, No. 4, 243–262.

Contensou, P.: 1965, 'Contribution à l'étude schématique des trajectoires semi-balistiques a grande Portée', Paper Communicated to the *Association Technique Maritime et Aeronautique*, Paris, France.

Fave, J.: 1968a, 'Approche analytique du problème du domain accessible à un planeur orbital', *La Recherche Aérospatiale*, No. 124.

Fave, J.: 1968b, 'Domain accessible au Sol par un planeur orbital rentrant dans l'atmosphère', Communication présentée au *6 Congrès International des Sciences Aéronautiques* (I.C.A.S.), Munich.

Griffin, J. W.: 1973, 'Optimal Three Dimensional Maneuvers of Rocket Powered Hypervelocity Vehicles With Lift, Bank Angle and Thrust Controls', Ph.D. Dissertation, Dept. of Aerospace Engineering, the University of Michigan, Ann Arbor, Michigan.

Griffin, J. and Vinh, N. X.: 1971, 'Three Dimensional Optimal Maneuvers of Hypervelocity Vehicles', AIAA paper No. 71-920 presented at the *AIAA Guidance, Control and Flight Mechanics Conference*, Hofstra University, Hempstead, New York.

Leitmann, G.: 1956, 'A Calculus of Variations Solution of Goddard's Problem', *Astronautica Acta* 2, No. 2.

Loh, W. (ed.): 1968, 'Entry Mechanics and Dynamics', in *Re-Entry and Planetary Entry Physics and Technology*, Springer-Verlag, New York Inc.

Miele, A.: 1958, 'General Variational Theory of the Flight Paths of Rocket-powered Aircraft, Missile, and Satellite Carriers', *Astronautica Acta* 4, No. 4.

Speyer, J. and Womble, M.: 1971, 'Approximate Optimal Re-Entry Trajectories', AIAA paper No. 71-919 presented at the *AIAA Guidance, Control and Flight Mechanics Conference*, Hofstra University, Hempstead, New York.

OPTIMIZATION OF ALTITUDE AND INCLINATION
CHANGE SCHEDULE DURING LOW THRUST ASCENT
TO GEOSYNCHRONOUS ORBIT

DONALD V. MCMILLEN

CTS Project, Communications Research Centre, Shirley Bay, Ottawa, Ontario K1N8T5, Canada

and

ROBERT E. LOHFELD

Sherman Fairchild Technology Center, Germantown, Md. 20767, U.S.A.

1. Statement of the Problem

Mission requirements for an advanced program* called for placement of a scientific satellite in a geosynchronous orbit with sufficient power to operate a 2 kW communications experiment for a minimum of 2 yr. Early tradeoff studies indicated significant cost savings and increased payload could be obtained if a low cost booster were used to place the spacecraft initially in a near Earth inclined orbit. The spacecraft would then spiral slowly out to geosynchronous orbit using low thrust ion engines.

The power requirements for ion engines (on the order of 15–25 kW N^{-1}) are large. Severe payload weight limitations for a Thor-Delta booster indicate the use of large solar arrays (on the order of 9–12 m^2 kW^{-1}) to provide this power. Radiation effects in the Van Allen Belt degrade the performance of the solar arrays requiring increased array area at launch to assure adequate end of life power.

The primary goals of the investigation were to: (1) minimize fuel consumption during transfer from near earth circular inclined (28.5°) orbits to a geosynchronous orbit using 30 cm mercury bombardment ion engines (2) minimize effects of solar array performance degradation in the Van Allen Belt; (3) minimize solar array area consistent with having approximately 5 kW of power available at the beginning of life in a geosynchronous orbit; (4) optimize initial orbit altitude and spacecraft weight and thus to maximize payload in a geosynchronous orbit.

2. Analytical Approach

Analysis of the spiral ascent trajectory was begun after comparing existing techniques (Cowell's method, Encke's method and Variation of Parameters method).

Cowell's method is based upon direct integration of the equations of motion. This approach utilizes rectangular coordinates and requires a small integration step to limit computer roundoff errors. It is unsatisfactory for modelling low thrust perturbations.

* NASA Contract NAS3-14360 for ATS Advanced Mission Study.

L. G. Napolitano et al. (eds.), Astronautical Research 1972, 55–67. All Rights Reserved
Copyright © 1973 by D. Reidel Publishing Company, Dordrecht-Holland

Encke's method also utilizes rectangular coordinates and integrates the low thrust perturbation accelerations to obtain deviations from the Keplerian orbit. This approach requires many integration steps and can build up significant errors caused by roundoff and truncation in the computer.

The Variation of Parameters method is the most satisfactory technique for modelling low thrust perturbations. Six orbital parameters are chosen to fully describe the motion and rectangular coordinates are not used. The parameters are:

Ω – longitude of ascending node
ω – argument of perigee
θ – true anomaly
i – orbital inclination
a – semimajor axis of elliptical orbit
e – orbital eccentricity.

The variations in the Keplerian elements with respect to the perturbing accelerations R, T and W are defined by the Lagrange Planetary equations. R, T and W are the components of acceleration along three mutually perpendicular coordinates. R is along the radius vector r from the Earth's center to the spacecraft, T is perpendicular to R in the osculating plane nominally in the direction of motion and W is perpendicular to the osculating plane forming a right hand set. (See Figure 1).

3. Planetary Equations

The Lagrange equations are:

$$\frac{da}{dt} = \frac{2}{n\sqrt{1-e^2}} \left(e \sin\theta \cdot R + \frac{p}{r} \cdot T \right)$$

$$\frac{de}{dt} = \frac{\sqrt{1-e^2}}{na} \left(\sin\theta \cdot R + (\cos\theta + \cos E) \cdot T \right)$$

$$\frac{di}{dt} = \frac{r}{na^2\sqrt{1-e^2}} \frac{(\omega+\theta)}{} \cdot W$$

$$\sin i = i\frac{d\Omega}{dt} = \frac{r \sin(\omega+\theta)}{na^2\sqrt{1-e^2}} W$$

$$\frac{d\tilde{\omega}}{dt} = 2 \sin^2\frac{i}{2}\frac{d\Omega}{dt} + \frac{\sqrt{1-e^2}}{nae}\left[-\cos\theta \cdot R + \sin\theta\left(1 + \frac{r}{p} \right) \cdot T \right]$$

$$\frac{d\varepsilon}{dt} = \frac{-2r}{na^2} \cdot R + \frac{e^2}{1+\sqrt{1-e^2}}\frac{d\tilde{\omega}}{dt} + 2\sqrt{1-e^2}\sin^2\frac{i}{2}\frac{d\Omega}{dt}.$$

Terms in the above equations, not previously defined, are:

n – mean motion $= \sqrt{\mu/a^3}$ (rad s^{-1})

p – semi-parameter $= a(1-e^2)$ (km)
E – eccentric anomaly $= \cos^{-1}\left[(e+\cos\theta)/(1+e\cos\theta)\right]$ (rad)
r – radius to the spacecraft $= p/(1+e\cos\theta)$ (km)
$\tilde{\omega}$ – longitude of the pericenter $= \omega + \Omega$
ϵ – mean longitude of epoch $= M + \tilde{\omega}$
M – mean anomaly $= E - e\sin E$ (rad)

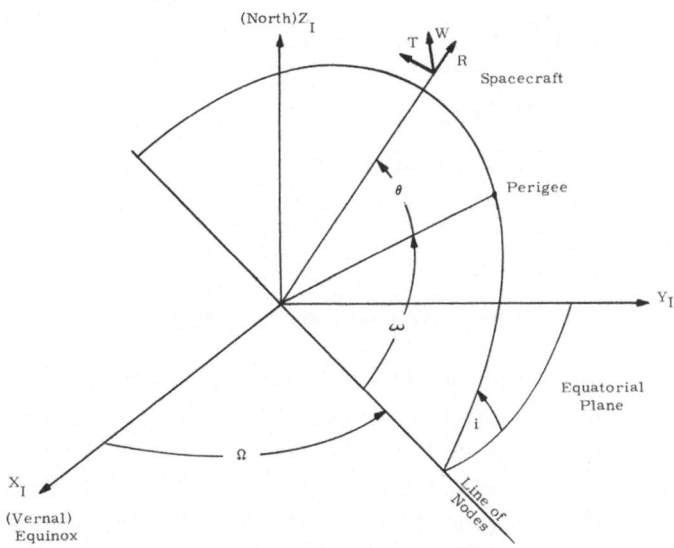

Fig. 1.   Definition of quantities.

For convenience, the planetary equations can be made functions of the true anomaly (θ) by multiplying them by: $dt/d\theta = r^2/na^2\sqrt{1-e^2}$ (s rad^{-1}). The resulting equations give the change in orbital elements with respect to the change in true anomaly.

Defining a nondimensional acceleration as:

$$A_c^* = Ac/g_a \ (-),$$

where A_c is the actual acceleration (F/m) and g_a is the gravitation acceleration at the distance corresponding to the semimajor axis,

$$g_a = \frac{\mu}{a^2} = n^2 a \ (\text{m s}^{-2})$$

thus $R = R^* n^2 a$, etc.

Substituting these relationships into the original planetary equations gives:

$$\frac{da}{d\theta} = \frac{2aeA\sin\theta}{B^2}R^* + \frac{2aA}{B}T^*$$

$$\frac{de}{d\theta} = A^2\frac{(\cos\theta + \cos E)}{B^2}T^* + \frac{A^2}{B^2}R^*$$

$$\frac{d\Omega}{d\theta} = \frac{A^2 \sin(\omega+\theta)}{\sin i B^3} W*$$

$$\frac{di}{d\theta} = \frac{(\omega+\theta)}{B^3} W*$$

$$\frac{d\tilde{\omega}}{d\theta} = \frac{-A^2 \cos\theta}{eB^2} R* + \frac{A^2(1+B)\sin\theta}{e}\frac{}{B^3} T* + 2\sin^2\frac{i}{2}\frac{d\Omega}{d\theta}$$

$$\frac{d\varepsilon}{d\theta} = \frac{e}{1+\sqrt{A}}\frac{A^2}{B^3} T* - \left[\frac{2A^{5/2}}{B^3} + \frac{eA^2}{(1+\sqrt{A})B^2}\frac{\cos\theta}{}\right] R* +$$

$$+ 2A^{1/2}\sin^2\frac{i}{2}\frac{d\Omega}{d\theta}.$$

A special short hand notation added above is:

$$A = 1 - e^2$$
$$B = 1 + e\cos\theta.$$

4. Integrated Planetary Equations

The Planetary Equations were programmed and several computer runs were made integrating them using a fourth order Runge-Kutta technique. To keep the errors small over many orbit revolutions (from 100–2000) a very small integration step was required resulting in very long computer times per run. To avoid this problem, the variational equations were integrated by hand and the closed form solution use in the simulation wherever possible. Three computation steps per orbit produced a Δe buildup of 1.0×10^{-5}. Δe did not decrease rapidly with an increase in number of steps reaching only about 3.5×10^{-6} at 50 steps. Simulation time was greatly reduced by using three steps per orbit at negligible sacrifice in accuracy.

5. Ascent Time, Δa and Δe only, no Van Allen Belt, no Eclipse

A series of sumulation runs were made using the closed form solution for several constant thrust and mass values calculating Δa and Δe only. These simulations gave an indication of ascent time required and number of spiral orbits for different thrust levels. The results for Δa vs time and thrust are shown in Figure 2.

A comparison case for constant thrust and mass was simulated which included a crude estimate of the Van Allen Belt degradation, over an altitude range of 5556 to 14816 km, thrust was decreased linearly with altitude over the Van Allen Belt until 50% of the starting thrust level remained. This thrust level was then held constant until geosynchronous altitude was reached. Ascent time and number of spiral orbits to reach geosynchronous altitude increased, as was expected. The percentage of time spent in the Van Allen Belt region (35%) was approximately equal to the percentage of time spent in the same altitude range without considering Van Allen Belt degradation. (See Figure 3).

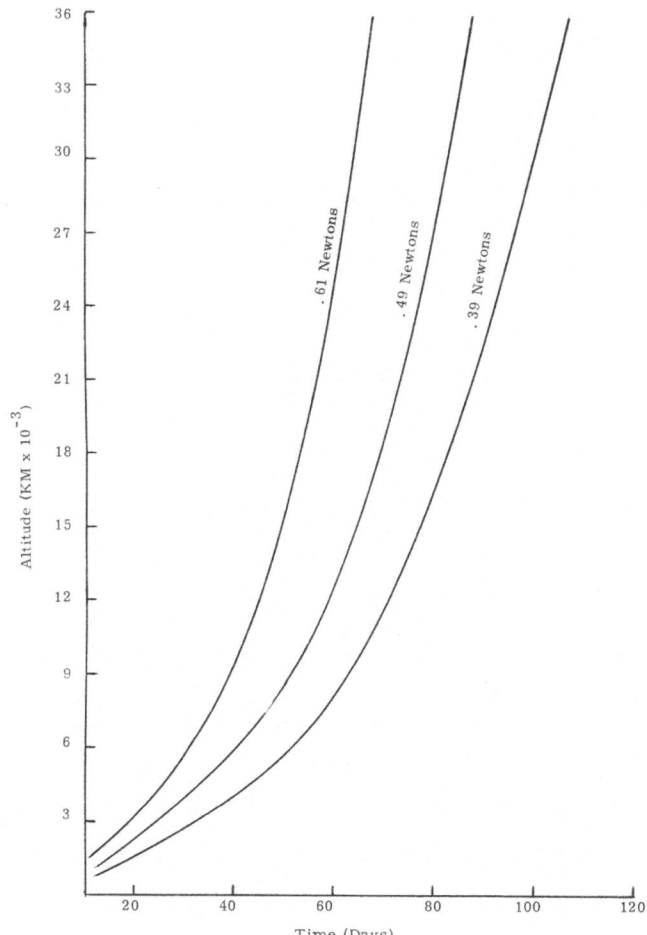

Fig. 2.   Ascent time, Δa and Δe only, no Van Allen Belt.

Fig. 2.   Ascent time, Δa and Δe only, no Van Allen Belt.

6. Block Diagram – Spiral Ascent Programm

The thrust to produce Δa and Δi comes from the ion engines. The power for the engines comes from the solar arrays which convert the solar energy into electrical energy at an efficiency (λ_{pc}) of about 8–12%. For this study, various initial power valves were chosen up to 15 kW. The performance of the solar cells is affected by bombardment of electrons and protons in the Van Allen Belt. The efficiency (λ_{PDVAB}) decreased with absorbed particles. Solar cell efficiency (λ_{pt}) increases with decreasing solar array temperature as the spacecraft altitude increases. The power actually produced by the solar cells is then divided between the spacecraft systems (Psc) and the power conditioning for the ion engines (λ_{pc}). The trust produced by the ion engines (F_t) is a function of the engine current (I_5) and voltage (V_5). Holding the voltage constant at 1000 V causes the thrust (F_t), electrical efficiency (λ_E), mechanical fuel usage efficiency

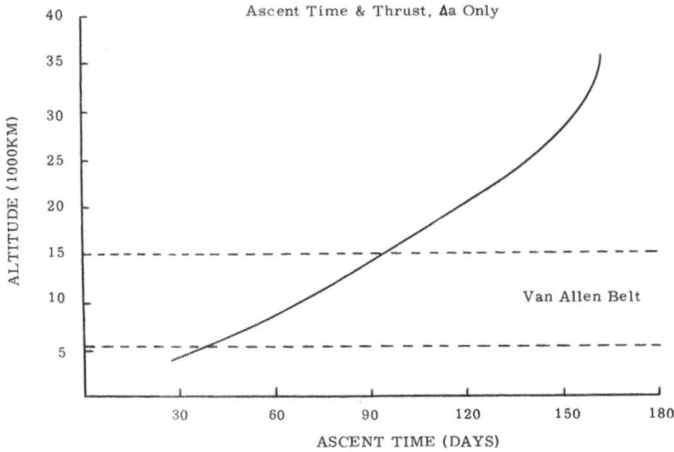

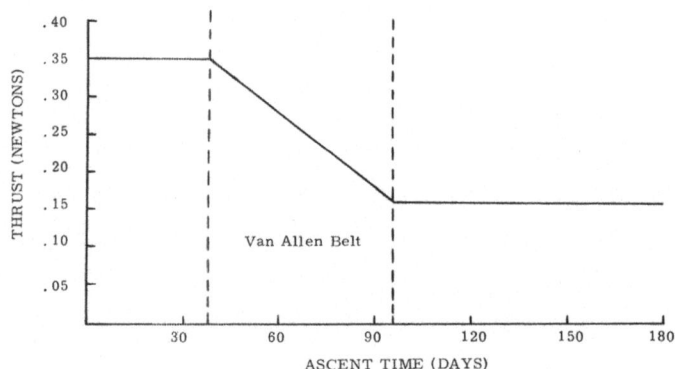

Fig. 3.   Ascent time as a function of thrust and altitude.

(λ_m) and specific impulse (I_{sp}) to vary with the current. The thrust is applied tangentially (T) to the orbit to produce Δa and normal to the orbit (W) to change Δi. The electron bombardment rate in the Van Allen Belt $(\dot{\Phi})$ is integrated (Φ) by the solar array decreasing efficiency (λ_{PDVAB}). (See Figure 4).

7. 1-MeV Equivalent Fluence

Data for the 1-MeV Equivalent Fluence $(\dot{\Phi})$ was obtained from tables listed in NASA SP-3024 (models of the trapped radiation environment). The worst case model for the magnetic equator was selected and programmed into the simulation. When the data is plotted on log-log paper with altitude increasing vertically and $\dot{\Phi}$ increasing to the right as in Figure 5, the graph has the appearance of an angry shark. This simplified model was chosen to reduce the simulation run time. The complete set of tabular data incorporated as an interpolation and table-look-up subroutine would have required more computer capacity than was available at the time the study was

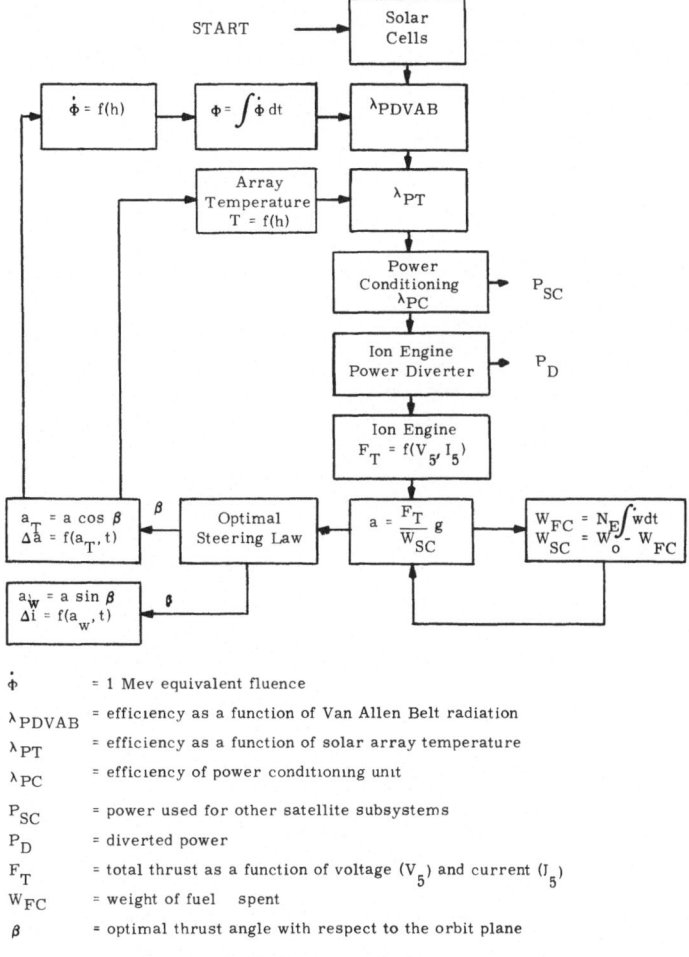

$\dot{\Phi}$ = 1 Mev equivalent fluence

λ_{PDVAB} = efficiency as a function of Van Allen Belt radiation

λ_{PT} = efficiency as a function of solar array temperature

λ_{PC} = efficiency of power conditioning unit

P_{SC} = power used for other satellite subsystems

P_D = diverted power

F_T = total thrust as a function of voltage (V_5) and current (I_5)

W_{FC} = weight of fuel spent

β = optimal thrust angle with respect to the orbit plane

Fig. 4.   Block diagram, spiral ascent program.

performed. The 'inner belt' peak $\dot{\Phi}$ is about 5×10^{14} at 2400 km and the 'outer belt' peak $\dot{\Phi}$ is about 6.6×10^{14} at 5200 km. (See Figure 5).

The effects of trapped radiation on solar cells varies with the cell design. Integrated fluence (Φ) data effects on the performance of the 10 Ω-cm cells as used on INTELSAT III were used during the early phases of the study. At $\Phi = 10^{15}$ electron cm^{-2} power is down by about 24% and at $\Phi = 10^{16}$ electron cm^{-2}, the power is down by about 44% for this cell. (See Figure 5).

Power output of the solar array increases as the altitude increases and the stable temperature decreases. A graph of the efficiency (λ_{pt}) for a solar array with equal front and back surface absorptivities (0.7) and emissivities (0.95) (representative numbers) shows an increase from about 0.82 at 1000 km to about 0.96 at 35 786 km (synchronous altitude). The data fits a curve of the form $\lambda_{pt} = ah^b$ ($<$) where h = altitude.

The efficiency of the power conditioner (λ_{pc}) was assumed to be 0.85. Its purpose is to convert 'raw' solar array power to the proper voltages required to operate the ion engine.

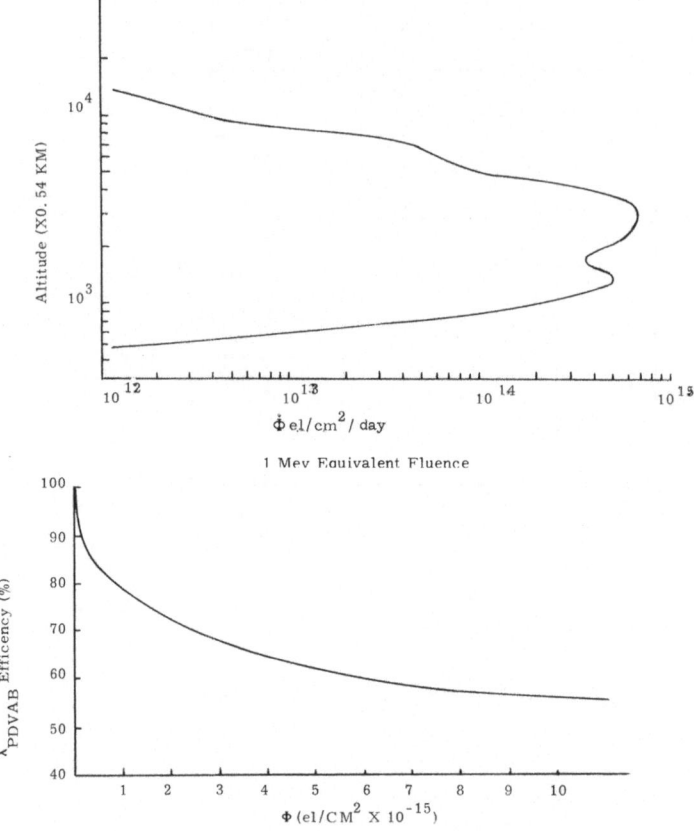

Fig. 5.   Solar cell efficiency vs integrated fluence.

8. Ion Engine Characteristics at Constant Voltage

Performance data for the 30 cm mercury bombardment ion engine is somewhat better at a constant voltage rather than at constant current. A nominal value of 1000 V (representative) was selected. The thrust (F_t) is linear with current ranging from 0.134 N at 2 A to 0.036 N at 0.5 A. Actually, $F_t = C\sqrt{V_5}\, I_5$ (N) where $C \approx 2 \times 10^{-3}$ N $V^{-1/1}$ A^{-1}. Electrical efficiency (λ_e) is approximately a quadratic function of current I as is the mechanical (mass utilization) efficiency (λ_m). The specific impulse (I_{sp}) is:

$$I_{sp} = C_1 \lambda_m \sqrt{V_5}\ \text{(s)},$$

The fuel consumption rate $(\dot{w})$ is determined by: $\dot{w} = C_2(F_t/I_{sp})$, where, for the ion engine, $C_2 = C/C_1 \approx 0.179$ for units of kg day^{-1}. (See Figure 6).

The maximum number of ion engines to be utilized is a function of available space as well as power and weight. For the design study, considering the Delta shroud and spacecraft configuration a maximum number of 5 engines could be utilized. Locating one engine in each corner of an imaginary 'square' and one engine in the center of the

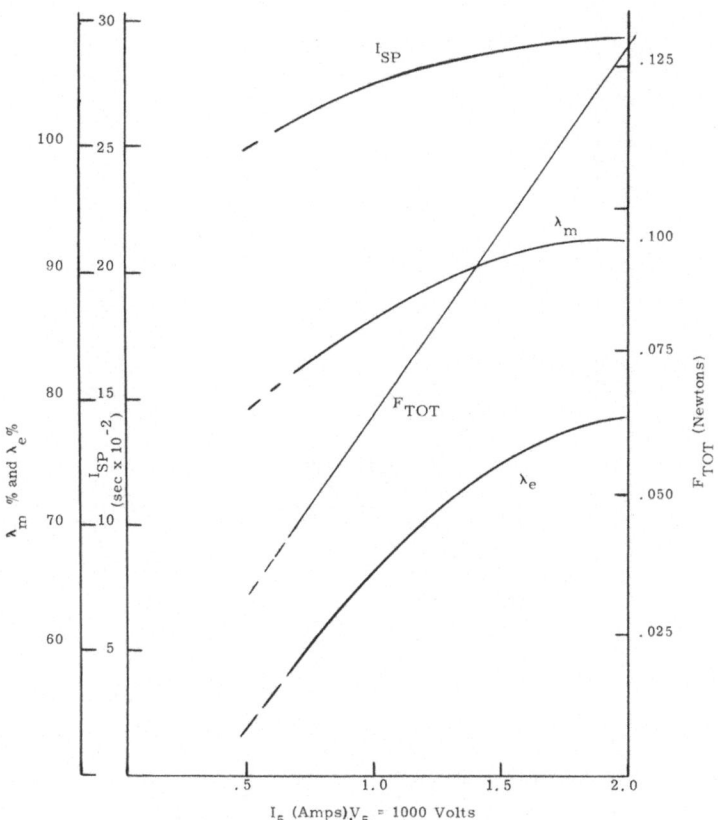

Fig. 6. Ion engine performance.

'square' allows switchover from N to $N-1$ engines in all cases. Switchover is required when the power available for N engines at reduced performance equals the power required by $N-1$ engines at maximum performance.

9. Switchover at Constant Power into the Engine (P_E)

It is more efficient, in terms of power usage, to switch at constant P_E. This results in a sudden jump in thrust and current (power) inside the engine (P_5). Switching at a constant P_5 would keep thrust constant but would require diverting power from P_E (on the order of 0.3 kW per engine and gradually returning it to P_E as the solar array power output degrades until the maximum available output is again being utilized. (See Figure 7).

10. Computer Results, Δa Only, No Eclipse

A series of simulation runs were made with all the previous mathematical models incorporated over a wide range of initial weights (408–1362 kg), powers (5.5–15 kW) and altitudes (1019–18920 km). These runs were for the Δa case only (no Δi, no

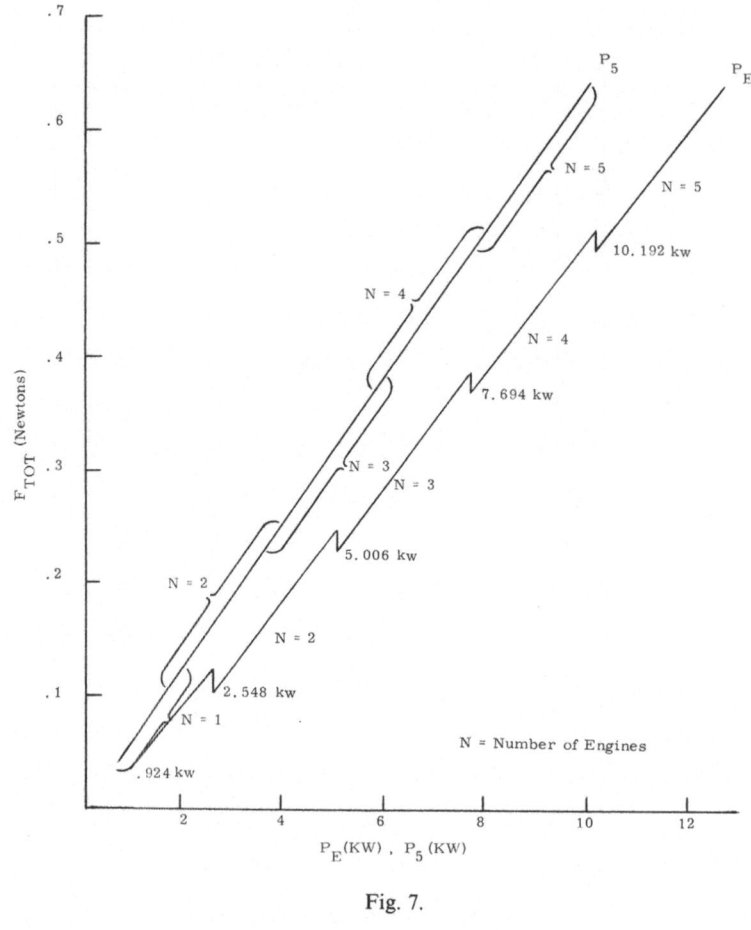

Fig. 7.

Fig. 7.

eclipse). The results indicated that the solar arrays degraded more than 50% for near Earth initial altitudes with the total degradation being about equal for all initial altitudes below 4000 km. The mass ratio of fuel consumed to initial weight is also nearly constant with initial altitude ranging from 0.141 at 1019 km to 0.033 at 18 520 km. There is a slight indication of minimum mass ratio with initial power of 13 kW over 9, 11 and 15 kW. This is probably due to the fact that at 13 kW, the initial engine complement is 4 which drops to 3 after a few orbits and the I_{sp} remains slightly higher for a longer period of time. The total fuel saving is on the order of 0.1% of initial weight. Ascent times range from 482 days (1019 km, 1362 kg, 9 kW) to 20 days (19 520 km, 408 kg, 5.5 kW). (See Figure 8.)

11. Optimization of Altitude and Inclination Charge

Low thrust optimization of simultaneous altitude and inclination changing has been treated by Edelbaum (1961) and others. Under the assumption that the orbits

COMPUTER RESULTS, Δa ONLY, NO ECLIPSE

Initial Power (KW)	Initial Altitude (KM)	Weight of Fuel Consumed (Kilograms) Initial Weight (KG)				Initial Altitude (KM)	Ratio Of Power Lost/Initial Power Initial Weight (KG)				Initial Altitude (KM)	Ascent Time (Days) Initial Weight (KG)			
		680	907	1134	1361		680	907	1134	1361		680	907	1134	1361
11 kw	4630	67.7	90.1	112.8	135.6	4630	.490	.515	.534	.548	4630	98.2	138.5	181.6	226.1
	3704	73.5	97.9	123.0	147.6	3704	.513	.537	.555	.569	3704	111.5	157.6	207.5	258.9
	2778	80.2	107.4	134.2	161.7	2778	.527	.550	.568	.582	2778	124.5	177.0	231.9	290.8
	1852	88.2	117.5	147.4	177.8	1852	.541	.564	.581	.595	1852	140.9	199.3	262.5	329.4
	1019	96.1	127.8	160.5	193.1	1019	.542	.565	.582	.595	1019	142.3	209.2	275.2	344.2
		680	907	1134	1361		680	907	1134	1361		680	907	1134	1361
13 kw	4630	69.7	91.4	111.2	133.9	4630	.478	.503	.520	.535	4630	86.4	117.5	144.2	179.4
	3704	74.8	97.0	121.4	145.4	3704	.501	.524	.541	.555	3704	95.2	125.6	163.7	202.9
	2778	80.1	106.3	[132.2]	158.6	2778	.514	.537	[.553]	.567	2778	101.4	140.5	[182.1]	226.5
	1852	87.2	116.2	144.8	173.4	1852	.527	.550	.556	.580	1852	112.3	157.8	205.0	254.9
	1019	95.6	126.7	158.1	189.9	1019	.529	.551	.568	.581	1019	119.3	166.2	215.9	268.1
		680	907	1134	1361		680	907	1134	1361		680	907	1134	1361
15 kw	4630	68.0	90.3	113.2	136.7	4630	.462	.488	.508	.524	4630	68.5	96.8	127.4	160.1
	3704	74.7	98.6	124.1	149.0	3704	.486	.511	.531	.546	3704	78.9	110.8	146.6	183.2
	2778	81.2	108.5	135.0	162.8	2778	.501	.526	.545	.559	2778	87.9	125.1	163.5	205.3
	1052	89.0	118.3	148.7	176.0	1852	.516	.541	.559	.571	1852	99.3	140.6	185.8	221.1
	1019	97.2	129.4	162.1	191.4	1019	.517	.542	.560	.572	1019	105.1	148.6	195.4	231.4

Fig. 8.   Computer simulation of Edelbaum's technique.

remain quasi-circular ($e \approx 0$) and that the thrust angle β (actually, the absolute magnitude of the thrust angle) with respect to the orbit plane remains constant for each orbit, Edelbaum found a simple formulation for optimization. His results are:

$$V \sin \beta = V_0 \sin \beta_0 \ (\mathrm{m\ s}^{-1})$$

and

$$\Delta i = \frac{2}{\pi}(\beta - \beta_0)\ (\mathrm{rad}),$$

where V = velocity in circular orbit.

β_0 can be determined by rearranging the Δi equation to read:

$$\beta_0 = \tan^{-1} \frac{\sin \frac{\pi}{2} \Delta i}{\dfrac{V_0}{Vf} - \cos \frac{\pi}{2} \Delta i} \ (\mathrm{rad}),$$

where $V_0/V_f = \sqrt{a/a_0}\ (-)$.

The subscripts '0' and 'f' denote initial conditions and final conditions respectively.

There are limitations to this technique which Edelbaum did not define. Two obvious ones are: $\beta_f \leqslant \pi/2$ rad (90°) and $h_f > h_0 > 0$.

By suitable mathematical manipulations the following can be defined:

$$(\beta_0)_{\beta_f = \pi/2} = \frac{\pi}{2}(1 - \Delta i)\ (\mathrm{rad})$$

$$= \sin^{-1}\left(\frac{V_f}{V_0}\right)\ (\mathrm{rad})$$

$$(a_0)_{\beta_f = \pi/2} = a_f \left(\cos \frac{\pi}{2} \Delta i\right)^2 \text{ (km)}$$

$$(h_{0\,\text{max}})_{\beta_f = \pi/2} = (Re + h_f)\left(\frac{\pi}{2} \Delta i\right)^2 - Re \text{ (km)}$$

$$(\Delta i_{\text{max}})_{\beta_f = \pi/2} = \frac{2}{\pi} \cos^{-1}\left(\frac{V_f}{V_0}\right) \text{ (rad)}.$$

For the case of $\Delta i = 28.5°$ (ETR launch): $h_{0\,\text{max}} = 14{,}875$ km and $\beta_{0\,\text{max}} = 45.232°$.

12. Computer Simulation of Edelbaum's Method

Simulation runs were made for Δa and Δi covering the same initial conditions as used for the Δa only cases. One sample case (2778 km, 1135 kg, 13 kW) is presented. β_0 was 26.1°, β_f was 71°, ascent time was 255.7 days and 179.8 kg of fuel were consumed. (See Figure 9.)

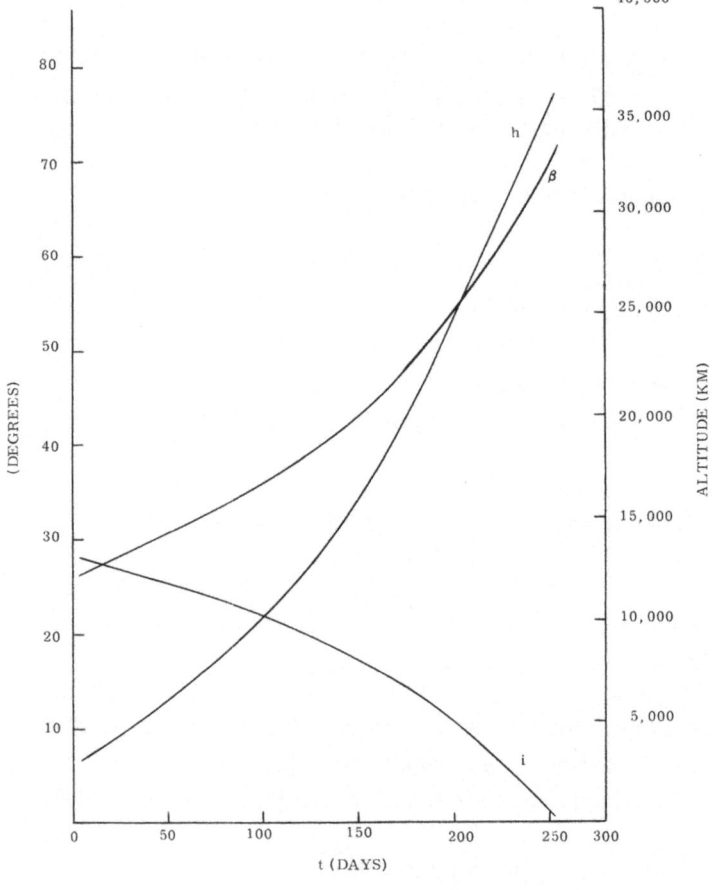

Fig. 9.

Fig. 9.

New data on the (0.0034 in.) 2 Ω-cm solar cells indicated superior resistance to radiation damage at high values of integrated fluence. They produce less power initially than thicker 10 Ω-cm cells (0.012 in.) but degrade less.

A simulation run was made using the 2 Ω-cm cells at the same initial conditions as the previous case (2778 km, 1135 kg, 13 kW). Results show less degradation, decreased ascent time and more thrust available. Fuel requirements are unchanged.

Reference

Edelbaum, T. N.: 1961, 'Propulsion Requirements for Controllable Satellites', *ARS Journal* August 1961, 1079.

COMPARATIVE STUDY OF REGIONAL HEMODYNAMICS DURING TILT TEST AND LOWER BODY NEGATIVE PRESSURE EXPOSURE

KH. KH. YARULLIN, T. N. KRUPINA, T. D. VASILYEVA, and D. A. ALEKSEYEV

Academy of Sciences of the U.S.S.R., Moscow, U.S.S.R.

One of the major adverse effects of weightlessness upon the cardiovascular system is a decline of orthostatic tolerance. In recent years (Berry, 1969, 1970; Suvorov *et al.*, 1972; Musgrave *et al.*, 1971) orthostatic tolerance has been estimated using both tilt tests and lower body negative pressure (LBNP) exposures. However, many problems regarding similarity or dissimilarity of the mechanisms and possible effects of these two tests upon the cardiovascular system still remain in doubt.

There is little literature clarifying variations in the regional, particularly cerebral and pulmonary hemodynamics. The present paper summarizes the results of rheographic studies of the cerebral, pulmonary and peripheral hemodynamics in 24 healthy male test subjects, aged 22–30, exposed to tilt tests and LBNP.

Orthostatic tolerance was evaluated in 20 min tilt tests (at an angle of $+70°$) and 10 min LBNP (at a pressure of -40 mm Hg) tests in the recumbent position. Before, during and after the tests rheoencephalographic data was recorded from frontomastoidal and bimastoidal leads (reflecting the hemodynamic condition in the basin of the internal carotid artery and the system of the basilar and vertebral arteries respectively); rheograms of the right lung, forearm and shin as well as photoplethysmograms of the hand and shin were also obtained. The measurements were made by two four-channel transistored rheographs based on a demodulation alternating current bridge. The working frequency of the generator of rheographs was 120 kHz, current strength 2.5 mA and voltage 3 V. Rheograms were recorded in the 10-channel electroencephalograph at a constant time of 1 s. Electrodes used to record right pulmonary rheograms were fixed to the chest in the region of the 2nd intercostal space along the mid-clavicular line and to the back at the level of the IV–VII thoracic vertebrae near the blade angle; electrodes to record rheograms of the forearm, hand and shin were fixed in the longitudinal leads. Pulmonary rheograms were recorded during breath retention between inhalation and exhalation.

The state of blood-filling, elasticity and tone of cerebral, pulmonary and peripheral vessels was estimated on the basis of visual data and quantitative analysis of main rheographic parameters. The relative value of pulse blood-filling of the regions tested was determined using the maximum value of rheographic waves measured as ohm fractions. The ratio of the anacrotic phase of the curve (α) to the cardiac cycle time (T), i.e. $\alpha/T\%$, was also measured. The outline analysis of rheograms allowed determination of the dicrotic index (the ratio of the amplitude at the incisure height to the maximal amplitude of the rheogram) indicating mainly the tone of arterioles (Donzelot *et al.*, 1950; Chlebus, 1962; Eninya, 1967; Yarullin, 1967). The diastolic index (the ratio of the amplitude at the dicrotic wave peak to the maximal amplitude of the rheoencephalogram) showing the state of the blood return and venous tone was also estimated (Yarullin and Levchenko, 1969).

Our studies revealed significant individual variations in the changes of regional hemodynamics. Despite these variations, the majority of the test subjects showed a common trend. Twenty one subject displayed a decline of the REG amplitude by 20–30% during tilt test and by 41% during LBNP as compared to the baseline values (Figure 1). This decline of the pulse blood-filling was followed by a decrease of the

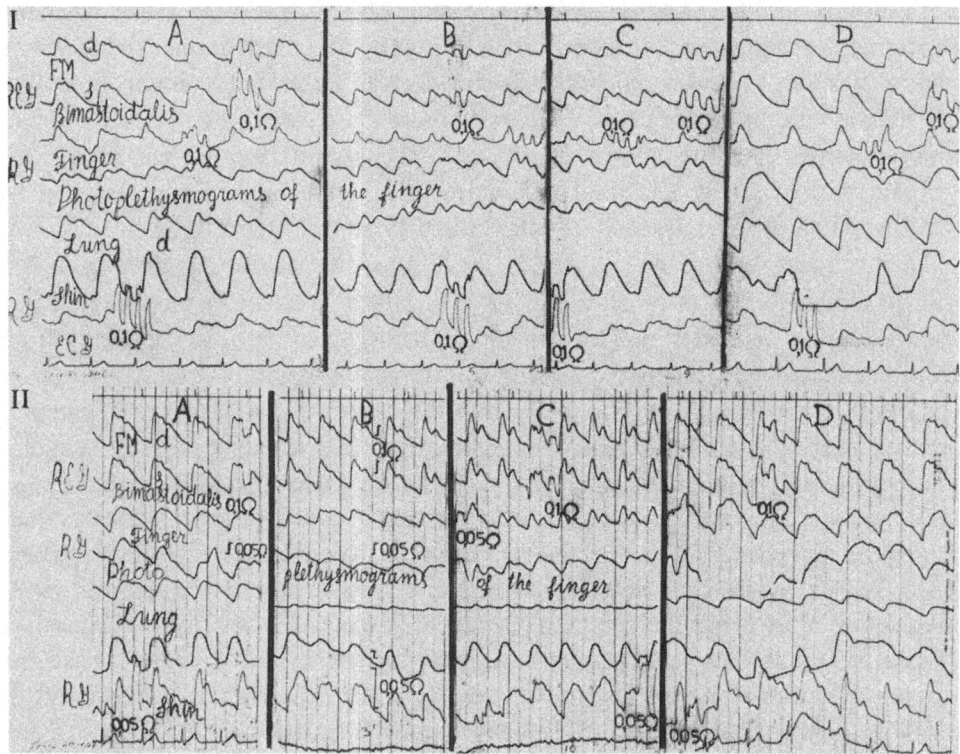

Fig. 1.   Changes in the rheogram of the test subject G, during lower body negative pressure (1) and during tilt (2): 1 – REG of the right hemisphere; 2 – REG of the left hemisphere; 3 – Bimastoidal REG; 4 – rheogram of the finger; 5 – Photoplethysmogram of the finger; 6 – rheogram of the right lung; 7 – rheogram of the shin; 8 – EKG in the 2nd lead. (I) A – reference level; B – the 5th min of LBNP; C – the 9th min of LBNP; D – the 1st min of the recovery. (II) A – reference level; B – the 3th min of tilt; C – the 10the min of tilt; D – the 1st min of the recovery.

dicrotic and diastolic indices during tilt test and LBNP by 33% and 48.9%, respectively. The reduction of the venous tone was noted in 21 subjects during tilt test and in 14 subjects during LBNP. It is obvious that a decrease of the blood-filling and tone of cerebral vessels, mainly arterioles and veins, was more distinct during LBNP than during tilt test. In 7 test subjects the pattern of responses of cerebral veins to

LBNP differed from the typical reduction of the vascular tone: they showed a noticeable increase of the venous tone.

As known, the level of reactive hyperemia developing in response to the deficient blood supply of this or that organ reflects the degree of its ischaemia. In comparison to the initial level an increase of the REG amplitude was not more than 24.3% after tilt table and reached 35.6% after LBNP (Figure 2); this also gives evidence for a higher decrease of blood supply during LBNP than during tilt table. It seems very likely that the feeling of the head-down position is related to the development of pronounced reactive hyperemia of the brain, which was more distinct in people with lowered orthostatic tolerance during test table. From the data of $\alpha/T\%$ it followed that the tone of large arteries increased as a compensatory reaction.

Most test subjects (20 people) showed a 33–40% (Figures 1 and 2) decrease of the amplitude of rheograms of the right lung, a decline of the tone of arterioles and an increase of the tone of veins and large arteries. The degree of the decline of the pulse blood-filling during tilt test and LBNP was almost identical; the reactive hyperemia noticed in the postischemic period was, however, greater after LBNP than after tilt test. An increase of the amplitude of rheograms following tilt test reached 19%

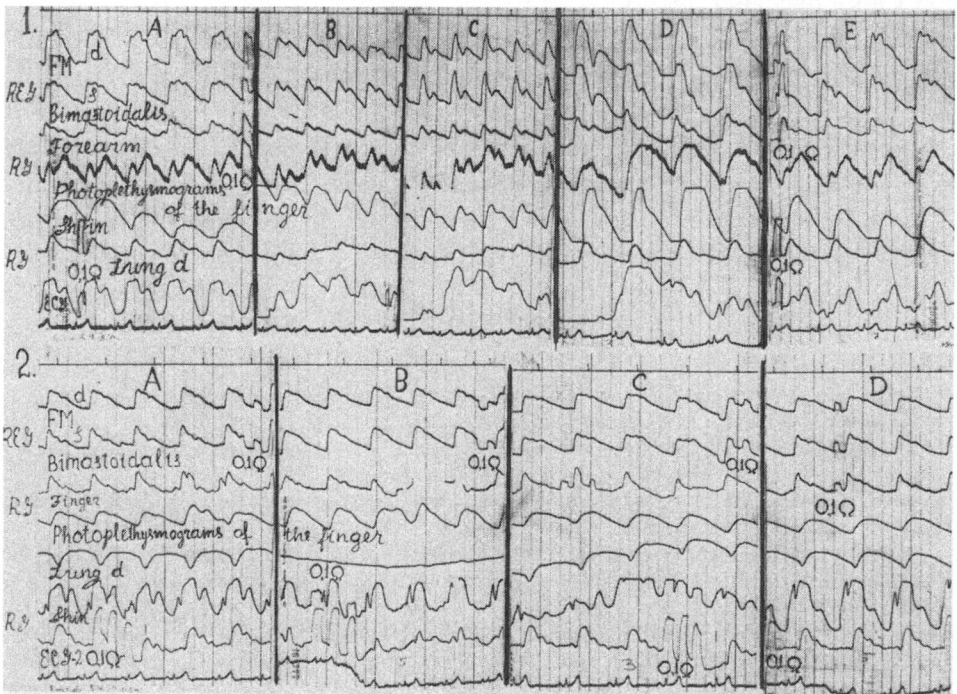

Fig. 2.   Changes in the rheogram of the test subject B, during lower body negative pressure (1) and during tilt (2): 1 – REG of the right hemisphere; 2 – REG of the left hemisphere; 3 – Bimastoidal REG; 4 – rheogram of the finger; 5 – Photoplethysmogram of the finger; 6 – rheogram of the right lung; 7 – rheogram of the shin; 8 – EKG in the 2nd lead. (I) A – reference level; B – the 4th min of LBNP; C – the 7th min of the LBNP; D – the 3th min of the recovery; E – the 7th min of the recovery. (II) A – reference level; B – the 5th min of tilt; C – the 3th min of the recovery; D – the 5th min of the recovery.

at the most and following LBNP it amounted to 26.7% (Figure 2); this is indicative of greater disturbances of pulmonary blood supply during LBNP than during tilt test. In case of LBNP the tone of pulmonary vessels increased mostly whereas during tilt tests it decreased.

As estimated from photoplethysmograms and rheograms, the tilt test and LBNP were followed by an increase of the total blood-filling of the shin and a simultaneous decline of the pulse arterial blood-filling. A substantial rise of the amplitude of the diastolic wave (diastolic index) by 80% during LBNP and by 65.8% during tilt test shows that an increase of the total blood-filling of the shin is related to venous congestion and blood-overfilling accompanied by a decrease of the arteriovenous difference, i.e. pulse blood-filling. The latter can be seen in rheograms of the shin. as a decrease of the systolic wave. (Figures 1 and 2).

Disturbances of the venous tone were more distinct during LBNP than during tilt test. In addition to a greater increase of the diastolic index this manifested itself in a more frequent occurrence of venous waves in rheograms of the shin during LBNP than during tilt test.

The initial level of amplitude and time parameters of rheograms recovered later after LBNP than after tilt test.

Thus, LBNP exerts a greater effect on the regional hemodynamics, particularly cerebral hemodynamics, than tilt test. A similar degree of changes in the regional hemodynamics during these two exposures was noted only in 3 test subjects. It seems probable that these differences in the changes of regional hemodynamics can be attributed not only to the varying level of the conditioning of test subjects and to the different mechanisms underlying the effect of the exposures but also to individual features of the cardiovascular system of each test subject.

References

Chlebus, H.: 1962, 'Value of Examination of Carotid Pulse by Means of Resonance Electrosphygmographs in Relation to Intracranial Pressure Tracings', *Am. Heart J.* **64**, 22–32.

Donzelot, E., Milovanovich, J. B., and Meyer-Heine, A.: 1950, 'Piezographie et diagraphie arterielles chez l'homme normal', *Arch. Mal. Cocur*, No. 11, 1013–1016.

Eninya, G. I.: 1967, 'The Principal Indices of Summarized Cranial *t* Rheograms in Apparently Healthy Persons of Diverse Age', *Clin. Med.*, *Moskow* **5**, 93–101.

Musgrave, F. Story, Zehmann, Fred W., and Mains, Richard C.: 1971, 'Comparison of the Effects of 70° Tilt and Several Levels of Lower Body Negative Pressure on Heart Rate and Blood Pressure in Man', *Aerospace Med.* **42**, No. 10, 1065–1069.

Suvorov, P. M., Voloshin, V. G., Dyachenko, S. N., and Krivets, V. F.: 1972, 'Hemodynamic Studies During Application Negative Pressure and Accelerations', *Kosmich. Biol. Med. (Moscow)* **6** (3), 59–64.

Yarullin, Kh. Kh. and Levchenko, N. I.: 1969, Paraclinical Methods of Research in the Neurobiological Clinic' (*All-Union Symposium*, November, 1969); 1969, II Issue. 'Up-to-date Methods of Studying Cerebral and Peripheral Circulation'. *Rheoencephalography during Brain Vascular Diseases*, Moscow, 1969, 93–108.

PROGRESS IN THE DEVELOPMENT
OF THE REVERSE OSMOSIS PROCESS FOR
SPACECRAFT WASH WATER RECOVERY

J. N. PECORARO

NASA, Washington, D.C., U.S.A.

J. M. SPURLOCK

T. A. Jonas, Consultant, Washington, D.C., U.S.A.

and

H. E. PODALL

U.S. Dept. of the Interior, Washington, D.C., U.S.A.

1. Introduction

The major expendable required by manned space missions is water. Drinking water requirements increase proportionately with mission duration; however, the requirements become even more important as missions become longer. The present Skylab water system requires over 4540 kg of stored water and tankage for supplying potable water, with minimal water for personal hygiene.

Wash water consists principally of the water and soap used for personal hygiene, including showering, as well as detergent and residue from washing clothes. In addition to the soap (which may contain a mild antiseptic) and detergent (which is a major component), the wash water contains various microorganisms and particulate matter. Dissolved solutes, in addition to those from the soap and detergent, include sodium and potassium chloride, ammonia, urea, lactic acid, etc. Upon recovery and reuse of water, the latter components may build up to appreciable concentrations unless they are actually removed in solid form, such as, for example, by absorption or ion exchange demineralization. In the latter case, the weights of expendables would be further increased.

In considering various processes for recovery and reuse of wash water, the reverse osmosis (RO) process was considered as a means for water recovery and concentration of the dissolved solids. Reverse osmosis is an extremely attractive process for recycling large volumes of wash water onboard manned spacecraft because its specific energy and volume requirements are substantially less than for other processes.

This paper reviews some of the work being done to develop RO membranes for wash water recovery, compatible cleansing agents, tentative wash water standards by the National Academy of Sciences and trade-off analyses for system design and optimization.

2. Brief Review of Process and Rationale for its Selection

The reverse osmosis process, first demonstrated by Prof. C. E. Reid of the University of Florida, in 1955, for desalting, is basically a pressure-driven membrane diffusion process. The process differs from ultrafiltration in that it pertains primarily to the removal of dissolved solutes (organic as well as inorganic). The key to the process resides, however, in the semi-permeability of the membrane; i.e., in the ability of the membrane to pass only water and to reject efficiently the solutes.

Specific attractive features of the process for use in wash water recovery are (1) its simplicity; (2) its general applicability for the removal of dissolved organic, as well as inorganic molecules; (3) its relatively low energy requirements when compared to evaporation or vapor diffusion processes; (4) its compactness; and perhaps most important, (5) the fact that no phase change is involved, thereby obviating the need for zero-G phase-separation devices.

3. Basic Considerations Relevant to its Application

Major requirements of the process for application to space missions include: (1) reliability of performance of the RO system, including membranes, pressure-seals, pumps, etc.; and, (2) assurance that the product water contains no harmful components to hinder its reuse.

In reference to the first requirement, it is apparent that for any RO system to be effective, adequate pretreatment is required to remove sufficient particulate matter to prevent serious fouling and consequent flux decline of the membrane. In addition, it is apparent that for the process to be useful, high water recovery efficiencies from the wash water are required, preferably above 90%. Indeed, even at this level of recovery, the concentrate would build up rapidly in the course of a space mission, thereby necessitating further recovery of water from it, possibly by other processes. An alternative would be to seek recoveries of 99% or higher so that further recovery of water from it would not be required, depending, of course, upon the initial amount of water, the actual water requirements per man-day, the other water sources available and the length of the mission between 'fresh' supplies.

In reference to the second requirement, concerning removal of all possible harmful components, the approach has been to aim at producing potable water from the RO process. Key components of concern, other than the soap and detergent, include various microorganisms and bacteria. To assure removal and prevention of growth of the latter, two approaches have been used, viz., (1) addition of biocide to maintain sterilization and/or (2) conduct of the process at pasteurization temperatures (such as at 75 °C) so as to eliminate the possibility of any contamination of the spacecraft atmosphere as well as the product water.

4. Ambient-Temperature RO Process

The ambient-temperature RO process, utilizing a biocide for destruction of any bacteria and sterilization of the wash water, was investigated initially [1]* because

---

* Numbers in brackets [ ] refer to literature citations at the end of the paper.

of the commercial availability of RO membranes for ambient-temperature use. However, it was found that (1) the available membranes were either unstable in the presence of the biocide and, in the case of cellulose acetate membranes, to alkaline detergents, and (2) the available membranes did not reject adequately these substances at the desired high water recoveries without significant loss in flux and rejection. Thus, it appeared necessary to either conduct the process in stages or to employ extensive recycling of the wash water feed to achieve a product water approaching acceptable quality for reuse.

Although there has been considerable recognition of the bacterial problems associated with long-duration space flights, there has been little effort to develop a general all-purpose sanition agent that will be effective and yet compatible with the spacecraft environment and materials. There are two fundamental paradoxes concerning antimicrobial agents and their use which are the cause of the problem; these are:

(1) In order to kill bacteria, a bactericide must, obviously, be toxic to bacteria. But since the biology of bacterial protoplasm is not substantially different from the biology of human protoplasm in many respects, bacterial toxicity must be accomplished, to some degree, with attendant human toxicity.

(2) In order to penetrate the living bacterial cell and destroy protoplasm, a bactericide must be chemically reactive. The same property causes it to react with substances other than bacteria, such as RO membranes.

Although there is no perfect sanitation agent, NASA conducted a study to perform the trade-offs that will select the best agent or agents for space use [2]. The problem is much more than a question of bacteriology. It involved the full spectrum of life support in space; the crew, the hardware, and all the interfaces that exist between them. Results of the definition and selection of the sanitation agents are shown on Table I.

Studies also showed that the available RO systems were not effective for processing the small volumes of wash water of interest to provide a product of acceptable water quality and with sufficient recovery for the process to be useful. This conclusion is not, however, entirely surprising in view of the fact that the available commercial RO modules are designed for processing either large volumes of brackish water or at lower water recoveries, generally below 85% (particularly for the smaller volumes).

A somewhat unexpected, though very significant, finding was that relatively large amounts ($\sim 1\%$) of certain biocides were required for a real wash water feed to assure adequate sterilization, and that at these concentrations and at the progressively high concentrations encountered in the course of the RO process, the available membranes gave poor rejections of the biocide. In this regard, a high rejection of the biocide is, of course, necessary in view of the fact mentioned earlier that bacterial toxicity is generally accompanied, to some degree, by human toxicity.

Accordingly, it is concluded that the ambient-temperature RO processes thus far investigated are not viable processes for wash water recycle in a spacecraft environment.

Sanitation Agent Requirements

THE SANITATION AGENTS MUST:

● **BE COMPATIBLE WITH THE HUMAN**

● **NOT PRODUCE CONTACT DERMATITIS**

● **BE NONTOXIC IF INGESTED IN SMALL QUANTITIES**

● **RETAIN EFFECTIVENESS IN THE PRESENCE OF ORGANIC MATTER**

● **BE EFFECTIVE IN REMOVING SOILS AND INHIBITING MICROORGANISMS**

● **BE BACTERIOSTATIC AGAINST A BROAD SPECTRUM OF MICROORGANISMS**

● **HAVE A RESIDUAL BIOCIDAL EFFECT**

● **BE EFFECTIVE IN THE REMOVAL OF POTENTIAL NUTRIENTS**

● **HAVE MODERATE GREASE CUTTING ABILITY**

● **REMOVE, BY ENTRAINMENT, A MINIMAL AMOUNT OF GRIT**

5. Pasteurization Temperature RO Process

It was soon recognized that in spite of the higher energy requirements associated with the use of a process conducted at pasteurization temperatures, conduct of the RO process at 75 °C is preferred in order to avoid the problems associated with the use, monitoring and control of the concentration of a biocide throughout the wash water-product water system. In brief, it appeared that the pasteurization approach would be the most reliable and practical approach.

It was found that there were no commercial RO membranes available which could operate effectively at 75 °C, particularly in the presence of various soaps or detergents, without appreciable loss in rejection (due to hydrolysis or swelling and dissolution) or flux (due to compaction). It was apparent that (1) the membranes had to be stable to water at 75 °C, and (2) they had to be compatible with, or essentially inert to, the soap and detergent employed at this temperature. In addition, the membranes had to give

Surfactant Classes Considered

CLASS I   ANIONICS

> ACCEPTABLE. WATER SOLUBLE. LOW FOAM. STABLE TO ACIDS AND ALCOHOLS. INCOMPATIBLE WITH CATIONIC COMPOUNDS.

CLASS II   CATIONICS

> UNACCEPTABLE. SHOW LITTLE VALUE AS DETERGENT OR WETTING AGENT.

CLASS III   NONIONICS

> ACCEPTABLE. DO NOT POSSESS AN IONIZABLE GROUP. COMPATIBLE WITH CATIONIC COMPOUNDS. LOW FOAMING.

CLASS IV   AMPHOTERICS

> UNACCEPTABLE. POSSESS ABILITY TO CHEMICALLY ACT BY A CATIONIC OR ANIONIC REACTION. INCOMPATIBLE WITH MOST GERMICIDAL AGENTS.

high rejections of the soap and detergent components to enable reuse of the product water, particularly when obtained at 90% water recovery or higher.

In initial studies, two types of membranes were examined: (1) porous-glass membranes (whose pores were reportedly less than 50 Å); and (2) cellulose acetate membranes modified by various feed additives to enhance their rejection and overall stability characteristics.

It was found that the porous-glass membranes in capillary tube form gave surprisingly good rejections of organics, as well as inorganic salts, but they had a reatively short lifetime (≤ 10 days). The latter effect was, however, significantly improved by the use of zirconia in the glass melt for preparing the porous-glass tubes. It was also found by NASA workers that the addition of a small amount of aluminum chloride to the feed water significantly increased the useful lifetime of the glass membranes at 75 °C. Thus, a minimum of a 30-day lifetime appears feasible.

The second approach, involving the use of feed additives for modifying the surface of cellulose acetate membranes, resulted in little improvement in the rejection of organics, though some improvement in salt rejection was noted.

Under a joint contract with OSW and NASA, three contractors are developing advanced membranes for recovery of wash water at sterilization temperatures. The objectives of these contracts are shown on Table II. The three membranes are sulfonated PPO polymer, cellulose acetable blend, and PBI hollow fibers.

It was found that certain modified cellulose acetate membranes developed for deslating seawater were effective at 75 °C, provided the tests were conducted at neutral

Biocide Classes Considered

PHENOLICS

UNACCEPTABLE. PRODUCE PUNGENT ODORS. INACTIVATED BY THE PRESENCE OF ORGANIC MATTER.

ALCOHOLS AND ALDEHYDES

UNACCEPTABLE. PRODUCE PUNGENT ODORS. HIGHLY FLAMMABLE.

SOAPS

UNACCEPTABLE. PROLONGED USE OF GERMICIDAL SOAPS CAN CAUSE CONTACT DERMATITIS.

DYES

UNACCEPTABLE. NOT A BROAD SPECTRUM BACTERICIDE.

QUATERNARY AMMONIUM COMPOUNDS

ACCEPTABLE. NONTOXIC AND NONIRRITATING. EFFECTIVE IN LOW CONCENTRATIONS AGAINST ALL TYPES OF BACTERIA EXCEPT PSEUDEMONAS GROUP.

HEAVY METALS

UNACCEPTABLE. NOT A BROAD SPECTRUM BACTERICIDE.

HALOGENS

CHLORIDES (ORGANIC)

ACCEPTABLE. RETAIN EFFECTIVENESS IN THE PRESENCE OF ORGANIC MATTER. BACTERICIDAL ACTION DOES NOT DEPEND UPON RELEASE OF FREE CHLORINE.

IODINE COMPOUNDS

UNACCEPTABLE. RENDERED INEFFECTIVE IN PRESENCE OF ORGANIC MATTER. LEAVE A STICKY RESIDUE. UNSTABLE WHEN DILUTED TO LOW CONCENTRATIONS. EFFECTIVENESS DEPENDS UPON RELEASE OF FREE IODINE.

ANTIBIOTICS AND ENZYMES

UNACCEPTABLE. NOT A BROAD SPECTRUM SANITATION AGENT.

OXIDIZING AGENTS

UNACCEPTABLE. NOT COMPATIBLE WITH MOST SURFACTANTS. STORAGE AND MATERIALS INCOMPATIBILITY.

GASEOUS COMPOUNDS

UNACCEPTABLE. TOXIC UPON CONTACT AND INHALATION.

Objectives of OSW-NASA Advanced Membranes

Flux $\geq 2.3 \times 10^{-4}$ cum/sec at .56 Kgf/mm^2 or less when operated 180 days
Rejection as follows for 90% recovery:

WASH WATER CONSTITUENTS	CONC. (PPm)	% R
Na Cl	1500	85.0
Urea	1000	99.3
Lactic Acid	700	99.0
Detergent or Soap	10,000	99.0
PH = 4.8		

Reverse osmosis purification of wash water at sterilization temperations (75°C)

or slightly acidic pH's. This imposes a limitation on the type of soap and detergent
that can be employed, but it is nevertheless a feasible approach.

It was concluded from these studies that with the commercially available cellulose
acetate membranes a preferred approach would be to seek a non-cellulosic membrane

Life Tests on Sulfonated PPO Membrane with 75° Wash Water

Feed: Wash Water with 10,000 ppm "Olive Leaf" Soap-Membrane Designation: RO-54

Time (Hours)	Flux (Cum/Sec)	NaCl	% Rejection Lactic Acid	% Rejection Soap
17	—	—	—	—
65	—	—	—	—
161	1.4×10^{-4}	91	—	—
185	2.3×10^{-4}	92	97	97
233	2.1×10^{-4}	94	—	—
279	1.9×10^{-4}	93	—	—
303	1.9×10^{-4}	92	—	—
380	1.9×10^{-4}	93	98	96
474	2.6×10^{-4}	93	—	—
543	2.1×10^{-4}	93	—	—
806	2.7×10^{-4}	95	—	—

Tentative Standards for Wash Water Specification

PHYSICAL PARAMETERS

COLOR, COBALT UNITS	15
CONDUCTIVITY, SPECIFIC, UMHOS-CM^{-1} AT 25°C	2,000
FOAMING	NONPERSISTENT MORE THAN 15s
ODOR	NONOBJECTIONABLE

CHEMICAL CONSTITUENTS

CARBON, TOTAL ORGANIC, MG/1	200
DETERGENTS	NOT SPECIFIED
LACTIC ACID, MG/1	50
NITROGEN, AMMONIA, MG/1	5.0
OXYGEN DEMAND, CHEMICAL, MG/1	NOT SPECIFIED
pH	5.0, MINIMUM 7.5, MAXIMUM
SODIUM CHLORIDE, MG/1	1,000
SOLIDS, DISSOLVED, RESIDUE-ON-EVAPORATION AT 180°C, MG/1	1,500
UREA, MG/1	50

MICROBIOLOGICAL

MICRO-ORGANISMS, NUMBER PER ML ON A STANDARD 48-H PLATE	10

which is intrinsically more stable thermally and hydrolytically. It was found that a sulfonated polyphenylene oxide (PPO) membrane, studied for desalting brackish water (particularly of the sodium sulfate type), could be modified to significantly improve its rejection characteristics for various solutes such as are present in wash water. Typical results obtained with the latter in flat-sheet form are shown in Table III. In addition, there was no evidence of any bacteria or viruses in the product water.

Current studies in this area are now aimed at developing modular designs incorporating the above membranes.

6. Wash Water Standards

At the request of NASA, the National Research Council, National Academy of Sciences-National Academy of Engineering, has undertaken the task of recommending a wash water standard. A tentative standard has been provided by that

Spacecraft Sanitation Agent Formulations

PERSONAL HYGIENE

PHASE A	GLYCEROL MONOSTEARATE		6.0 %
	P.E.G. 400 DISTEARATE		4.0 %
	TWEEN 60		5.0 %
	CETYL ALCOHOL		1.5 %
	ISOPROPYL PALMITATE		4.5 %
	CETOL		0.5 %
PHASE B	NONISOL 250		5.0 %
	PROPYLENE GLYCOL		5.0 %
	DISTILLED WATER	G.S.	100 CC

EQUIPMENT MAINTENANCE

PHASE A	CETYL ALCOHOL		1.7%
	STEARYL ALCOHOL		2.25 %
PHASE B	SODIUM LAURYL SULFATE		5.0 %
	PROPYLENE GLYCOL		7.0 %
	CHLORAMINE T		0.3 %
	DISTILLED WATER	G.S.	100 CC

study group, primarily as guidance for application in the early design of a system for further testing and evaluation, under the control of medical personnel. As indicated earlier, the major problem has been the soap and detergents. In their report, NAS indicated that a specification for these constituents could not be made in the absence of information on the specific detergent formulation and the composition of the degradation products of the detergent materials.

7. Selection of Soaps and Detergents

Another NASA study has centered on formulating criteria for selecting or specifying cleansing agents, for crew personal hygiene and for laundry purposes, which will be compatible both with human skin and with RO membrane materials [3]. Results of this study provide an important input to an overall systems design tradeoff analysis. The objectives of the soap-effects study are (1) the identification of effects of candidate cleansing-agent products on human skin and its 'normal' microflora; and (2) the definition of potential hazards to the crew from any ecological changes that might be produced by cleansing-agent candidates.

Candidate cleansing agents considered and tested to date, for either or both personal hygiene and laundry use, include:

(1) 'Neutrogena' – a proprietary formulation of a 'superfatted' soap, widely used by persons with dermatological and allergic problems.

(2) 'Miranol JEM' concentrate – a dicarboxylate having a general structure which includes a mixture of $\frac{2}{3}$ caprylic acid 2d $\frac{1}{3}$ ethylhexoic acid. The pH of the concentrate is 10.0 to 10.5 at 30 °C. The 'Miranols' are amphoteric surface active agents having

various types of cationic and anionic groups of equal strength (isoelectric point at pH 7.0).

(3) Sodium dodecylbenzen sulfonate – a surface active detergent.

(4) 'Olive Leaf' – a saponified oil an detergent.

For each soap candidate, twenty subjects were selected as a representative sampling for statistical purposes, while small enough to permit a manageable number of bacteriological samples. The chosen subjects represented in each case a rather wide selection of skin types (color, sex and sensitivity differences). The twenty-subject group was subdivided such that some participated in skin-patch tests as well as bath-regimen tests (acute and chronic exposure) and others participated only in the latter (chronic exposure) tests.

In the chronic-exposure tests, the bathing regimen for each soap candidate was a compromise based upon several possible personal-hygiene protocols for long-duration space missions. Subjects were swab-sampled initially during a period when they used their normal personal-hygiene procedures; this established a subjective 'normal' skin-flora baseline for each subject. Then the subjects used the prescribed bathing regimen with the candidate test soap for periods up to six weeks. Six body sites were selected for swab sampling (in the laboratory); back of right ear, left axilla, back of right hand, upper left thigh at croth, upper right thigh at anus, bottom of left foot. After carefully controlled sampling, the samples were plated on blood agar, eosin methylene blue agar, mannitol, thioglycollate broth and a Sabouraud medium. These were then incubated and read (counted) at 24 and 48 h, with data reported as general types of organisms and approximate numbers of each. Baseline ('normal') and regimen-produced results were compared for each subject to establish changes in skin microflora characteristics.

The second category of testing involved skin-patch tests on several of the subjects of each group to identify any overt dermatological effects associated with acute exposure to candidate cleansing agents. A 2″ by 2″ sterile gauze patch, moistened in the center with 0.25 cc of a 2% cleansing-agent solution (2 g of agent per 100 ml of distilled water), was affixed to the subject's back and left arm. These patches were allowed to remain in position for five days. On removal of the patches, the subject was examined for signs of reaction. Testing criteria also called for careful examination of other regions of the subject's body, before and after the skin-patch tests, to permit determination of causes of any dermatological responses that might produce artifact reactions and interpretations.

8. System Design and Optimization

A detailed specification has been established [4] for a reverse-osmosis wash water recovery unit to be used in an integrated wash water recovery system. Figure 1 shows a schematic diagram of the integrated wash water system, and serves to define the location of the RO unit with respect to recovery interfaces.

The design process rate of the unit will be such that 2.6×10^{-4} m^3 of waste water

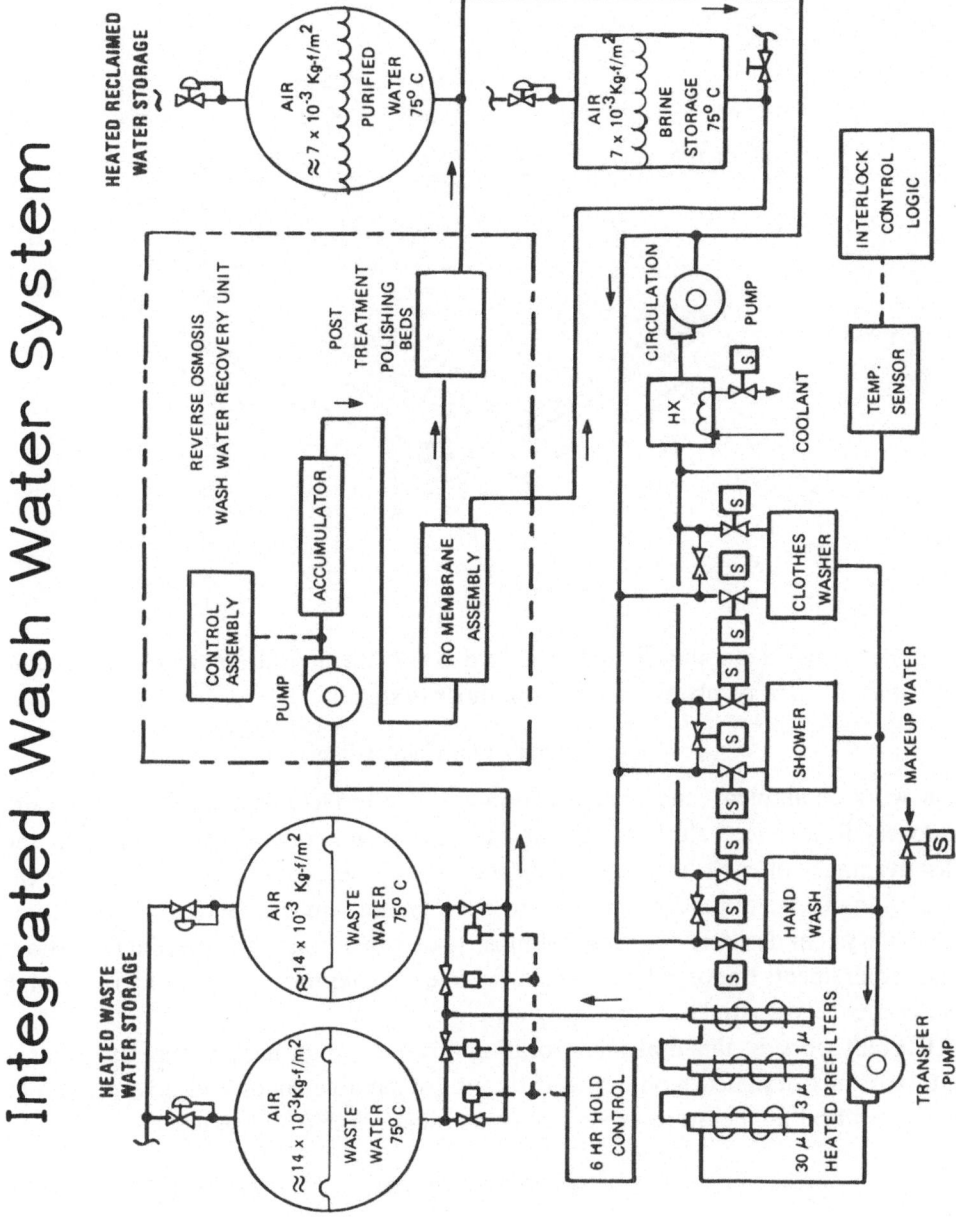

Integrated Wash Water System

per minute may be processed. The feed water suplied to the RO unit will be at a temperature of $75 \pm 2°C$ and a pressure of 4.22×10^5 kg m^{-2}. Feed water will be prefiltered before supply to the RO unit.

The RO unit must discharge the product water and brine at the system interface at a temperature of $75° \pm 2°C$. The unit also must be capable of unimpaired operation with product and brine back pressures at the unit interface of 0 to 2.1×10^{22} kg m^{-2}.

Nomenclature Used for Spiral Wound Module

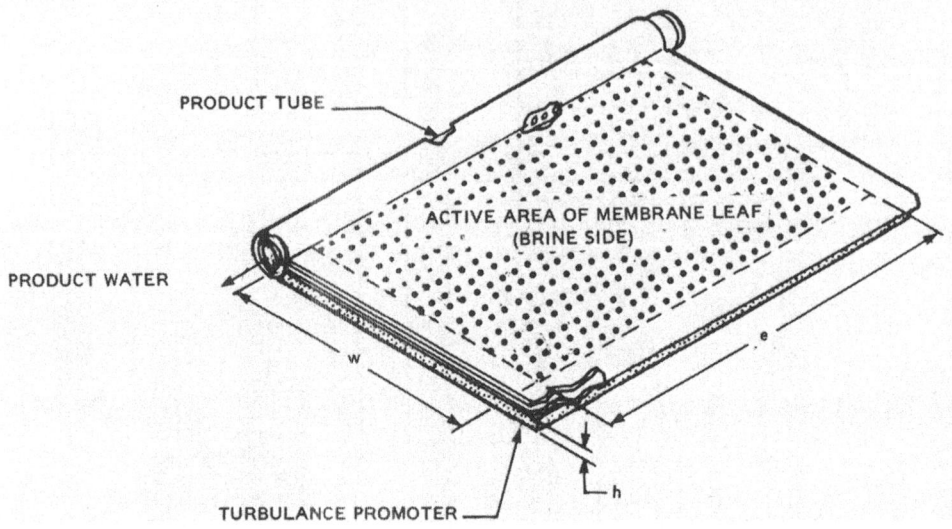

Figure 2 shows the nomenclature used to describe the module geometry, applicable to a spiral-wound or plate-and-frame modular design.

9. Summary and Conclusions

The work on ambient- and pasteurization-temperature RO processes for wash water recovery in a spacecraft environment has been reviewed, and the advantages and disadvantages of each have been indicated. A key requirement in each case is the provision of a membrane of appropriate stability and semi-permeability. In addition, the RO system designed for such use must take into account the specific limitations and requirements imposed by the small volume of water to be processed and the high water-recovery desired.

Current work is aimed at incorporating advanced high temperature membranes into specially designed modules, and the use of suitable soaps and detergents. The pasteurization temperature process appears most reliable and practical.

Acknowledgements

The authors wish to acknowledge data contributions from David F. Putnam, of McDonnell Douglas Astronautics Co., and Robert L. Durfee of Versar, Inc.

References

[1] *NASA Contract NAS 9-9199*, w/Chemtric, Inc., 'Water-Waste Management Test and Evaluation.'
[2] *NASA Contract NAS 9-12205*, w/Fairchild (Republic Division), 'Spacecraft Sanitation Agent.'
[3] *NASA Contract NAS 9-12672*, w/Versar, Inc., 'Evaluation of Proposed Skylab and SSP Soap Products.'
[4] *OSW Contract 14-30-3062*, w/McDonnell Douglas Astronautics Co., 'Definition of Reverse Osmosis Requirements for Space-craft Wash Water Recycling.'

THE SIGNIFICANCE OF EXPIRED AIR AS A SOURCE
OF THE SPACECRAFT ATMOSPHERIC CONTAMINATION
WITH CARBON MONOXIDE

YU. G. NEFEDOV, V. P. SAVINA, and N. L. SOKOLOV

Academy of Sciences of the U.S.S.R., Moscow, U.S.S.R.

It is a widespread opinion that carbon monoxide (CO) is among those trace contaminants which to a greater or lesser extent can occur continuously in the spacecraft atmosphere and, due to their high toxicity, can produce an adverse effect on the health and performance of a spacecraft crew.

Detailed studies and identification of the potential sources of carbon monoxide in space cabin simulators have been made. The first essential is to eliminate the more important sources of CO production and to develop effective methods of removing carbon monoxide when elimination of a CO source in a spacecraft environment is not feasible. The metabolic products of the human body are among constant sources of CO production.

As has been shown by T. Sjöstrand, R. Cobburn, V. Kustov, and L. Tiunov *et al.*, carbon monoxide is formed in the body by endogenous processes during the catabolism of hemoglobin. According to R. Cobburn, the rate of CO production is $0.42 \, \text{ml h}^{-1}$.

In addition to the catabolism of hemoglobin there also exist other processes by which CO production in the human body may occur, in particular carbohydrate oxidation, which is at present poorly understood. Carbon monoxide produced in the body is removed through the lungs.

The carbon monoxide concentration in expired air of humans, according to our data, is on average about $10 \, \text{mg m}^{-3}$. Studies carried out with the aid of different techniques (among them chromatographic ones) have indicated that the CO concentration in expired air of non-smokers is 3–4 times lower than that of smokers. Indeed, an increased content of carbon monoxide in expired air of the smokers during their stay in a sealed environment is maintained for several days despite the fact that the test subjects do not smoke during this period. The CO production by a man enclosed in a sealed chamber in the absence of an effective system of carbon monoxide removal in the first 10–15 days is always associated with a rise of CO production in the chamber atmosphere. It should also be noted that the rate of increase of CO concentration in the chamber air falls gradually, and almost stops in 10–15 days. An initial rise of CO concentration in a sealed environment occurs due to an increase of CO production in expired air of men, at the cost of differences in CO partial pressures between chamber atmosphere and alveoli. This process is later stabilized at a constant value.

This assumption is supported by experimental results obtained during examina-

L. G. Napolitano et al. (eds.), Astronautical Research 1972, 85–87. All Rights Reserved
Copyright © 1973 by D. Reidel Publishing Company, Dordrecht-Holland

tion of expired air of test subjects participating in a 1-yr medico-engineering experiment.

If before the onset of experiment the CO concentration in expired air of the smokers was considerably higher than that of the non-smokers, during the experiment itself, when smoking was prohibited, the CO concentration in expired air of all test subjects levels off step by step and falls to $1-2$ mg m^{-3}. In this case the content of carbon monoxide in a chamber atmosphere was practically negligible because of the presence of a highly effective system of air cleaning. It appears that the CO concentration in expired air of men depends not only on the rate of its endogenous production but also on its exogenous entering in the body.

The gradual removal of exogenous carbon monoxide from the body during enclosure of man in a carbon monoxide-free environment promoted a reduction of CO concentration in expired air down to the level corresponding to the rate of endogenous CO production. The results from the 1-yr test are in close agreement with calculated values of R. Cobburn (0.42 ml h^{-1}).

The CO removal from the body is a very long-term process because the affinity of hemoglobin for carbon monoxide is some hundreds times higher than that for oxygen.

It should be also noted that human blood can retain carbon monoxide in a hemoglobin-unbound form. Consideration must be given to this form of binding in the cases of chronic CO poisoning.

It is reported in some papers that when chronic CO poisoning is encountered the carboxyhemoglobin content of the blood may not vary much. Therefore, measurements of carboxyhemoglobin do not always correlate with true CO concentration in the blood.

Our experiments with humans exposed to stresses indicated that CO removal from the body increases significantly during long-term fasting, also in very hot ($+35-40\,^{\circ}$C) and humid environments, and during hypokinesia as well.

It is supposed that during fasting, hypokinesia, and a stay of man in a sealed chamber under conditions of high air temperature and humidity, changes occur in the rate of carboxyhemoglobin decay, associated with increased CO production.

It is also possible that extra CO production is due to an increase of the oxidative processes in the body, in particular during fasting. However, the mechanism of this CO production is as yet imperfectly understood and special studies are needed.

Thus, expired air of a man may contain both an endogenous and exogenous carbon monoxide. Its concentration to a great extent depends on the microclimatic parameters of atmosphere, the duration of an enclosure of man in an environment containing carbon monoxide (tobacco smoke as well), his motor activity, nutrition, and, finally, individual peculiarities of the body.

The studies performed have shown that CO entering into a spacecraft atmosphere due to metabolic processes is definitely significant. According to calculation, daily CO production by a healthy man consists approximately 200 mg (at rest, without exposure to stress factor effects).

The rate of CO production significantly increases during exposure of man to stress factors. This must be taken into account when making recommendations for the systems of spacecraft air cleaning.

Numerous chamber experiments carried out before 1964, made it possible to recommend a level of 5 mg m^{-3} as the provisional maximum acceptable limit for air contamination of a sealed chamber with carbon monoxide during 30-day exposures. In order to improve these recommendations issued at an early date concerning the maximum permissible limits for carbon monoxide, we have studied the long-term and continuous exposure of man to carbon monoxide in a sealed environment. The preliminary data from experiments lasting up to 30 days indicates that concentrations up to 15 mg m^{-3} do not produce changes in the state of the main physiological systems of the body. Specific changes at this concentration characteristic of CO (carboxyhemoglobin content, non-hemoglobin iron, catalase, peroxidase, etc), return to the norm after cessation of its action.

Thus, summarizing the above data, it can be said that man himself is a constant source of CO production in the environment, and this factor acquires special importance when a man is enclosed in a sealed chamber. However, when an effective system of air cleaning is used, i.e. in the absence of CO in the inspired air, the CO content in expired air is within the limits caused by its endogenous production, i.e. about 1–2 mg m^{-3}.

PART II

ENGINEERING AND MANAGEMENT ASPECTS
OF SPACE TECHNOLOGY

ANALYTICAL INVESTIGATION OF THE
DUAL PROPELLANT MODE

HARRY O. RUPPE*

Technical University Munich, Germany

A publication by Salkeld [1] suggests that the performance of a single stage rocket vehicle can be improved using the dual propellant mode, also called mixed propellant mode. This can be described as follows:

Given the performance of a liquid hydrogen/liquid oxygen stage, we assume for the simple case that without any increase of propulsion system mass and without any mass increase for the additional bulkhead, feedlines etc., we can replace part of the hydrogen by for example kerosine, adjusting the oxygen volume as required by the different mixture ratio. Engine performance as measured by exhaust speed shall not be degraded due to the dual propellant capability. This propellant replacement keeps the total propellant mass constant. Tankage mass is corrected for the different propellants. By optimizing the ratio of LOX/Kerosine to LOX/Hydrogen the performance of the dual propellant stage can be better than the performance of the original LOX/Hydrogen version.

What I call the complex case, differs from the above simple case, in so far as the dual propellant vehicle is penalized by the additional bulkhead mass and by two sets of engines for the different propellants. I will not go into this complex case here any further, since my own investigation and the results of most other researchers in this field seem to agree, that there is little practical utility in this complex case. Furthermore, I will not consider drop engines or drop tanks, since the main application of the dual mode propellant vehicle seems to be a reusable design; and throwing away engines or tankage is of course in conflict with the goal of full reusability of a launch vehicle.

My attention was drawn to this problem by Mr Val Cleaver of Rolls-Royce. One of his coworkers, Mr Allan Bond, has also studied this field. His somewhat different conclusions are due to differences in assumptions concerning the dry masses of the stages. Therefore, these assumptions are very critical to the results obtained. Since I doubt that a valid, simple, analytical model of stage dry masses can be developed, I followed a pragmatic procedure. Figure 1 gives data for actual stages which have been built. Of course, those stages have been built at different times and therefore, they reflect a different state of the art. Nevertheless, these are real numbers. If we want to consider a single stage from Earth surface to low orbit, then these masses need correction, for two main reasons: Firstly their thrust when used as a first stage is either too high, and we have to reduce engine mass; or as upper stages, their thrust is too low, and we have to increase the propulsion system mass. Secondly, a correction is required, if we have to add a guidance and control system to the stage.

* *AIAA AF, BIS F, DGLR SM, IAA CM.*

L. G. Napolitano et al. (eds.), Astronautical Research 1972, 91–101. All Rights Reserved
Copyright © 1973 by D. Reidel Publishing Company, Dordrecht-Holland

bulk density. 1.02 g/cm³
jet speed.   3.1 km/s   (average surface-orbit)

Vehicle	Propellant Mass [kg]	Dry Mass [kg]	Corrected Dry Mass [kg]
Long Tank Thor	66000	4200	4500
EUROPA I / Ist stage	82000	6500	
EUROPA II / Ist stage	88000	6000	
ATLAS	111000	7000	6800
SATURN IB / Ist stage	401000	39000	
SATURN V / Ist stage (IC)	2000000	125000	120000

Fig. 1.   Stage masses of LOX/Kerosine stages.

bulk density. 0.28 g/cm³
jet speed   4.3 km/s   (average surface-orbit)

Vehicle	Propellant Mass [kg]	Dry Mass [kg]	Corrected Dry Mass [kg]
Centaur	13500	1500	1593
S IV B	105000	10000	10654
S II	416000	37500	40000

Fig. 2.   Stage masses of LOX/LH$_2$ stages.

Two winged fully reusable stages,
all LOX/LH$_2$

First stage: $m_{payl.}$ = 300 t (= 2nd stage)

m_p = 1000 t

m_{dry} = 200 t

c = 4.25 km/s

Second stage: $m_{payl.}$ = 30 t

m_p = 200 t

m_{dry} = 70 t

c = 4.55 km/s

Fig. 3.   Space shuttle masses.

Masses so corrected are also shown in Figure 1. Figure 2 gives corresponding information for LOX/Hydrogen stages. The next figure (Figure 3) contains some data on a fully reusable, winged, two-stage shuttle design. Figure 4 is a plot of all this information, from which the equations shown in Figure 5 are derived. These equations seem to be valid for propellant loadings between 10^5 and 10^7 kg.

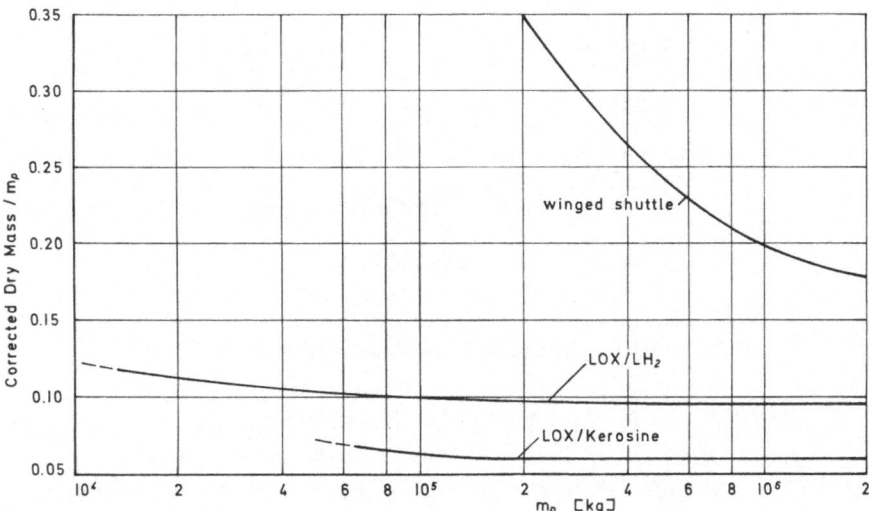

Fig. 4.   Corrected dry mass as a function of the propellant mass m_p for 3 different stage design types.

Type of equation :	$m_d = B + A\, m_p$	(B, A: const.)
Ballistic, LOX/RP :	$m_d = 300 + 0.0597\, m_p$	[kg]
Ballistic, LOX/LH$_2$:	$m_d = 600 + 0.0954\, m_p$	[kg]
Winged, LOX/LH$_2$:	$m_d = 37500 + 0.1630\, m_p$	[kg]

Fig. 5.   Approximate equation for dry mass m_d.

We now introduce the symbol x so that $x=0$ means we have a LOX/Kerosine vehicle, and $x=1$ describes a LOX/Hydrogen design. Then we write the first two equations (valid for the present state of the art of ballistic vehicles) in the form shown in Figure 6. Next we see the corresponding equation for winged designs. I have transferred the 0.6 from the ballistic case to this case. I find no good reasons why this 0.6 should change, though Mr Bond has argued that it might increase a little for the winged design. This would of course then change the 0.1 also. To cover this possibility, I include the choice $\beta=1$ (and A changed thus that $A\beta=$const.) in this investigation.

Below we see two more equations, which are of similar form, but having smaller coefficients. They describe an advanced state of the art, in order to permit an assess-

$$x = 0 : \text{LOX / Kerosine}$$
$$x = 1 : \text{LOX / LH}_2$$

Ballistic : $m_d =$ $450 + 0.06(1 + 0.6x) \, m_p$

Winged : $m_d = 37500 + 0.10(1 + 0.6x) \, m_p$

Advanced structures (assumption!) :

Ballistic : $m_d = 0.05(1 + 0.6x) \, m_p$

Winged : $m_d = 25000 + 0.08(1 + 0.6x) \, m_p$

Fig. 6.   Transformed equations for dry mass m_d.

ment of whether our results will change for structurally advanced designs. Figure 7 compares results from these equations with 'real' numbers. So far the evaluation of our data was straightforward. Now comes the decisive thought: We use these equations for mixed mode designs, but now interpret x differently: x will now be the hydrogen/oxygen fraction of the total propellant loading. It is easily seen, that

	Corrected Dry Mass [kg]	Dry Mass from equation fig.6 [kg]	
		today	advanced
Long Tank Thor	4500	4410	3300
ATLAS	6800	7110	5550
SATURN V	120000	120450	100000
Centaur	1593	1746	1080
S IV B	10654	10530	8400
S II	40000	40386	33280
Winged 1st stage	200000	197500	153000
2nd stage	70000	69500	50600

Fig. 7.   Comparison of dry mass values of different stages with approximative values from equation of Figure 6.

this is in agreement with our prior definition; but it is an extension of it. I think it is a very reasonable assumption, because the limiting cases (which are pure LOX/ Kerosine or pure LOX/Hydrogen designs) are described correctly.

For ballistic recovery and reusable vehicles (as proposed e.g. by Mr P. Bono and Dr D. Koelle), we will increase B and A by 36%; this factor 1.36 stems from a small private study, yet unpublished.

Consider one mission only: The establishment of a low Earth orbit. The velocity requirement (from Cape Kennedy, inclination 28.3°, altitude 200 km) is (with 1% reserve) given in Figure 8. Taking the Ziolkovsky-equation, we can write for a

ballistic / one shot	9314 [m/s]
ballistic / reusable	9460 [m/s]
Shuttle / 1st stage	4600 [m/s]
Shuttle / 2nd stage	4800 [m/s]

Fig. 8.   Speed requirements for establishment of low Earth orbit
(Cape Kennedy, $i = 28.3°$, altitude 200 km).

mixed-mode stage, burning the LOX/Kero portion first (burning the LOX/Hyd. first is always a disadvantage), as shown in Figure 9. Statement of problem and solution:

$$v = 3100 \ln \frac{m_p + B + A\,(1 + \beta x)\,m_p + P/100\,m_p}{x\,m_p + B + A\,(1 + \beta x)\,m_p + P/100\,m_p} + 4300 \ln \frac{x\,m_p + B + A\,(1 + \beta x)\,m_p + P/100\,m_p}{B + A\,(1 + \beta x)\,m_p + P/100\,m_p}$$

$$\text{Payload Mass} = \frac{P}{100}\,m_p$$

For 2-stage shuttle: $\left(\dfrac{P}{100}\,m_p\right)_{I.Stage}$ = initial mass of orbiter stage;

jet speeds for this case:

first stage: 3100 / 4300 [m/s]

second stage: 3350 / 4550 [m/s]

Fig. 9.   Ziolkovsky equation for dual propellant mode (1 stage).

given m_p, B, A, β, p
select $x = x_0$ so, that $v = v_{max}$
select $v_{max} = 9314$ m s^{-1} etc.
Read $p = p_0$ for this case
Plot p_0 or total payload mass and x_0 (where $0 \leqslant x_0 \leqslant 1$) vs m_p or total lift-off mass M_s.
From $dv/dx = 0$ we can find an analytical expression for x_{opt}.
This is shown in the next figure (Figure 10).

$$x_{opt} = \left[\frac{43}{31 + 31 A\beta - 12q} - 1 \right] \frac{q}{A\beta}$$

$$q = \frac{B}{m_p} + A + \frac{p}{100} \; ; \; \text{of course,}$$

$$\text{if} \quad x_0 \leq 0 : \quad \text{use} \quad x_0 = 0$$

$$x_0 \geq 1 : \quad \text{use} \quad x_0 = 1$$

For shuttle, 2nd stage: 43 to be replaced by 45.5

31 to be replaced by 33.5

Fig. 10.   Optimality condition for the dual propellant parameter x.

By investigation of d^2v/dx^2 it can be shown, that the value x_{opt} will indeed lead to a maximum of v.

1. Results

Numerical investigations of the model as explained lead to following results:
(1) Figure 11 shows the payload ratio $\bar{p}$ (in % of lift-off Mass M_s) for various

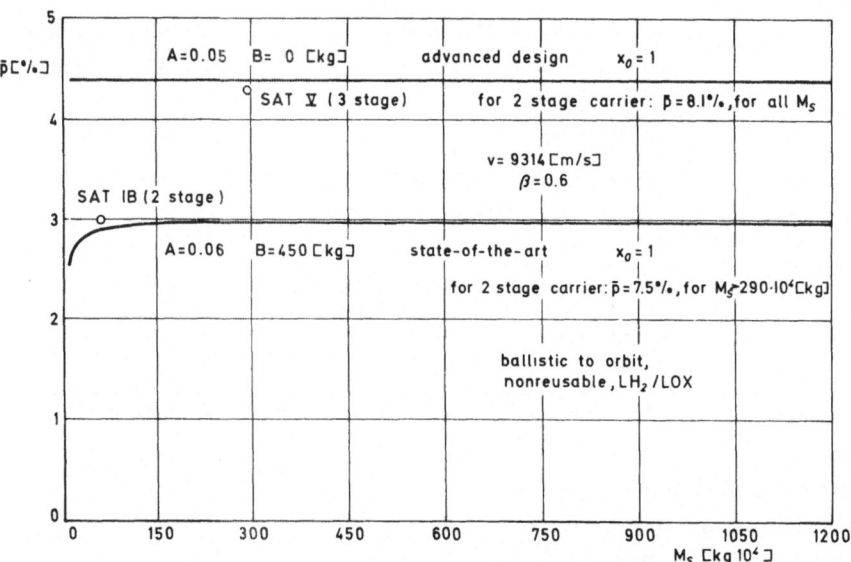

Fig. 11.   Payload ratio $\bar{p}$ (% of M_s) vs stage lift-off mass M_s.

lift-off masses, for ballistic non-reusable single-stage rocket vehicles. Both the 'state of the art'-design and the structurally advanced design are included. The optimization procedure leads to $x_0 = 1$, i.e.: the pure LOX/LH$_2$ design is superior to any mixed-mode application. Numerically, the 'state of the art' results in 2.95% of payload for vehicles of $M_s \geqslant 1500\ t$. The advanced version will carry 4.4%, for all sizes. – A 2-stage design using the 'advanced' figures results in $\bar{p} = 8.1\%$. We get $\bar{p} = 7.5\%$ for large ballistic nonreusable two stagers, using state-of-the-art structures. These figures demonstrate the comparatively low sensitivity of two-stagers for this mission. – Pure LOX/LH$_2$ operation is optimum for 2 stagers. – For comparison, some SATURN-figures are given.

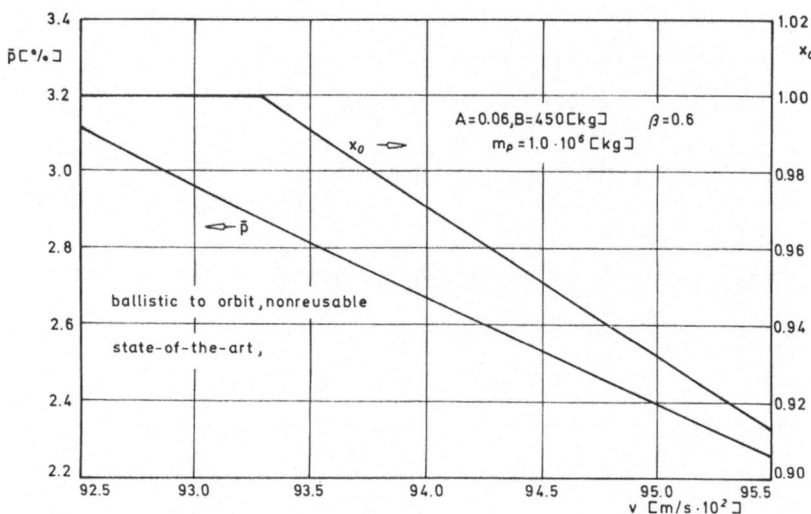

Fig. 12.   Payload ratio $\bar{p}$ and dual propellant parameter x vs velocity requirement v.

(2) In order to get some feel for the sensitivity of the results of Figure 11, we have picked the same problem as in Figure 11 (single-stage ballistic to orbit, nonreusable), one specific propellant mass ($m_p = 1000\ t$), state-of-art-design, but varied the speed requirement v, in Figure 12.

The mixed-mode design becomes superior to the all LOX/LH$_2$-vehicle for $v > 9340$ m s^{-1}. But the actual performance gain (measured in improvement of $\bar{p}$) compared to a LOX/LH$_2$-vehicle is too small to be shown in this figure, up to $v = 9550$ m s^{-1}.

(3) Figure 13 is another attempt to show sensitivity: we take the case of Figure 12, keeping this time $v = 9314$ m s^{-1} constant, but vary A:

For $A > 0.06$ we get a mixed-mode design. For about $A > 0.083$, $\bar{p}$ becomes negative. For all positive $\bar{p}$-values, there is but little $\bar{p}$-difference between the optimized mixed-mode vehicle and the pure LOX/LH$_2$-design. But for rather large A-values, the $\bar{p}$-difference between those two designs becomes quite significant (albeit all $\bar{p}$-values are negative!)

(4) Figure 14 depicts results for the single-stage to orbit ballistic reusable vehicle. Four cases are shown:

(a) state-of-the-art, LOX/LH$_2$:

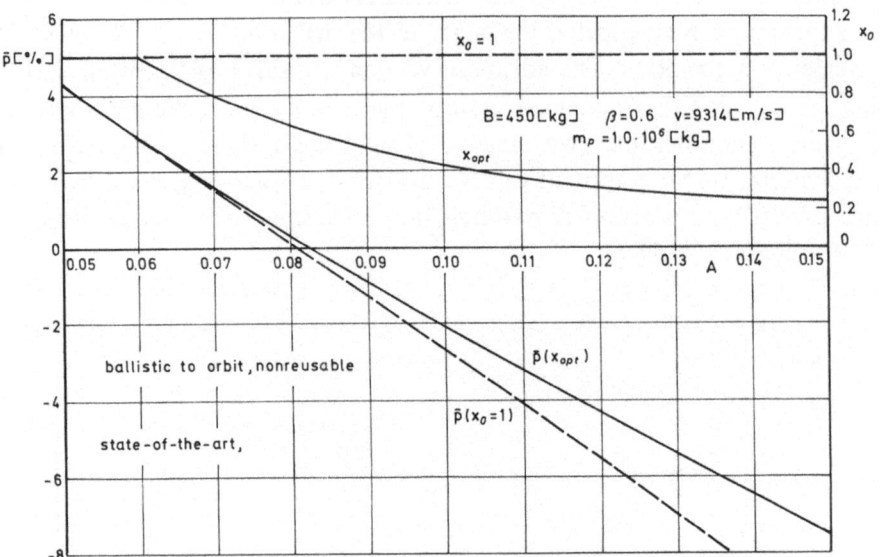

Fig. 13.   Payload ratio $\bar{p}$ and dual propellant parameter x_0 vs structure parameter A.

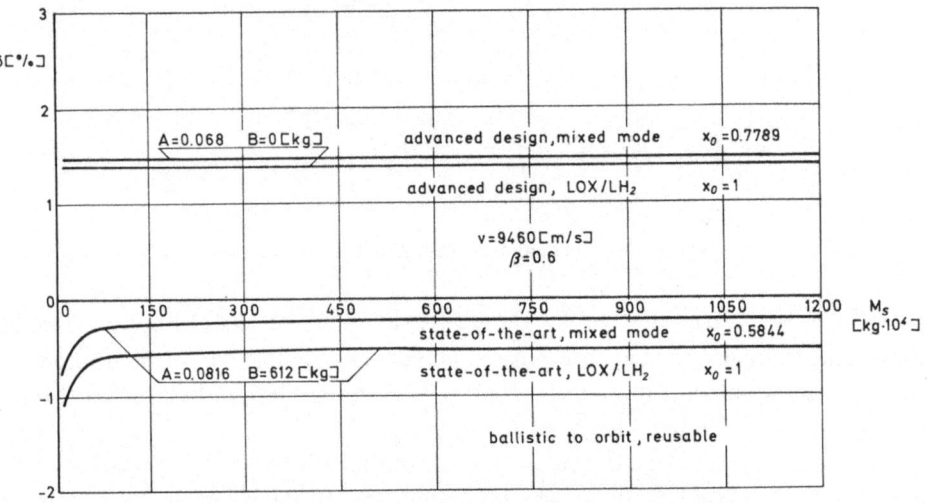

Fig. 14.   Payload ratio $\bar{p}$ (% of M_s) vs stage lift-off mass M_s.

Fig. 14.   Payload ratio $\bar{p}$ (% of M_s) vs stage lift-off mass M_s.

for vehicles of $M_s > 1500\ t$, $\bar{p} \approx -0.5\%$ of M_s.

(b) state-of-the-art, optimized mixed-mode:

$$x_0 = 0.5844$$

for vehicles of $M_s > 1500\ t$ $\bar{p} \approx -0.22\%$ of M_s.

(c) advanced structural design, LOX/LH$_2$:

for all M_s-values, $\bar{p} \approx 1.4\%$ of M_s.

(d) advanced structural design, optimized mixed mode:

$$x_0 = 0.7789$$
for all M_s-values, $\bar{p} \approx 1.5\%$ of M_s.

(5) In Figure 15 the first stage of the winged shuttle has been optimized for both present and advanced state-of-the-art for $x_0 = 1$, i.e. pure LOX/LH$_2$-usage. This

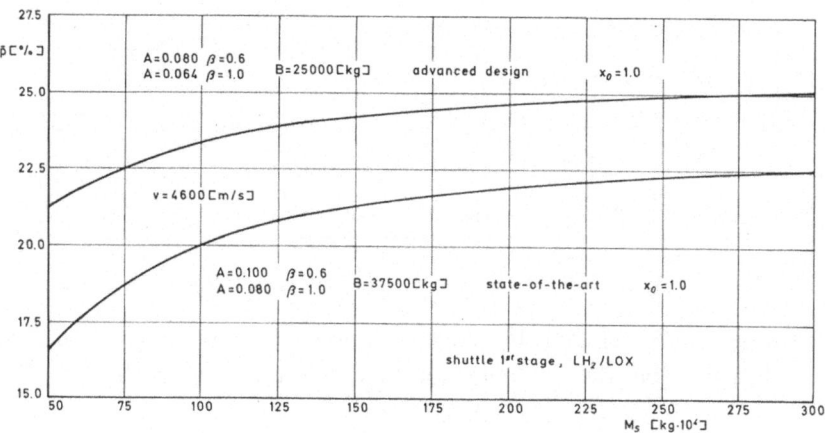

Fig. 15. Payload ratio $\bar{p}$(% of M_s) vs stage lift-off mass M_s.

is true for both the 'standard' choices of $A = 0.10$, $\beta = 0.6$ (respectively $A = 0.08$, $\beta = 0.6$) and the 'β-increased' choices of $A = 0.08$, $\beta = 1$ (respectively $A = 0.064$, $\beta = 1$).

(6) The same observations as for the preceding figure hold for Figure 16, depicting the upper stage of the winged shuttle.

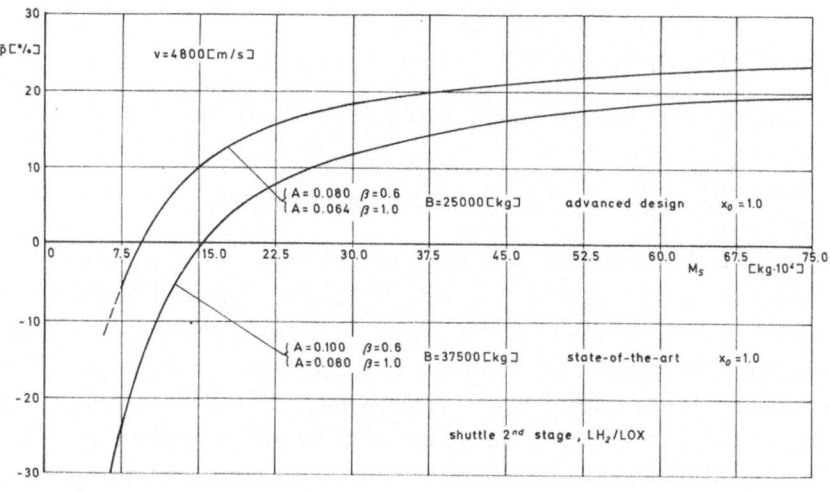

Fig. 16. Payload ratio $\bar{p}$(% of M_s) vs stage lift-off mass M_s.

(7) From the two previous figures, it is easy to put together two-stage orbital winged carriers as in Figure 17. Staging losses, separation devices etc. are neglected;

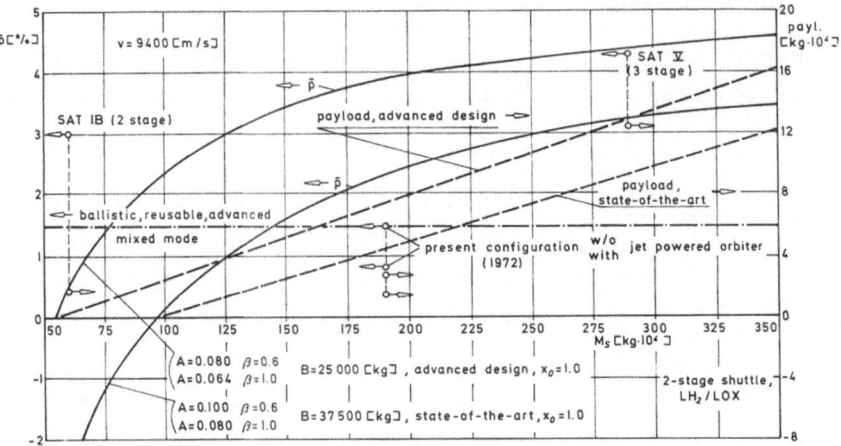

Fig. 17.   Payload ratio $\bar{p}$(% of M_s) vs total lift-off mass M_s for a 2-stage shuttle.

on the other hand, velocity distribution between the two stages is not optimized (though I expect, that the optimum is not far from the figure chosen).

2. Main Results – Summary

(a) In no case enters mixed-mode operation.
  (b) Payload is negative for

$M_s < 950\ t$ (present design) or
$M_s < 570\ t$ (advanced design).

(c) The minimum practical carrier (payl. $\approx 10\ t$) has

$M_s = 1200\ t$ (present design) or
$M_s =\ \ 750\ t$ (advanced design).

(d) Large vehicles will have quite reasonable payload fractions:

present state-of-the-art: $\bar{p} > 3\%$ for $M_s > 2550\ t$
advanced design: $\bar{p} > 4\%$ for $M_s > 2100\ t$.

Tentative performance data for the present (August 1972) design of the shuttle (2 solid boosters, external orbiter propellant tank) are indicated. Compared to these, even the 'present-state-of-the-art' data appear to be optimistic. This is at least in part due to lower jet speed of the shuttle design as it stands at present.

For further illustration, some SATURN-figures and results from Figure 14, advanced structures, mixed-mode are included, as well as absolute payload masses.

For comparison also: 2-stage ballistic nonrecoverable orbital transport, advanced structure: $\bar{p} = 8.1\%$; present state-of-the-art structure: $\bar{p} = 7.5\%$, for large vehicles.

3. Conclusions

(1) Dual propellant Mode ('Mixed Mode') operation will indeed in some cases lead to improved stage performance, compared to LOX/LH$_2$-use only.

(2) Whether this is or is not true, depends quite critically upon vehicle and mission parameters. If these lead to a low payload, dual-propellant-mode may give improved performance.

(3) In none of the cases investigated here, dual-propellant-mode will be significantly better than pure LOX/LH$_2$-use, if we have a positive payload at all.

(4) Single stage ballistic reusable orbital carriers become practical only for structually quite advanced designs; but then there is again little performance difference between mixed-mode and LOX/LH$_2$ only.

(5) From Figure 17, the advanced ballistic vehicle appears to be attractive as compared to the 2-stage advanced winged design for launch mass $M_s < 1000$ t or so; i.e. for payloads below about 20 t. A nonrecoverable 2 stage ballistic orbital carrier would carry more than 5 times as much payload! This is obviously detrimental to the economy of the ballistic single-stage reusable vehicle.

(6) Two-stage winged orbital shuttles will not use mixed mode operation profitably. Minimum useful launch mass will be about $M_s > 1000$ t. For large designs well above $M_s > 2000$ t, attractive payload ratios can be obtained, being of the order of 50% of this value for the ballistic nonrecoverable two stager.

Acknowledgements

I would like to thank the following gentlemen, all from my office:

Dr O. Reidelhuber, who did all numerical work;

Mr Lehmann, who prepared the figures;

and Mr H. Scheffold, who made the slides.

References

[1] Salkeld, R., 'Mixed-Mode Propulsion for the Space Shuttle', p. 52–58, *Astronautics and Aeronautics*, August 1971.

[2] Ruppe, H. O., *Introduction to Astronautics*, Vol. 1, Academic Press 1966.

[3] Ruppe, H. O., *Introduction to Astronautics*, Vol. 2, Academic Press 1967.

[4] *Raumfahrt-Typenblätter of the DGLR*, from 1964 on.

[5] Bond, A., Preliminary Assessment of 'Mixed-Mode Propulsion for the Space Shuttle', Rolls Royce, Ltd., Nov. 1971.

[6] Bono, Ph. and Gatland, K., *Frontiers of Space*, Blanford Press, London, 1969.

[7] Salkeld, R., 'Mixed-Mode Propulsion: Optimum Burn Profile for Two-Mode Systems', *J. of Spacecraft*, June 1972.

[8] Bond, A., personal letter, dated 27.4.72.

[9] Ruppe, H. O., 'Gedanken zum Raumtransporter: Vertikal- oder Horizontalstart' (unpublished).

STUDY OF SHUTTLE-BASED SYSTEMS FOR HIGH-ENERGY PLANETARY MISSIONS*

D. J. SHAPLAND

European Space Research Organisation

and

W. MÜLLER

Messerschmitt-Bölkow-Blohm, F.R.G.

Abstract. The work described in this paper was conceived at a time when Europe was considering the development of the space tug as a possible contribution to the post-Apollo Programme. In this context, it was natural to ask what role the tug might play in the performance of planetary missions. Other injection stages, such as the expendable Agena or Centaur, will also be available and were considered in the study. Data are presented for 10 sample missions representing a spectrum of high-energy, solar system exploration possibilities. Principal mission requirements are established, and their compatibility with the performance of selected post-Apollo Programme transportation elements is examined. Two particularly promising missions emerge Jupiter orbiter and out-of-ecliptic probe.

1. Introduction

Advanced planetary missions can be very demanding in terms of injection-ΔV, and, often, heavy on-board systems are required for the fulfillment of exacting scientific mission requirements. Employing conventional, expendable launch systems can lead to prohibitive weight and volume constraints for the scientist and spacecraft designer, and, sometimes, high programme costs.

The great promise of the space shuttle is a reduction in payload transportation costs, thus providing a potential for economically utilizing low-Earth orbit as a staging point for injecting relatively large payloads on energetic transfer trajectories. The subject study was initiated to explore this potential.

2. Objectives and Approach

The overall purpose of the study was to determine the compatability of the requirements for advanced planetary missions with the capability of the post-Apollo Programme transportation elements, and to determine the associated operational and cost implications. Only feasibility level analyses were performed, thus no detailed evaluation was to be expected; rather, the study provided information on mission feasibility and a first indication of how planetary objectives might be accomplished in the framework of the post-Apollo Programme.

Ten example high-energy missions, covering a range of scientific objectives through-

* This work was performed for the European Space Research Organisation by Messerschmitt-Bölkow-Blohm, under Contract No. 1515/71 EL. More detailed information is provided in the Final Report, URV-52(72), June 1972.

out the solar system were chosen for the establishment of critical mission requirements. The targets ranged from the Sun itself to the outer reaches of the solar system and included one lunar surface application. The missions considered were:

(1) Near-Solar Probe   (6) Asteroid Rendezvous Probe
(2) Mercury Orbiter/Lander   (7) Cometary Probe
(3) Lunar Hopper   (8) Jupiter Satellite
(4) Mars Retriever   (9) Saturn Satellite
(5) Out-of-Ecliptic Probe   (10) Outer-Solar System Satellite

Practical means of performing these missions, utilizing post-Apollo Programme hardware, were then examined in terms of spacecraft concepts and launch configurations. A particularly relevant parameter derived was the number of tugs (i.e. shuttle flights) required to effect the mission.

3. Study Assumptions

The space shuttle was assumed to be the first stage of the transportation system for delivering the planetary system elements viz. spacecraft and injection stage to low-Earth orbit. Some relevant performance and cost data for the assumed injection stages are presented in Figure 1.

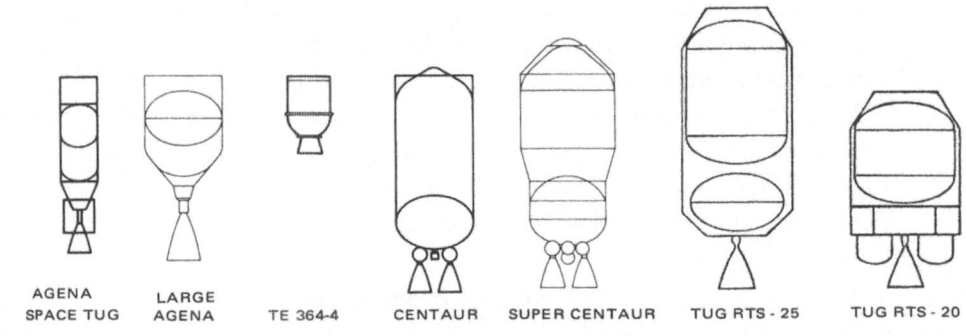

SHUTTLE +	M_P	λ'	I_{SP}	RECURRING PROD. COST	OPERATION COST	NRC
Agena Tug	6.1	0.91	310	3.41 Mio $	0.78	40.5 Mio $
Large Agena Tug	23.5	0.96	310	3.86 Mio $	0.85	46.8 Mio $
Centaur Tug	13.6	0.87	444	8.7 Mio $	1.26 (est.)	31 Mio $
Super Centaur	25	0.91	444		1.3 (est.)	86 Mio $
TE – 364 – 4	1.04	0.92	286	0.7 Mio $ (est.)		
Tug RTS – 20	20	0.87	460	7.65	1.7	500 Mio $
Advanced Tug (25 t)	25	0.90	470	9	1.9	700 M $

Fig. 1.   Shuttle based systems.

The particular reusable tug assumed was the MBB RTS-20, a single stage vehicle of 20 tonnes propellant loading, and 9.4 m length whose performance and cost

TABLE I

Summary of principal mission requirements

Mission name	Scientific objectives	ΔV injection (km s^{-1})	ΔV at target (km s^{-1})	Max. temp. (°K) $\alpha/\varepsilon=1$	Range from Earth at target (AU)	Number of tugs for injection	Mass of spacecraft at target (kg)
(1) Near Solar Probe	Determination of Sun's gravitational potential (Harmonics)	12.3 0.1/1.0 AU orbit		1300	1.96	3	700
(2) Mercury Lander	Direct determination of chemical composition of surface and atmosphere, gravity tides, seismicity	6.7	9.8 (ideal)	650	1.5	6	700
Mercury Orbiter	Surface imagery, geometrical figure, mass distribution by orbit tracking	6.7	6.3	650	1.5	2	300
(3) Lunar Hopper	Geological exploration, geophysical observation, physical processes on surface	3.14	2.38	393	384 000 km	2	3600
(4) Mars Retriever	Sampling of surface material, determination of chemical composition	3.8	7.8 (ideal) for landing/return	300	1.35	3–7	100 (to Earth)
(5) Out-of-Ecliptic Probe	Magnetic field-solar wind interactions, cosmic dust and zodiacal light, latitudinal dependencies	11.5 (30° inclination Earth-Synchromous)		400	0.52	1	350
(6) Asteroid Rendezvous Probe	Surface imagery, mass distribution, investigations of plasma and radiation	CERES 6.2 EROS 3.3	4.4 3.6	250 500	2.3 0.61	1 1	450 500
(7) Cometary Probe	Chemical composition of nucleus and coma, interactions solar-wind-'rocket effect', ionization process of tail gases	KOPFF 3.9 (one impulse transfer)	5.9	400	~1	1	350
(8) Jupiter Satellite	Investigation of magnetosphere, radiation belts, atmosphere, red spot, interaction of magnetosphere-atmosphere, solar wind radiation belts, mapping of planet and satellites	~6.6	2–3.5	160	4.9	1	1000
(9) Saturn Satellite	Investigation of rings composition, existence of magnetic fields, radiation belts, mapping of planet and satellites	7.4 (Jupiter swing-by)	1.6 (10-day orbit)	130	9.98	1	1000
(10) Outer Solar System Satellite (at 15 AU)	Distribution of stars, galactic matter, particle radiation (celestial background), observation of space beyond Saturn	7.4 (Jupiter swing-by)	6.14 (insertion into circular orbit)	100	~14	4	3000

estimates were derived in recent European studies of the reusable space tug concept [1]. The actual specification for a space tug is extremely nebulous at this time, and, to reflect more recent trends towards a larger tug concept, a version containing 25 tonnes of propellant was also included in the study. As such it is representative of an improved performance tug and is referred to in this document as 'advanced tug'. Single and clustered launch configurations were assumed for both vehicles, in the reusable and expendable modes. Data on the expendable systems were taken from NASA-published data.

As regards the space shuttle, the principal design features assumed were a payload capability of 29 500 kg to a 185 km orbit on a due East launch, and payload bay dimensions of 4.56 (dia) × 18.3 m. The launch cost was taken as 10 MAU.

4. General Mission Feasibility

The results of the mission analysis for the ten sample missions are summarized in Table I. These results provide an overall review of the mission requirements, and representative data rather than fully optimized solutions are presented. Six analytical steps were involved in generating the data:

(1) compilation of planetary data (mass, gravity, etc);

(2) identification of scientific objectives and corresponding orbits;

(3) calculation of delta-V requirements for injection from low-Earth orbit on to transfer orbit;

(4) estimation of retro- and rendezvous-manoeuvre requirements and definition of on-board propulsion system;

(5) establishment of total injected mass, assuming a given spacecraft mass; and,

(6) evaluation of the required launch configuration.

The two parameters – maximum temperature (for a surface with an absorptivity/emissivity ratio of unity) and spacecraft-to-Earth range – were derived from typical transfer trajactories; they typify the magnitude of the thermal and communication problems, respectively.

Column 7 of Table I gives the number of tugs required for injection of the planetary system from Earth orbit and resulting in the total mass of spacecraft at the target shown in the final column. The latter mass is considered to be compatible with the stated scientific objectives. It is interesting to note that five of the example missions can be performed using single tug launches. These are:

Out-of-Ecliptic Probe   Jupiter Satellite
Asteroid Rendezvous Probe   Saturn Satellite
Cometary Probe.

The remainder are considered feasible, but the requirement for multiple shuttle launches with the attendant operational and cost implications, render these missions implausible.

Two of the missions – Jupiter Satellite and ex-ecliptic – are considered interesting enough for further discussion here.

5. Jupiter Satellite

The great planet Jupiter, by virtue of its massive size, dense turbulent atmosphere, strong radio emissions, associated Jovian moons, and its primordial state of development, promises rich scientific returns. Fortunately its exploration requirements are generally compatible with a single shuttle launch.

Two different types of orbit were considered in the analysis. The first type comprises ellipitical orbits in Jupiter's equatorial plane well suited for the exploration of the planets surface, magnetosphere, and inner satellites. If a Niehoff trajectory [2] is adopted, encounters are possible with Io, Europa, and Ganymede during one orbital period of the spacecraft. This type of orbit is illustrated by the solid line of Figure 2. The second type includes near-polar orbits with inclinations between 60 and 90 deg to Jupiter's equatorial plane as indicated by the dashed line of Figure 2.

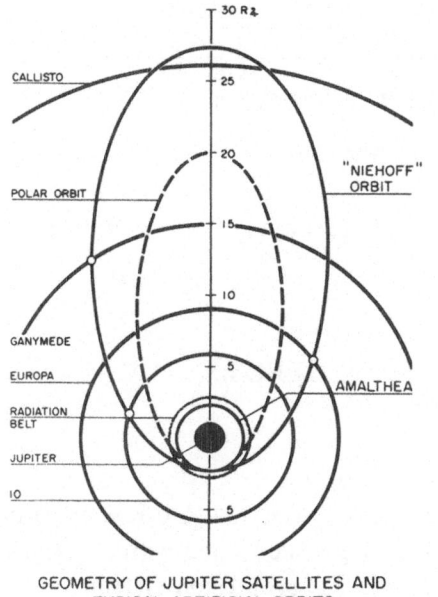

GEOMETRY OF JUPITER SATELLITES AND
TYPICAL ARTIFICIAL ORBITS

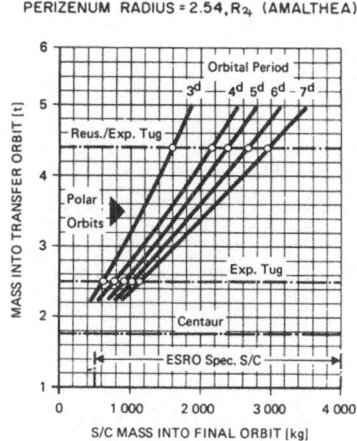

Fig. 2.   Jupiter orbiter-typical orbits and performance characteristics.

Further data on the 1986 mission are given in the figure where the relationship between the transfer mass and final orbit-mass for various orbital periods is illustrated. Quite heavy spacecraft can be orbited by a single shuttle-tug launch.

Because of its scientific promise, this mission merits additional analysis to define a spacecraft concept that is possibly compatible with more readily available stages such as the Shuttle/Centaur or L. T. Agena.

6. Out-of-Ecliptic Mission

Because of its flexibility and current interest, the out-of-ecliptic mission was analysed

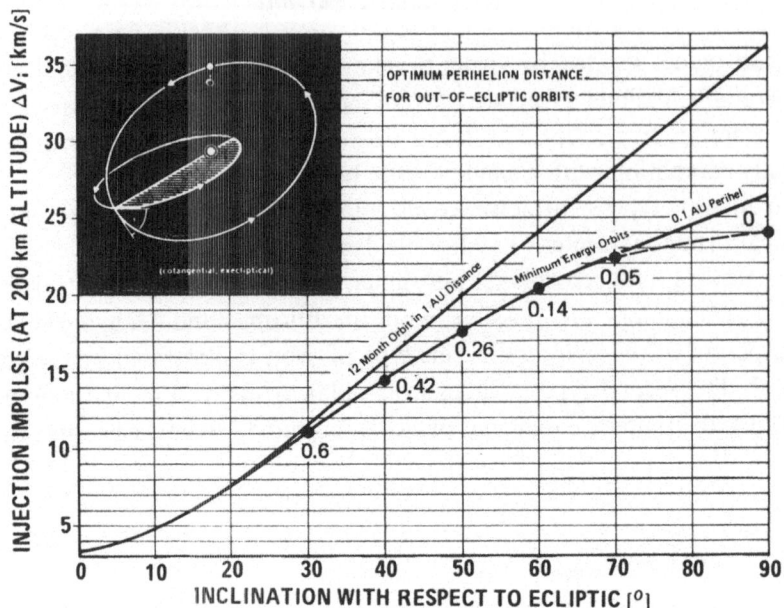

Fig. 3.   Out-of-ecliptic probe injection energy requirements.

further for the following launch modes, to provide preliminary information on a spectrum of possible approaches:

(1) Shuttle + Tug + Kick Stage;
(2) Shuttle + Centaur + Kick Stage;
(3) Shuttle + Tug + Solar-Electric Propulsion System (SEPOS);
(4) Shuttle + Centaur + Kick Stage + SEPOS;
(5) Shuttle + Tug (using Jupiter Swingby);
(6) Titan IIIE + Centaur + Kick Stage (for comparison).

The assumed kick-stage was the solid propellant stage TE 364. A 6 KW stage of total weight 780 kg, utilizing solar power, with an Isp = 5000 s and with a burning time of 10000 h was taken as representative of an electric propulsion system.

The required direct injection impulses (from a 200 km base orbit) for out-of-ecliptic orbits of various inclination to the ecliptic plane are shown in Figure 3. The upper line represents the requirements for Earth-synchronous missions. In fact minimum energy missions exist for certain perihelion distances as shown by the lower curve. However, Earth-synchronous orbits are preferred from the point of view of reduced mission complexity regarding attitude control, communication distance, and thermal problems.

6.1. SCIENCE PAYLOAD

The model payload package assumed for the out-of-ecliptic mission is shown in Table II. Its composition is based on a model payload derived by an ESRO Preliminary Mission Definition Group.

The package includes instrumentation for the measurement of particles, X-rays, and plasma together with instruments for studying the upper latitudes of the Sun and deep space. The total weight is approximately 50 kg.

TABLE II

Model payload package for out-of-ecliptic probe

Experiment	Mass (kg)	Power (W)	Data flux (bps)	Orientation
Plasma probe	15	10	200	to Sun
Magnetometer (2)	6	10	150	mounted on booms
Proton/electron detector	3	3	60	to Sun
X-ray detector	1.5	1.5	15	to Sun
Cosmic ray detector	4	4	90	to Sun
Dust photometer + detector	2.5	6	15	in flight direction
Interstellar H/He detector	5	5	70	$\perp$ to orbital plane
Solar spectrometer	12	10	200	to Sun

6.2. SPACECRAFT CONCEPTS

Utilizing the requirements provided by the model payload, various spacecraft concepts were generated corresponding to the possible mission modes viz. direct injection (spinning and 3-axis stabilized spacecraft), Jupiter Swingby, and electric propulsion. Data regarding these spacecraft are summarized in Table III.

TABLE III   TYPICAL OUT-OF-ECLIPTIC PROBE WEIGHTS

ELEMENT \ WEIGHT (Kg)	Jupiter Swingby	Direct Injection		Solar Electric
	Spin Stabilized	Spin Stabilized	3 Axis Stabilized	3 Axis Stabilized
PAYLOAD	20	20	50	50
STRUCTURE	85	63	75	105
POWER (including solar panels)	78	39	38	160
ATTITUDE CONTROL	13	10	10	13
COMMUNICATIONS DATA	40	26	26	38
PROPULSION PROPELLANT	25	-	-	396
CONTINGENCY + 15%	39	25	33	72
TOTALS	300	183	232	834

The direct injection missions can be initiated any year and a June launch (when the ascending Earth crosses the nodal line of the Sun's equator) results in the maximum heliographic latitude. Utilizing a Jupiter swingby, the perihelion of the final trajectory can be controlled by the distance of closest approach at Jupiter and the orientation of the hyperbola. Typically, for a 1984 mission with an injection delta-V of 7.8 km s^{-1}, an approach to 5 Jupiter radii results in a perihelion of 1 AU and a total flight time of 4 yr.

The solar electric mode is a special case and is the only way of satisfying the large heliographic requirements for a fully 3-dimensional exploration of the solar system. In the example taken, escape from Earth was provided by the Shuttle/Tug or Shuttle/Centaur, resulting in an initial orbit inclination of about 26 deg to the ecliptic. The low thrust from the SEPOS was applied at the ascending and descending nodes (with respect to the ecliptic plane). After a period of 3 yr (7 burns) a heliographic latitude in excess of 60 deg could theoretically be obtained.

6.3. LAUNCHER PERFORMANCE COMPARISON

The performance of the launchers assumed for the study is summarized in Figure 4 with the Titan IIIE – Centaur – Kick-Stage shown for comparison. For the derived

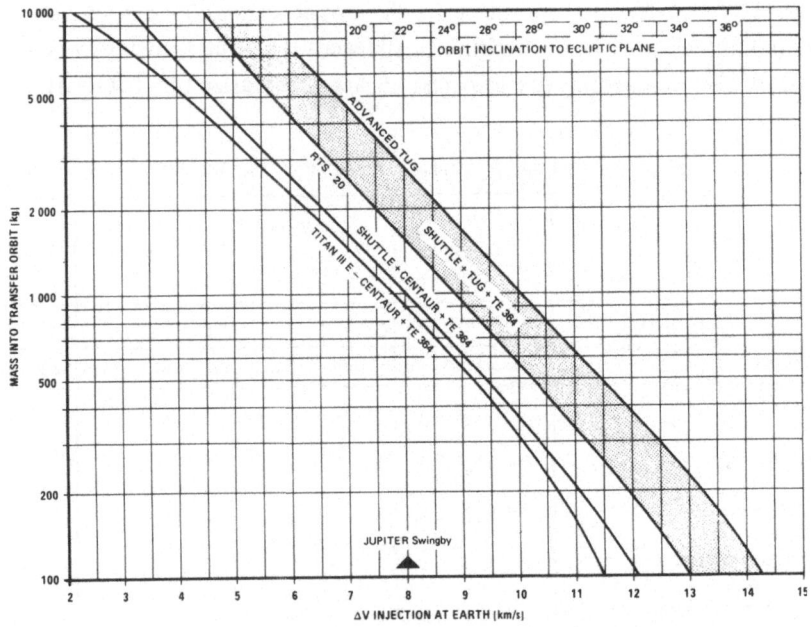

Fig. 4.   Laimcher performance comparison for out-of-ecliptic probe.

spacecraft, the Titan/Centaur can achieve about 28 deg ecliptic inclination compared with 31 deg for the Shuttle/Tug, and even 34 deg if the advanced tug is assumed.

6.4. Programme Cost

Due to the large difference in the launch hardware associated with the various modes studied for accomplishing the out-of-ecliptic mission, programme costs vary from mode to mode. These costs are summarized in Table IV but it is emphasized that they are very tentative and are presented for comparative purposes only.

TABLE IV

Out-of-ecliptic mission cost summary

	Titan III E + Centaur + TE 364	Shuttle + Centaur + TE 364	Shuttle + Tug + TE 364	Shuttle + Tug + SEPOS
Inclination to Ecliptic	28°	23°	31–34°	∼60°
Basic Launch Cost (MAU)	27	10.5 + 8.7 = 19.2	10.5 + 1.5 = 12.0	12.0
Kick Stage (Unit Cost)	0.5	0.5	0.5	0.5
Kick Stage Development	–	–	–	(15.0)
Spacecraft Development Cost	30.5	26	26	35
Spacecraft Flight Unit Cost	12	10.5	10.5	15
Program Cost without experiments and operations (MAU)	70	56.2	49	62.5 (77.5)

Regarding the performance of the out-of-ecliptic mission, it can be concluded that: The mission can be performed in the shuttle era using either a Centaur-class or space tug as a second stage, augmented by a third kick-stage; the resulting mission performance would be generally better, and program costs somewhat less, than could be achieved with an existing Titan IIIE-Centaur. However, an electric propulsion stage is desirable for full achievement of out-of-ecliptic mission objectives.

References

[1] *European Space Tug System Study*, Messerschmitt-Bölkow-Blohm, Final Report on ELDO Contract CTR 17/4/31.
[2] Niehoff, J. C., 'Touring the Galilean Satellites', AIAA Paper No. 70–1070 presented at Astrodynamics Conference, Santa Barbara, Aug., 1970.

ANALYSIS OF POGO STABILITY

S. RUBIN

The Aerospace Corporation, El Segundo, Calif., U.S.A.

1. Introduction

Many liquid propellant rocket vehicles have experienced self-excited longitudinal vibration. The self-excitation or instability arises from interaction of the vehicle structure with the propulsion system. The early experiences with this phenomenon involved vibrations in the first longitudinal mode of the vehicle; the nickname 'pogo' was coined on the basis of a superficial similarity of the out-of-phase motion of the ends of the vehicle with the motion of the pogo stick (child's jumping stick). Figure 1 illustrates a typical occurrence of instability during powered flight in terms of longitudinal acceleration at a point on the vehicle structure. Corresponding oscillations occur within the liquid propellant and in the combustion chamber. The oscillation begins spontaneously, intensifies, and then dies away – typically in a period of 10 to 40 s. The frequency of vibration tracks that of the first structural mode which increases as propellant depletes. Less often, pogo vibrations have occurred in higher modes of longitudinal vibration; also multiple periods of instability, each involving a different mode of vibration, have occurred during operation of a single vehicle stage.

A summary of vibration amplitudes resulting from pogo instabilities on various space vehicles appears in [1]; the frequency of flight instabilities has ranged from 5 to 50 Hz, while acceleration amplitudes (zero to peak) have reached 330 m s^{-2} ($34g$) at engines and 170 m s^{-2} ($17g$) at spacecraft. The structural vibrations can produce an intolerable environment for equipment and astronauts, and can overload vehicle structure. In addition, the attendant pressure and flow fluctuations in the

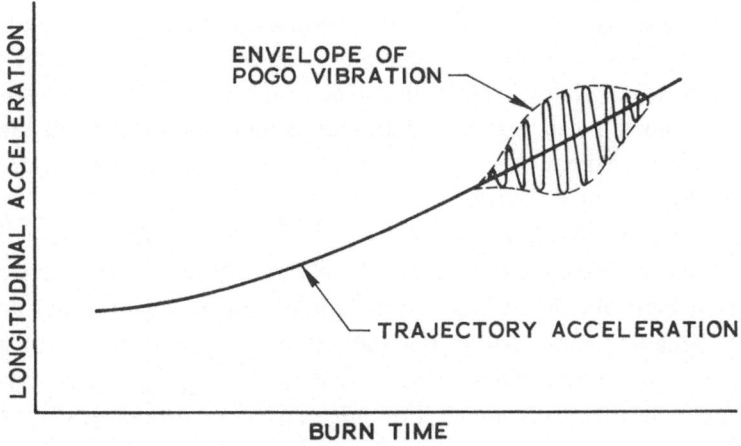

Fig. 1. Typical occurrence of pogo vibration.

L. G. Napolitano et al. (eds.), Astronautical Research 1972, 113–125. All Rights Reserved
Copyright © 1973 by D. Reidel Publishing Company, Dordrecht-Holland

propulsion system may produce various deleterious effects, one of which has been premature engine shutdown. Hydraulic accumulators were incorporated into the feedlines near engine inlets on the Gemini, Titan III, and Saturn V vehicles, and were successful in suppressing or eliminating pogo instability. Thor-Delta vehicles are currently undergoing such a development.

A block diagram of the positive feedback process which can lead to instability is shown in Figure 2. When the structural vibratory accelerations induce the propul-

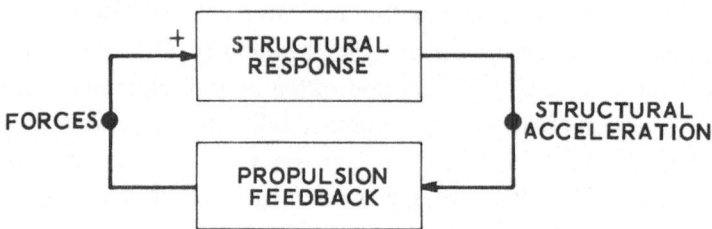

Fig. 2.   Block diagram of pogo feedback process.

sion system to generate forces which then act to intensify the original vibration, the system becomes negatively damped or unstable. The system is time-varying during a flight, with properties affected primarily by propellant depletion and changing hydraulic conditions at engine inlets. At the time when the system properties yield an instability, oscillations will appear spontaneously. At a later time, the oscillations disappear when the system again becomes stable.

The common form of pogo, which can be called 'engine-coupled pogo', results from the action of the tank-to-engine propellant feedlines and the engine itself (see Figure 3).* When the vehicle vibrates longitudinally due to some excitation, the engine and the propellant in the flexible tank undergo oscillatory motions. These two motions produce oscillating pressure and flow in the feedline and within the engine. These hydraulic oscillations produce forcing functions on the vehicle structure, the prime example being engine thrust oscillations. Thus the propulsion system acts to regenerate forces on the structural system, thereby closing the feedback loop as indicated in Figure 2. Although a pump is included in Figure 3, it is not an essential ingredient and pogo has occurred in pressure-fed systems, an example is the French Diamant B vehicle.

In most cases, the variation of the amplitude of oscillations with time during a pogo instability is believed to be the result of a slowly varying limit cycle controlled by nonlinear behavior. Modeling of the coupled system for accurate prediction of limit-cycle behavior is not within the state-of-the-art due to a lack of quantitative definitions of the nonlinearities, such as structural damping, pump cavitation and dynamic gain, and flow resistance. For this reason, we do not attempt to design a

---

* A much less common form of pogo results from the pneumatic behavior of an active pressurization system for the propellant tank ullage. This form is known as 'ullage-coupled pogo' and has been experienced only on Atlas vehicles immediately after liftoff. In this paper we shall restrict discussion to the 'engine coupled' form.

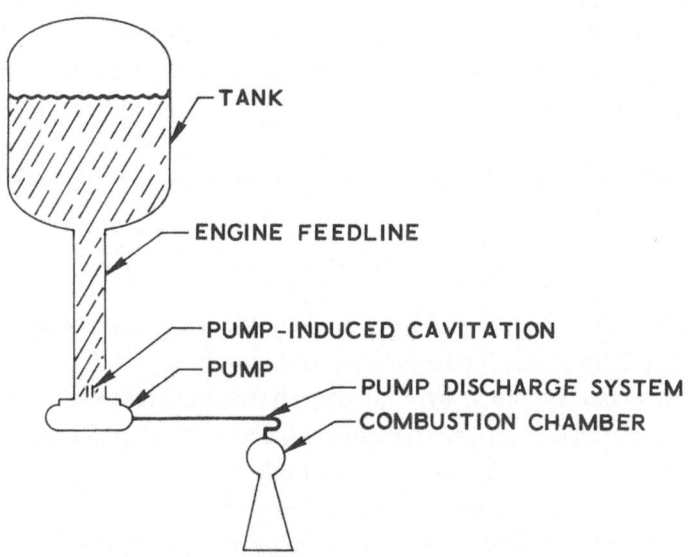

Fig. 3.   Schematic for engine-coupled pogo.

vehicle to meet specified limits of amplitude during a pogo instability. Instead we require that the system remain stable in a linear sense, so that perturbations of the system do not diverge. This objective requires only an ability to predict the damping associated with perturbations of the system variables relative to their steady-state values. As a further simplification we do not, in general, analyze the time varying system. The rate of change of the system properties is usually sufficiently slow that a sequence of fixed-parameter stability analyses at successive times of flight provides an adequate description of the time variation of system stability. Therefore, stability of a vehicle can be determined in most cases through analysis of a time-invariant, linearized-parameter mathematical model. Experience has shown this approach to be valid.

Guidance for the design of space vehicles relative to pogo is provided in a NASA design criteria document [2]. Its statement of a general design criterion is as follows:

The coupled structural and liquid-propulsion system of a space vehicle shall be stable, as determined by a suitable combination of analysis and test. The analysis shall be performed with linearized mathematical models for engine-coupled and ullage-coupled pogo, and prior to flight a space vehicle shall be considered stable when the analysis demonstrates an adequate margin of stability. Instability shall be eliminated by appropriate modification of the coupled system, when necessary. The accuracy of the mathematical models shall be substantiated by ground test and flights of suitably instrumented vehicles.

The ground tests provide experimental data to establish values for the most significant model characteristics, particularly resonant frequencies, dampings, and gains associated with modes of the structural and propulsion systems. Initial flight tests are specially instrumented to permit verification of vehicle stability and to identify dynamic characteristics of the coupled system for overall verification of the mathematical model.

This paper abstracts and expands on certain material from [2] regarding (1) the general character of the mathematical model, (2) the methods and criteria for the stability analysis, and (3) the utility of eigenvectors in evaluating the mathematical model through analysis of test data. [2] contains additional information on these and other aspects of the design for pogo stability, including a comprehensive list of references. A recent overview of the pogo phenomenon is given by [3].

2. Character of Mathematical Model

The structural response to forces generated by the propulsion system is described most conveniently in terms of natural modes of vibration of the overall space-vehicle structural system (see Figure 2). Most analysts have found it advantageous to exclude the tank-to-engine feedlines from the model of the structural system, relegating them to the propulsion system. The tanks are then taken to be closed off at each entrance to a feedline; that is, no propellant is permitted to leave the tank during oscillations in the structural modes. When the propulsion system is coupled to the structural system, special contributions to the generalized forces acting on the structural modes are employed to remove these artificial constraints at the tank bottoms [4]. Other 'external' dynamic forces acting on the structural system are engine thrust, the force acting on the engine inlet due to hydraulic pressure, and the forces resulting from such items as bends, area changes, and resistances in hydraulic flow. The possibility of instability of the closed-loop system stems from the fact that the engines are energy sources. Thrust is the only necessary force for production of an instability. The other forces cannot by themselves lead to instability, but they can influence system stability in concert with engine thrust.

The propulsion feedback indicated in Figure 2 represents the generation of forces by the propulsion system in response to the oscillations of the structural system. The propulsion system is described mathematically in terms of the dynamics of pressure and flow for the feedlines and for the engines. The 'external' excitation of the propulsion system is provided by the motion of the feedline pipes and of the engines, and by oscillatory pressures and motions of the propellant in the tanks. The 'outputs' of the propulsion system are the aforementioned 'external' forces acting on the structural system.

Free oscillations of the closed-loop system are governed by a homogeneous system of equations in the Laplace variable s. This presumes a time variation of the form $\exp(st)$, where $s = \sigma + i\omega$ can be viewed as the 'complex frequency' of an oscillation. Thus, we consider the jth state variable of the coupled system (that is, an acceleration, force, pressure, or flow) to be represented by a complex amplitude $v_j(s)$ for an oscillation varying in time as $\exp(\sigma t) \cdot \exp(i\omega t)$. All this is quite standard: the frequency of the oscillation is ω, the imaginary part of s; the rate of decay or growth of the amplitude of oscillation with time is established by σ, the real part of s; the associated amplitude and phase are represented by $v_j(s)$. Then the system of equations can be put into the matrix form

$$A(s) v(s) = (A_k s^k + A_{k-1} s^{k-1} + \cdots + A_1 s + A_0) v(s) = 0, \tag{1}$$

where $v(s)$ is a vector containing the complex amplitudes of all the state variables and the A_k, $A_{k-1}, \ldots, A_1$, A_0 are real, square, unsymmetrical matrices (usually $k = 2$ to 4). Numerical analysts refer to the solution of Equation (1), with possibly complex coefficient matrices, as the generalized eigenvalue problem.

The eigenvalues or natural complex frequencies of the coupled system are given by those values $s_n = \sigma_n + i\omega_n$ for which the determinant of $A(s)$ vanishes. For each eigenvalue s_n there is a corresponding eigenvector v_n which satisfies the equation $A(s_n) v_n(s_n) = 0$; the eigenvector describes the 'shape' of the natural mode. Thus each eigenmode (i.e., eigenvalue plus associated eigenvector) provides us with a complete physical description of a mode of free oscillation of the coupled system: oscillation frequency ω_n, convergence or divergence rate $\exp(\sigma_n t)$, and the relative amplitudes and phases of the state variables v_n.

In the next section we discuss two basic approaches to the practical conduct of stability analysis. First, we appraise the open-loop technique, which historically has been the most widely employed, and identify certain shortcomings. Then we delve more deeply into the eigensolution (or closed-loop) approach, which is gaining increasing popularity.

3. Stability Analysis

3.1. MULTI-LOOP CHARACTER

Figure 2 shows the closed-loop system in its most elementary schematic form. Actually, a multi-loop block diagram is required to indicate all details of the system's actions and interactions. The multiple loops result from such things as (1) the consideration of a number of natural modes of the structural system, (2) the presence of the hydraulic paths for both propellants, (3) the need to consider a multiplicity of engines due, for example, to differences in their feedlines or in their motion in a structural mode, and (4) the several contributions to the generalized force on a structural mode as mentioned in the previous section.

A less obvious internal loop of the system is the one involved with a purely 'chugging' (or 'buzzing') type of instability of the system [5]. This type of instability results when an oscillation in combustion chamber pressure induces the propellant feed system to generate flows into the chamber, which then cause intensification of the original chamber pressure oscillation. No participation of any total vehicle mode is involved, although some local structural characteristics of the propellant lines or the thrust chamber may be of significance. An elementary block diagram displaying such a loop within the overall pogo loop appears in Figure 4. The diagram is kept simple by restricting consideration to a single structural mode and to the hydraulics of a single propellant. The lower loop by itself could be the source of a chugging instability, ordinarily at a higher frequency than is involved in a pogo instability.

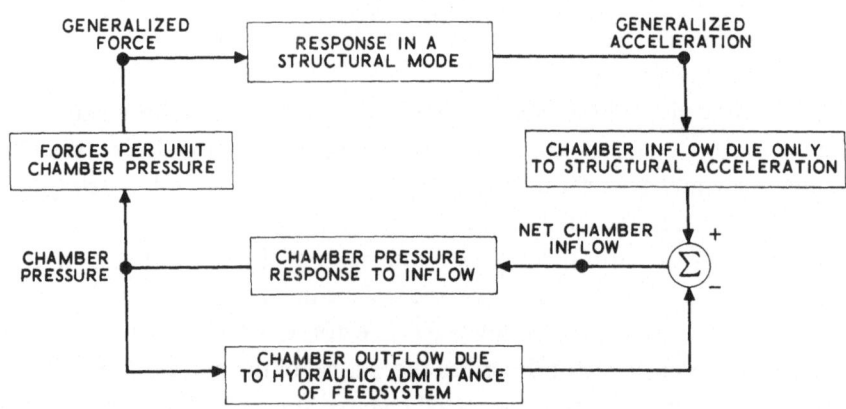

Fig. 4.   Block diagram of chugging loop within pogo loop.

3.2. OPEN-LOOP ANALYSIS

The most popular form of stability analyses has been the application of the Nyquist stability criterion (e.g., [6]). In its simplest concept, application of this criterion involves the opening of the loop at a single point and then determining the frequency response across the opening (that is, output per unit input for $s = i\omega$). This establishes gain and phase margins; these define the minimum gain or phase change which, when instituted at the point of loop opening, would produce a neutrally stable system. Conceptually, the frequency response could also be employed to design a change to the system at the loop-opening point to achieve stability, but this has not been a practical approach for pogo.

Unfortunately, this simple concept is necessarily applicable only for a system involving a single loop. The open-loop approach increases in complexity when the 'complete Nyquist criterion' is applied to a multi-loop system (discussed for the pogo problem in [7] and [8]. Since the open-loop system can itself be unstable, the determination of closed-loop stability makes it necessary either (1) to determine frequency response for a number of configurations of loop openings, or (2) to determine the number of unstable open-loop eigenvalues in addition to the single open-loop frequency response. In one instance of a multi-engine vehicle, the loop was opened by the thrust of one engine and the simple Nyquist criterion did not indicate the actual instability of the system because that particular engine was involved to a minor extent; in this case the open loop was itself unstable.

The system formed by opening the loop of the coupled structure/propulsion system is merely a mathematical artifice used for convenience in the conduct of stability analysis. Its popularity is based on the fact that frequency-response determination is readily accomplished in practice on either an electrical analog or digital computer. However, it is questionable that the application of the complete Nyquist criterion holds any computational advantage relative to an eigensolution approach which is discussed next. Furthermore, the results of the open-loop analysis do not have the clear physical significance of the eigensolutions.

Additionally, a drawback to open-loop analysis of a multi-loop system has been that the resulting frequency response depends on the location of the loop opening; that is, the derived gain and phase margins are not unique. This has led to difficulties in comparing results of different analyses and in relating results to prescribed stability goals. This aspect is correctable, as will be shown later (under *Stability Criteria*) in a discussion of the special margins defined in [2].

3.3. CLOSED-LOOP ANALYSIS

The direct approach to stability analysis is the determination of eigensolutions. This has been less widely employed for pogo because of the lack of recognition on the part of analysts of (1) the practical importance of eigensolutions and (2) the existence of refined computer programs for the general nonsymmetric form of the pogo equations.

Mathematically, the system for pogo is nonself-adjoint (see Section 9-3 of [9]); that is, it is governed by equations of unsymmetrical form. Because the system coefficient matrices in Equation (1) are real, the system characteristically possesses eigenvalues and associated eigenvectors in complex-conjugate pairs, along with real eigenvalues and eigenvectors. The structural dynamicist is generally familiar with a simpler type of eigenvalue problem which involves only real eigenvectors. The more general eigenvalue problem is common to physical systems which can give rise to self-excited oscillations. Indeed, the powerful computer programs now being employed for the general eigenvalue problem are an outgrowth of work begun some twenty years ago to analyze flutter stability of aircraft (for example, [10] and [11] describe the method employed at the author's organization; see also [12] and [13]). A digital-computer program for pogo stability analysis, using the closed-loop approach, is described in [14]; this approach was begun in 1966.

3.4. STABILITY CRITERIA

Experience with pogo instabilities has shown that large flight-to-flight dispersions in peak vibration levels can occur for presumably (but actually not completely) identical vehicles. This indicates a great sensitivity to small deviations in parameters of the coupled system. Although in some cases it is possible to tolerate a coupled structure-propulsion vibration, there is a danger that seemingly minor modification of the vehicle, or of its flight conditions, will increase substantially the severity of the vibration. It is therefore deemed necessary to design for stable operation of the coupled system over the range of uncertainties in system parameters.

Preflight stability-margin requirements were first prescribed for the Titan IIIB vehicle. The requirement for a nominal set of parameter values was 6 dB of gain margin and 30 deg of phase margin (when the loop is opened at each structural modal coordinate). These requirements were based on known practice for control/structural system stability (e.g., Table II of [15]). In addition, a stable vehicle was required for an off-nominal set of parameter values. With the background of favorable Titan IIIB flight experience, these same requirements were instituted for correction of pogo which occurred during first-stage operation of the Saturn V/Apollo vehicle. On subsequent flights of the corrected vehicle, this stage was free of pogo. The requirement for stability of Titan III vehicles for off-nominal parameters was later replaced by one stating that the probability of an instability during vehicle flight shall be less than 0.001 35 [this value corresponds to a onesided three-standard-deviation level for a normal (Gaussian) distribution]. As a result of this experience, it is believed that requirements for pogo prevention are best stated by a combination of margin requirements for nominal parameters plus an acceptably small probability of instability during vehicle flight.

Two criteria are stated in [2]. The basic criterion is that

... uncertainties in parametric values shall be accounted for by appropriate statistical means for establishing that the probability of a pogo instability during space-vehicle flight is sufficiently small.

The additional criterion states that

... as a minimum requirement, the nominal coupled system shall be stable at all times of flight for the following two conditions imposed separately: (1) the damping of all structural modes is halved simultaneously (this corresponds to a *damping gain margin* of at least 6 dB), and (2) any phase shift up to ±30 deg is applied simultaneously to each of the structural modes (this corresponds to a *structural phase margin* of 30 deg).

These margin values are appropriate for a mature design; that is, one whose dynamics are well verified by test. Higher values of the margins can be employed for purposes of preliminary design at the discretion of the analyst.

In connection with the quantitative specification of margins in [2], it was felt desirable not to place a restriction on the method of stability analysis; either open- or closed-loop analysis is permitted. However, it was necessary to define margins in such a way as to avoid any lack of correspondence between an open-loop margin with one based on closed-loop damping. The key concept is to use the stability analysis only to establish whether the least damped system mode is stable, unstable, or neutrally stable. Any appropriate method must give this degree of information. A unique margin can then be based on the degree of change to some parameter of the mathematical model which just causes the system to become neutrally stable in the least damped mode.

An appropriate parameter for a gain margin is felt to be the structural damping ζ (fraction of critical). Thus, a *damping gain margin* (Δ, in decibels) is defined as

$$\Delta = 20 \log_{10}(\zeta/\zeta_N), \tag{2}$$

where, for any structural mode, ζ_N is the fraction-of-critical structural damping for which the system is neutrally stable and ζ is the actual value of the structural damping. The damping of every structural mode employed for the system model is changed proportionately to achieve neutral stability of the coupled system; so damping in any mode can be used for the margin. As an example of the use of Equation (2), when the actual structural damping ratio equals 0.01

$\zeta_N = 0.01$ yields $\Delta = 0$ dB (neutrally stable)

$\zeta_N = 0.005$ yields $\Delta = +6$ dB (6 dB stable)

$\zeta_N = 0.02$ yields $\Delta = -6$ dB (6 dB unstable).

Note that for stability, Δ has a positive value. If the value of $\zeta_N < 0$, there is no concern about the stability of the system. That is, the system is stable even when all damping is removed from the structural modes. In effect, the propulsion-system feedback provides a stabilizing influence rather than a destabilizing one.

In addition to the *damping gain margin*, which is a measure of the sensitivity of coupled-system stability to variation in damping of the structural modes, another measure which should be used in evaluating stability is the sensitivity of the stable coupled system to phase shifts introduced into the structural modes. It is recommended that this sensitivity be measured in terms of a *structural phase margin*, defined as that minimum value of pure phase shift introduced simultaneously into the response of each structural mode that will barely cause the coupled system to become neutrally stable in the least damped mode.

3.5. QUALITATIVE ANALYSIS

The mathematical model for pogo analysis of a vehicle is quite complex. It is therefore useful for the analyst to have a more manageable analytical approach for purposes of improving his understanding of the system and for conducting qualitative analysis. Such an approach has been developed by first simplifying the system to one involving one equivalent engine, one propellant, and one dominant structural mode. Except for the possibility of pure chugging, the system can be treated in a single-loop fashion; chugging can be considered as a separate matter. Next an expression is derived in closed form for the net propulsion-system frequence response in terms of the generalized force per unit modal acceleration. Such an expression appears in Equation (25) of [16]; [4] presents a more general expression which, among other things, includes the effect of dynamic outflow from the tank, an effect incorrectly considered to be negligible in earlier studies. It is then possible to express simply the structural damping ζ_N for neutral stability

$$\zeta_N \simeq -\frac{\bar{H}_I G_p}{2}, \qquad (3)$$

where $\bar{H}_I$ is the imaginary part of the propulsion-system frequency response evaluated at the natural frequency of the structural mode, and G_p is the structural gain (square of the modal displacement divided by generalized mass) for a reference position on the vehicle. Equation (3) is based on Equations (36) through (39) of [16] (note that Equation (39) has a minus sign missing). Thus, we have a qualitative analytical method for determining the damping gain margin defined in Equation (2). The author has found this approach to be highly useful, particularly during the preliminary design of a new vehicle such as the Space Shuttle.

4. Utility of Eigenvectors

The utility of the coupled-system eigenvectors has been poorly recognized in the study of pogo stability. In one case, a computer program [7] derived eigenvalues, but stopped short of determining the associated eigenvectors, although the additional computation is relatively small. As pointed out in [2], the eigenvectors provide a vital link between the mathematical model and observations of the physical system. Failure to consider the eigenvectors is comparable to ignoring mode shapes in studies of natural modes of structural vibration, both from the standpoint of physical insight into the system behavior and correlation with experimental observation.

The only opportunity to observe the dynamic behavior of the overall structure/propulsion system has been during flight of the vehicle. It has not been possible to perform ground testing while simulating the essential conditions of free flight (primarily, the lack of structural constraints and the steady acceleration effects on propellant pressures). Consequently, our ability to observe the dynamics of the overall system resides in two possibilities. First, we can observe the normal random oscillations of the system due to its own normal operation, recognizing that the sources of the excitation are undefined. Or, we may observe the system when it oscillates in a regular manner during an instability. In either case, we observe the frequency of one or more free modes of oscillation of the coupled system, but unfortunately not the associated damping. During instability, nonlinearities in combination with system changes with time prevent the detection of the system's linear damping. During stable operation, we are prevented from extracting damping information by the lack of definition of the random disturbances, and also by the nonstationary character of the data resulting from the system changes with time. However, we have the opportunity to observe the 'shape' of free modes of oscillation; namely, the relationships of amplitude and phase among the system variables. Of course, we are only able to correlate such relationships of the data with the eigenvectors from the mathematical model when the oscillations are small enough to be in the linear range. In the case of random oscillations, we employ spectral-density analysis to determine relationships between variables when the oscillations are relatively coherent (that is, well correlated); examples are given in [17]. Thus, inflight data provides us with a unique opportunity to observe the free modes of oscillation of the overall coupled system and corroborate to some extent the mathematical model.

For example, it has been possible from random flight data to relate pressure oscillations at engine inlets and in a combustion chamber with an acceleration oscillation on the vehicle structure. Figure 5 shows an example of correlation of the flight data with eigenvector information from the mathematical model. The ratio of amplitudes of a pressure oscillation to an acceleration oscillation is plotted as a function of the frequency of a particular mode of the coupled system. The solid line is based on eigensolutions of the mathematical model at a sequence of times of flight during which the frequency of the system mode varies from ω_1 to ω_2. The circled points are derived from spectral analysis of random data at various times of flight for

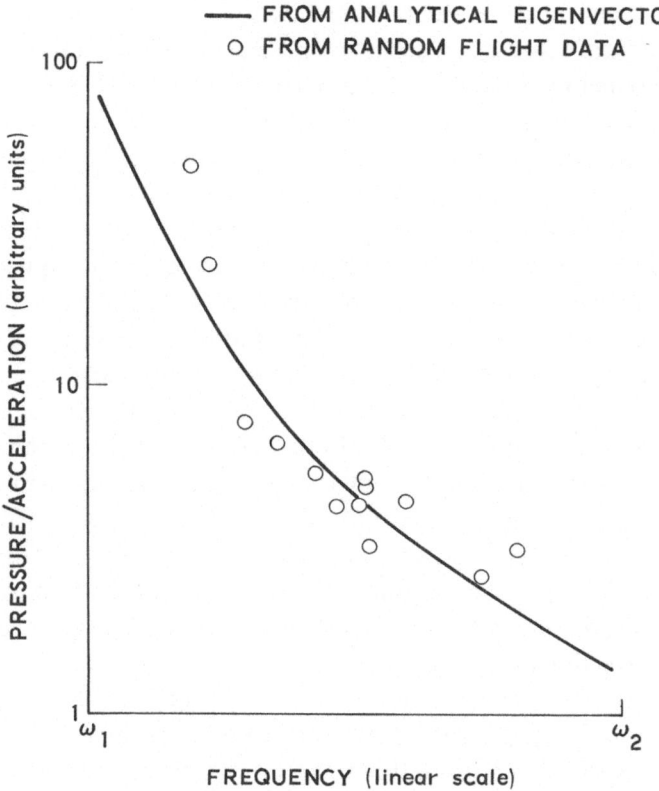

Fig. 5.   Example of correlation of flight data with analytical eigenvector.

two vehicles. Such correlation studies have been helpful in evaluating uncertainties in mathematical modeling.

The comparison of test data with mathematical eigenvector information can also be applied to ground tests of a portion of the flight coupled system. A ground-firing test of an engine is an example. The test is mathematically modeled for coupling of the propulsive portion with the test-stand structure. Eigensolutions for the test system are then compared with amplitude and phase relationships derived from test data (which is generally random in nature). In this manner, there can be an evaluation of the portion of the system which is common to the flight-vehicle system.

5. Summary

Pogo stability is an important consideration for the design of liquid propellant space vehicles. The design criterion established by NASA [2] states that the vehicle shall be designed to be stable under all conditions of flight, taking into account uncertainties in parameters. Experience has shown that stability can be established through analysis of a time-invariant, linearized-parameter mathematical model of the coupled structur-

al and propulsion systems. Special stability margins have been defined which permit
the use of either open- or closed-loop forms of stability analysis without any ambi-
guity in the numerical results. The accuracy of elements of the mathematical model
must be substantiated by ground tests. An overall evaluation of the model must be
obtained from vehicle flights.

Complete information about the stability character of the coupled system is contained in a description of its modes of free oscillation. Mathematically, these modes
are the eigenmodes of the governing system of nonsymmetrical equations which
describe the closed-loop system. Powerful and efficient digital-computer programs
for such a general type of eigenvalue problem do exist, but have not yet been widely
employed for pogo stability analysis. In the hope of stimulating the use of this analytical approach, the paper has identified the attributes of an eigensolution in terms of its
physical significance, and in terms of its utility in evaluating the mathematical model
using test data. A major supply of applicable data can be obtained from propulsion
development tests and from vehicle flights. The data are generally in the form of
random self-oscillations occurring naturally due to normal system operation. When
coherent, these data provide relative amplitude and phase information among the
system variables for comparison to analytical prediction in the form of the system
eigenvectors. Such correlation has been successfully accomplished, in actual practice,
and has led to correction of a mathematical model.

The most widely employed form of stability analysis has been the elementary form
of the Nyquist criterion applied to open-loop frequency response. This approach is
not necessarily applicable for a multi-loop system, which is inherently the case for
the pogo problem. The complete Nyquist criterion is quite complex in application and
any computational advantage over an eigenvalue approach is therefore questionable.
Moreover, the open-loop information does not have the direct physical significance
and the practical utility that the eigenmodes possess. Consequently, the closed-loop or
eigensolution approach is rising in popularity.

A method of qualitative stability analysis has been outlined. The method employs
an approximate closed-form expression for the structural damping required for
neutral stability. The coupled system is simplified to one involving one engine, one
propellant, and one structural mode. The method is proving to be highly useful during
current studies of the Space Shuttle vehicle.

References

[1] Rubin, S., 'Prevention of Coupled Structure – Propulsion Instability (Pogo) on the Space Shuttle.
    Space Transportation System Technology Symposium (Cleveland), Vol. II – Dynamics and Aero-
    elasticity', NASA TM X-52876 (1970), pp. 249–262.
[2] Prevention of Coupled Structure – Propulsion Instability (Pogo)', NASA Space Vehicle Design
    Criteria (Structures), NASA SP-8055 (1970).
[3] Rasumoff, A. and Winje, R. A.: 'The Pogo phenomenon: Its Causes and Cure', Proc. XXIInd Inter-
    national Astronautical Congress (Brussels) (September 1971).
[4] Rubin, S., Wagner, R. G., and Payne, J. G.: 'Pogo Suppression on Space Shuttle-Early Studies',
    NASA CR-2210 (March 1973).

[5] Harrje, D. T. and Reardon, F. H., (eds.): 'Liquid Propellant Rocket Combustion Instability', Chapters 5 and 6, NASA SP-194 (1972).

[6] Chestnut, H. and Mayer, R. W.: *Servomechanisms and Regulating System Design*, Vol. I, John Wiley and Sons, Inc., New York (1959).

[7] Rose, R. F., Staley, J. A. and Simson, A. K.: 'A Study of System-Coupled Longitudinal Instabilities in Liquid Rockets', AFRPL-TR-65-163, Parts I and II, Air Force Rocket Propulsion Lab (September 1965).

[8] Goldman, R. L. and Reis, G. C.: 'A Method for Determining the Pogo Stability of Large Launch Vehicles', Report No. TR-69-7G, Research Institute of Advanced Studies (June 1969).

[9] Halfman, R. L.: *Dynamics, Vol. II – Systems, Variational Methods, and Relativity*, Addison-Wesley Publishing Co., Inc. (1962).

[10] Holt, J. F.: 'Zeros of Arbitrary Functions Using Arbitrary Guesses', Proc. SIAM-SIGNUM 1972 Symposia on Numerical Analysis and Computation, Austin, Texas (October 1972).

[11] Holt, J. F.: 'ASC MULE, General Root Finding Subroutine', Report No. TOR-0073 (9320)-8, The Aerospace Corporation (March 1973).

[12] Peters, G. and Wilkinson, J. H.: '$Ax = \lambda Bx$ and the Generalized Eigenproblem', *SIAM J. of Numerical Analysis*, **7** (4) (December 1970), pp. 479–492.

[13] Wilkinson, J. H.: *The Algebraic Eigenvalue Problem*, Clarendon Press, Oxford (1965).

[14] Payne, J. G.: 'Pogo Stability Analysis Program', Report No. TOR-0059(6122-30)-9, The Aerospace Corporation (1970).

[15] 'Effects of Structural Flexibility on Launch Vehicle Control Systems', NASA Space Vehicle Design Criteria (Guidance and Control), NASA SP-8036 (1970).

[16] Rubin, S.: 'Longitudinal Instability of Liquid Rockets Due to Propulsion Feedback (POGO)', *J. Spacecraft Rockets*, **3** (8) (August 1966), pp. 1188–1195.

[17] Wagner, R. G. and Rubin, S.: 'Detection of Pogo Characteristics by Analysis of Random Data', Proc. ASME Symposium on Stochastic Processes in Dynamical Problems (Los Angeles) (November 1969), pp. 51–62.

DEVELOPMENT AND IN-FLIGHT PERFORMANCE OF THE MARINER 9 SPACECRAFT PROPULSION SYSTEM*

D. D. EVANS**, R. D. CANNOVA[†], and M. J. CORK[†]

Jet Propulsion Laboratory, Pasadena, Calif., U.S.A.

Abstract. On November 14, 1971, Mariner 9 was decelerated into orbit about Mars by a 1334 N (300 lbf) liquid bipropellant propulsion system. This paper describes and summarizes the development and in-flight performance of this pressure-fed, nitrogen tetroxide/monomethyl hydrazine bipropellant system. The design of all Mariner propulsion subsystems has been predicted upon the premise that simplicity of approach, coupled with thorough qualification and margin-limits testing, is the key to cost-effective reliability. The Mariner 9 subsystem design illustrates this approach in that little functional redundancy is employed. This paper summarizes the design and test rationale employed in the Mariner 9 design and development program.

The qualification test program and analytical modeling are also discussed. Since the propulsion subsystem is modular in nature, it was completely checked, serviced, and tested independent of the spacecraft. Proper prediction of in-flight performance required the development of three significant modeling tools to predict and account for nitrogen saturation of the propellant during the six-month coast period and to predict and statistically analyze in-flight data. The flight performance of the subsystem was excellent, as were the performance prediction correlations. These correlations are presented.

1. Introduction

The Mariner 9 Mars-orbiter spacecraft was the sixth in a series of Mariner spacecraft which have explored the planets Mars and Venus since 1962. The previous five spacecraft completed flyby missions which provided only brief encounters with the target planet. The Mariner 9 spacecraft, however, was placed in a 12-h orbit about Mars, thereby allowing a repeat of the close encounter sequence twice a day for an extended period of time. Scientific instruments included a wide-angle television camera for surface mapping, a narrow-angle television camera for close-up studies, two infrared instruments, and one ultraviolet instrument for surface and atmospheric properties measurements. Behavior of the spacecraft radio signal at the entrance and exit of Earth occultation phases provided additional atmospheric information, and the orbital characteristics allowed study of the Mars gravitational field.

The earlier Mars and Venus spacecraft utilized small monopropellant hydrazine spacecraft propulsion systems designed to accomplish up to two interplanetary trajectory correction maneuvers. The Mariner 9 spacecraft, illustrated in Figure 1, was designed to use the basic Mariner 6 and 7 (Mariner 1969) spacecraft with the incorporation of a new and larger propulsion subsystem. This bipropellant subsystem was designed to accomplish in-transit trajectory corrections, to decelerate the

---

* This paper presents the results of one phase of research carried out at the Jet Propulsion Laboratory, California Institute of Technology, under Contract NAS 7-100, sponsored by the National Aeronautics and Space Administration.

** Assistant Manager for Flight Systems Development, Liquid Propulsion Section.

[†] Group Supervisor, Liquid Propulsion Section.

[†] Group Supervisor, Propulsion Systems Analysis and Advanced Engineering Section.

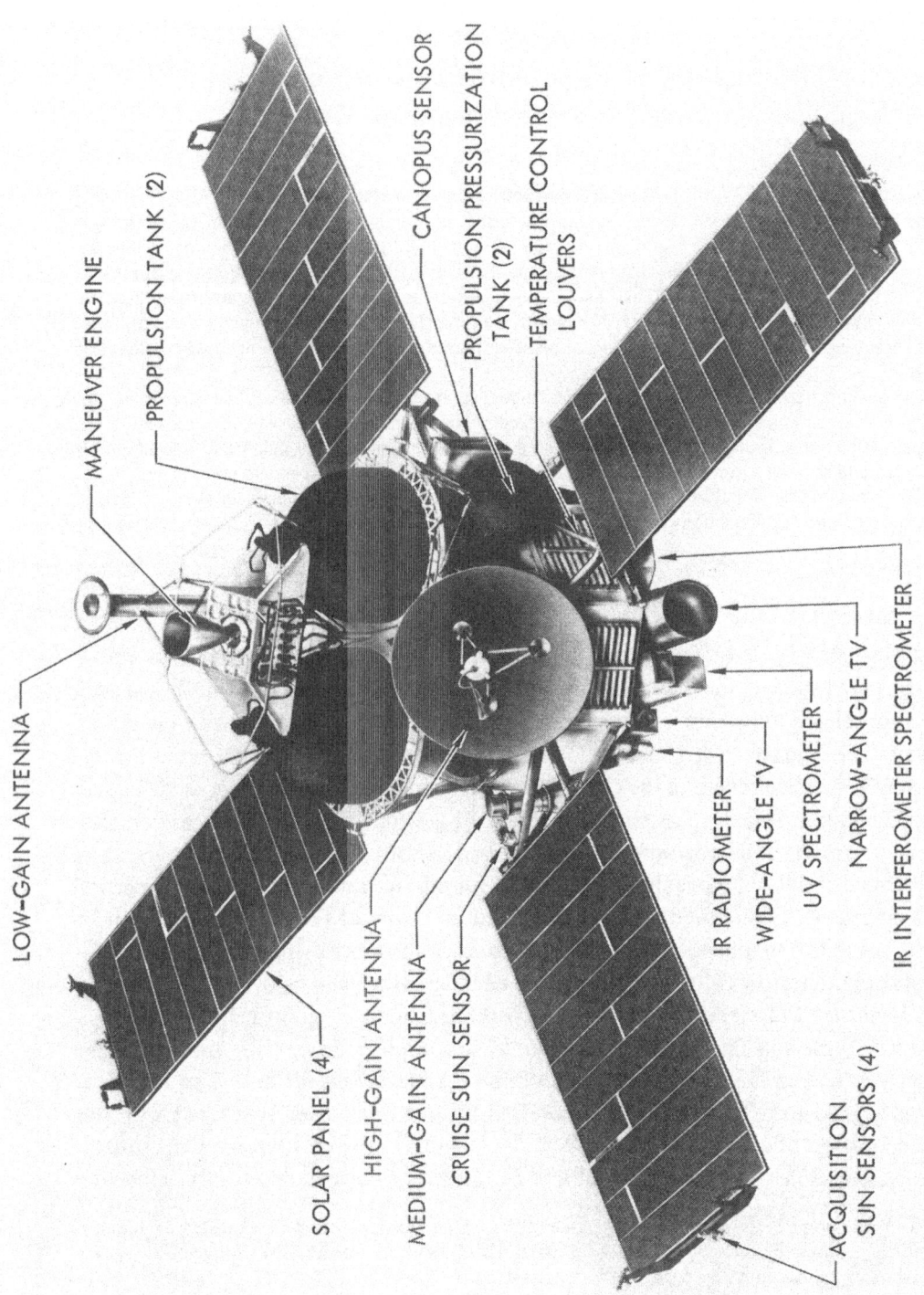

spacecraft from a hyperbolic approach trajectory into an elliptical orbit about Mars, and to perform subsequent orbit trim maneuvers as required.

The basic Mariner propulsion philosophy embodies three key principles:

(1) Provide cost and weight effectiveness by simplicity of design coupled with thorough margin limit testing. For unmanned spacecraft, this approach, rather than that of redundancy, has proven successful.

(2) Design the propulsion subsystem to be modular and man-rated when fueled and pressurized. This approach allows subsystem fabrication and propulsion testing independent of the spacecraft, thus decoupling expensive spacecraft operations from propulsion operations. This proves to be extremely valuable during launch preparations where the tested, fueled, and pressurized propulsion system can be independently checked and fueled and later delivered to the spacecraft for mating and encapsulation.

(3) Provide 'pathfinders' for all critical operations. Prior to assembly, test, or other operations on flight hardware, 'pathfinder' operations are conducted wherein all personnel, procedures, and equipment undergo a dress rehearsal before hazarding the flight hardware.

These principles were applied to the Mariner 9 design and development as will be discussed subsequently.

2. Description and Operation

2.1. SUBSYSTEM

The propulsion subsystem is shown in Figures 2 and 3. The subsystem pressurization is by gaseous nitrogen. Pressurant is isolated from the remainder of the subsystem by the commandable pyrotechnic valves of the pressurant control assembly (PCA). Upon actuation of one of the PCA normally closed valves, pressurant flows from the pressurant tanks through the pressurant filter and the regulator, whose outlet pressure is controlled to $1741 \times 10^3 \mathrm{N} \mathrm{m}^{-2}$ (253 lbf in.$^{-2}$). After flowing through the regulator, pressurant flows into the pressurant check and relief assembly (PCRA) and into the propellant tanks.

Once in the propellant tank, the pressurant causes the bladder to collapse about the standpipe and expel propellant through the gas separation device and into the propellant isolation assembly (PIA). The PIA controls propellant flow to the rocket engine with three normally closed and two normally open pyrotechnic valves and a filter. After leaving the PIA, propellant flows through the flex lines, which permit gimballing of the rocket engine, and to the rocket engine solenoid valve. The rocket engine operates with N_2O_4 and MMH at a mixture ratio of 1.57:1; the hot gases are expelled through a nozzle with an expansion ratio of 40:1.

Servicing valves are used to provide access to the inlet and outlet sides of the pyrotechnic valves in the PCA and PIAs, to the downstream side of the check valves in the PCRA, and to the propellant tank side of each standpipe. Pressure transducers provide pressure information at the PCA inlet, downstream of the check valve in the PCRA, and at the PIA outlet as well as providing rocket engine combustion chamber pressure.

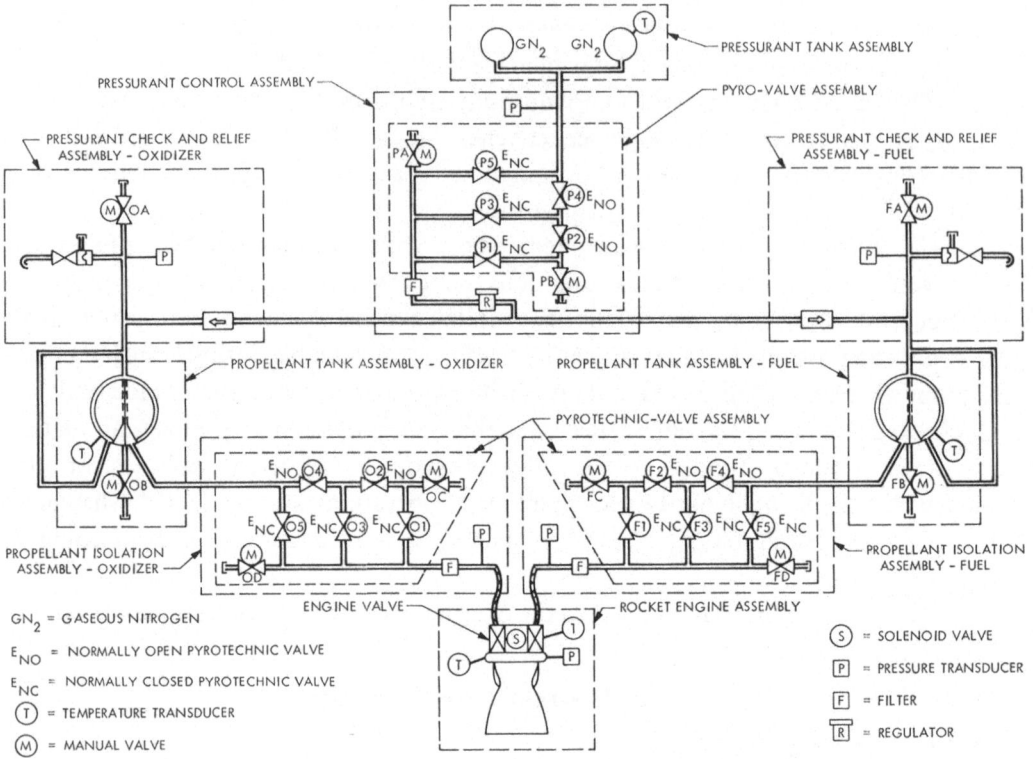

Fig. 2.   Schematic diagram of Mariner Mars 1971 propulsion subsystem.

Fig. 2.   Schematic diagram of Mariner Mars 1971 propulsion subsystem.

The propulsion support structure, a beryllium tube truss with magnesium and steel fittings, is attached to the upper octagonal spacecraft frame and supports the propulsion equipment, the high-gain antenna, and the lowgain antenna.

The following operating sequence was that upon which the subsystem design was based. A later section describes the sequence actually followed. Before the first trajectory correction, the engine valve must be opened to bleed the air trapped between the normally closed propellant pyrotechnic valves and the engine valve. Actuation of the first set of pyrotechnic valves, P-1, O-1, F-1, pressurizes the propellant tanks and allows propellant flow down to the engine valve. The trajectory-correction maneuver is performed by opening the engine valve; this causes the propellant to flow into the thrust chamber, undergo hyperbolic ignition, and continue to burn until such time as the desired velocity increment is obtained as determined by an on-board integrating accelerometer. At this time, the engine valve is closed by removing its electrical power. Later, the propellant and pressurant lines are closed by actuation of the second set of pyrotechnic valves, P-2, O-2, F-2, to guard against leakage after tracking data confirm that no more propulsion maneuvers will be required before the nominal time of the second trajectory correction. The pressurant and propellant lines are reopened, by the third set of valves, P-3, O-3, F-3, just before the second trajectory

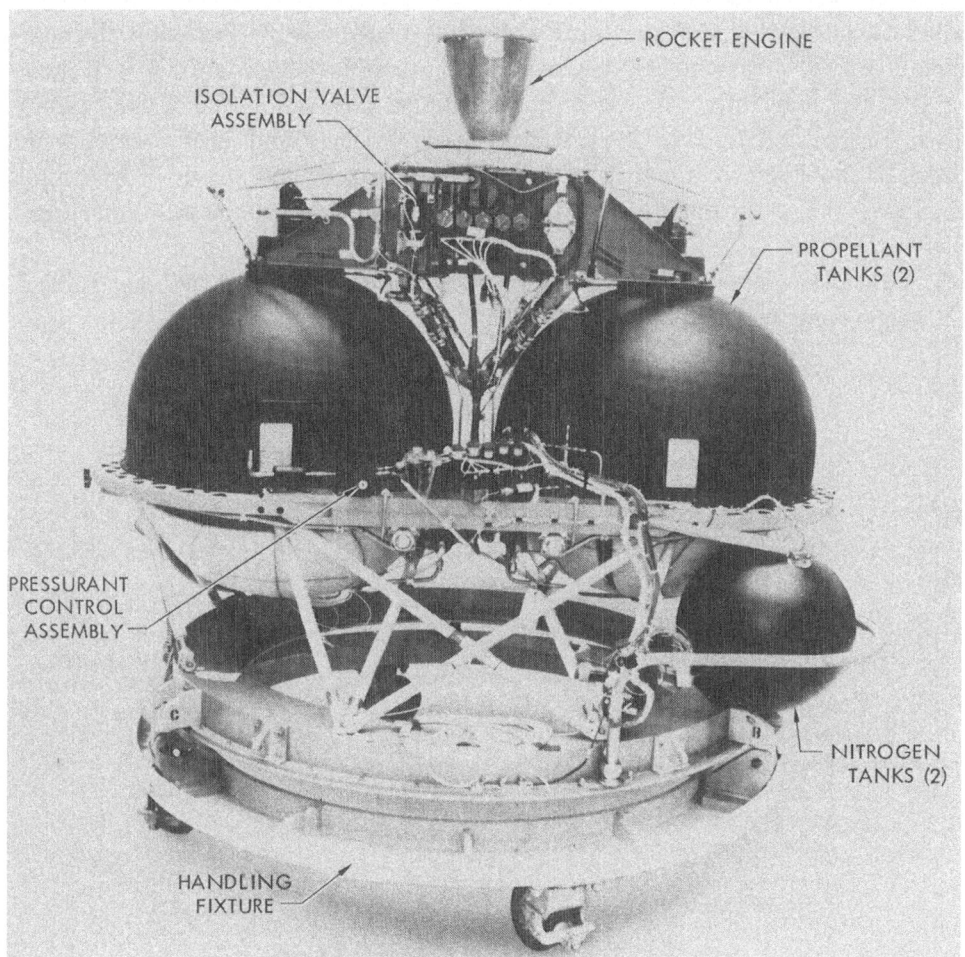

Fig. 3. Mariner 9 propulsion subsystem.

correction, if such a correction is needed. The valves remain open for the orbit insertion maneuver. The orbit insertion maneuver was expected to involve an approximate 840-s duration burn to place the spacecraft into the initial orbit. Within two days after orbit insertion, one or two orbit trim maneuvers were anticipated to place the spacecraft in a precision 12-hour period orbit. After tracking data confirm correct orbital characteristics, operation of the fourth set of valves, P-4, O-4, F-4, is possible to isolate the propulsion fluids for the rest of the mission. An additional set of valves, P-5, O-5, F-5, is available for subsequent maneuvers if needed.

Actuation of the pyrovalves and management of solenoid power for the engine valve is accomplished by necessary power switching in the pyrotechnics subsystem. Thrust vector control during engine firing is provided by the use of gimbal actuators for pitch and yaw control and cold gas jets for roll control.

Referring again to Figure 2, note how the components are arranged into identifiable subassemblies. Each subassembly contains a group of components that can be physically located together and functionally tested as a subassembly. Also note the commonality of the subassemblies. The pyro valve assembly, common to three of these, was designed to be interchangeable, thus allowing a 'production' run of this building block. Furthermore, the fuel and oxidizer pressurant check and relief assemblies are identical, allowing economies in design, production, test, and spares provisioning. This also is the case with the fuel tank assemblies and the propellant isolation assemblies.

Table I summarizes the propulsion subsystem performance characteristics; Table II is a weight summary.

TABLE I

Propulsion subsystem performance characteristics

Parameter	Value	
Vacuum thrust	1334 ± 89 N	$(300 \pm 20$ lbf$)$
Vacuum specific impulse	2775 ± 49 N-s kg^{-1}	$(283 \pm 5$ lbf-s lbm$^{-1})$
Thrust chamber expansion ratio		40:1
Thrust chamber pressure	$806.7 \pm 55 (\times 10^3)$ N m^{-2}	$(117 \pm 8$ lbf in.$^{-2})$
Propellant loaded mixture ratio, O/F by weight[a]		$1.50 ^{+0.05}_{-0.03}$
Nominal oxidizer flow rate	0.289 kg s^{-1}	$(0.637$ lbm s$^{-1})$
Nominal fuel flow rate	0.192 kg s^{-1}	$(0.424$ lbm s$^{-1})$
Propellant load capacity	462.7 kg	$(1020$ lbm$)$
Usable propellant load capacity	440 kg	$(970$ lbm$)$
Propellant loading accuracy	± 0.45 kg	$(\pm 1.0$ lbm$)$
Minimum burn duration		0.4 s
Shutdown impulse variation, 3σ	± 22.2 N-s	$(\pm 5$ lbf-s$)$

[a] O = oxidizer; F = fuel.

2.2. PROPELLANT FEED SYSTEM

The fabrication of the propellant feed system major subassemblies was performed by the Martin Marietta Corp. (MMC), Denver Division, under contract to JPL. This responsibility included the procurement of the components and their acceptance and qualification testing. The only components not purchased by MMC were the propellant tank shells and the flex lines, which were procured by JPL. The components were incorporated with detail parts machined by MMC to form the subassemblies, which were then acceptance tested and provided to JPL. Figure 4 shows a completed subassembly.

Upon their receipt at JPL, the subassemblies were mounted on the subsystem structure and joined to their interconnecting plumbing. When assembly of the propulsion subsystem was completed, it was then subjected to the subsystem flight acceptance test.

The connection of components within subassemblies and the interconnection of

TABLE II

Mariner Mars 1971 propulsion weight summary (dry mass)

Description	Dry mass	
	kg	lbm
Pressurant tank assembly	24.9	55.0
Pressurant control assembly	5.6	12.4
Pressurant check-relief assembly (2)	2.1	4.6
Propellant tank assembly (2)	30.4	67.0
Propellant isolation assembly (2)	10.2	22.4
Rocket engine assembly	7.8	17.1
Tubings and fittings	5.9	13.1
Thrust plate assembly	5.7	12.6
Truss and ring assembly	9.1	20.1
Propellant tank thermal covers (2)	2.5	5.5
Cable harnesses (2)	2.2	4.9
Gimbal actuators (2)	2.5	5.6
Squibs (15)	1.0	2.3
Total propulsion subsystem	110.0	242.6

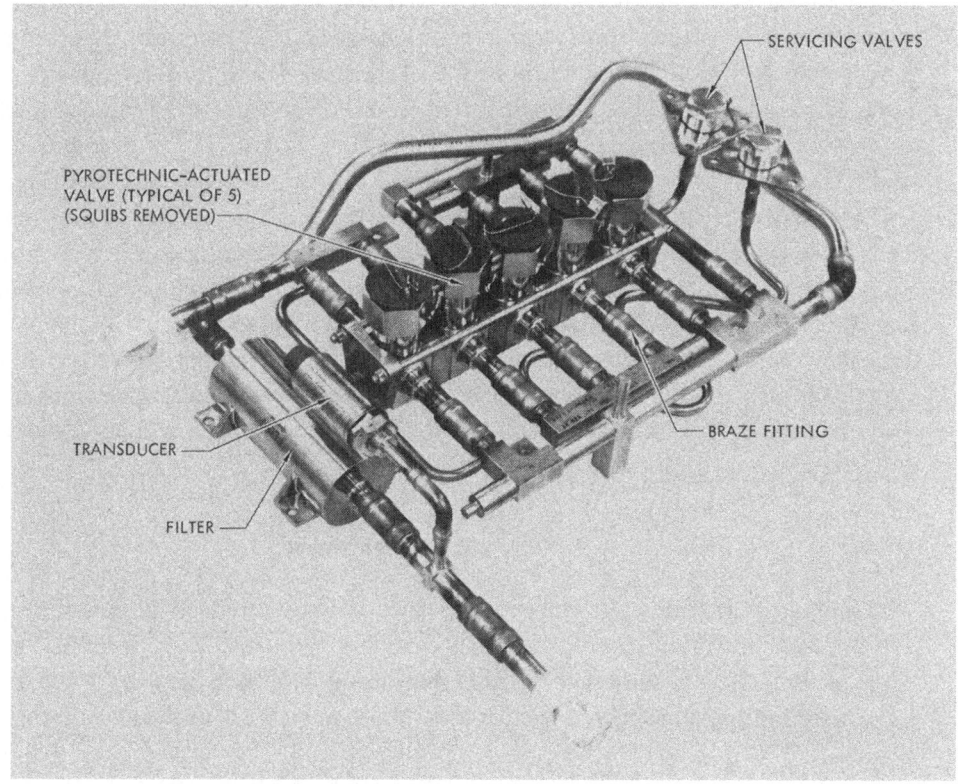

Fig. 4. Propelland isolation assembly.

subassemblies within the propulsion subsystem was accomplished by in-place induction brazing. With this technique the number of mechanical external seals on the subsystem was reduced to 16: 10 service valves, each with a primary and a redundant seal, two tank flanges with aluminimum crush gasket seals, and four 'AN-type' fittings, two on each flex hose, with crushable aluminum seals. This fabrication technique resulted in a subsystem external leakage rate of less than 1×10^{-5} STP cm^3 s^{-1} when the subsystem was pressurized to its operating pressures with helium.

Components from existing programs were selected wherever possible to minimize development and qualification. Some minor changes and improvements were incorporated in several components due to performance requirements and the need for long-term exposure to propellants.

After manufacture, the subassemblies were flight-acceptance-tested before being integrated into the subsystem. The sequence of FA testing was to ensure proper assembly, functional operation, and cleanliness verification.

2.3. ROCKET ENGINE ASSEMBLY

The Mariner 9 rocket engine, shown in Figure 5, was manufactured by theRocketdyne Division of North American Rockwell Corporation. It is a two-piece conductively cooled combustion chamber and radiation-cooled nozzle extension. The engine is equipped with a torque-motor-operated, mechanically linked bipropellant control valve produced by the Moog Corporation, Aerospace Division, East Aurora, New York. The combustion chamber, fabricated from hot-pressed beryllium, is attached to the 40: 1 cobalt alloy nozzle extension by a Renè-41 nut. The engine employs a unique method of thermal control developed by its manufacturer, and termed 'INTEREGEN'. Heat transferred convectively to the engine is conducted through the thick, highly conductive chamber walls and transferred, again convectively, to the boundary layer coolant (BLC) covering the thrust chamber walls near the injector. The BLC covering is also convectively heated from the hot gas side. In this manner the engine can run for long periods with a near-steady temperature distribution. Success of this cooling technique depends on the heat absorption capabilities of the BLC and the proper thermal management in the metal walls so that adequate protection from the hot combustion gases is afforded.

3. Subsystem Development

The subsystem development sequence is shown in Figure 6. Early design of the subsystem made maximum use of previously qualified hardware. The conceptual design was first evaluated in the breadboard system using surplus or prototype hardware. Tests were conducted using this system to evaluate general operation and characteristics.

The engineering test model (ETM) was the initial subsystem with fully operational components and subassemblies. The ETM was used to evaluate operation and performance of the subsystem over a wide range of conditions and environments. It

Fig. 5.   Mariner Mars 1971 rocket engine assembly.

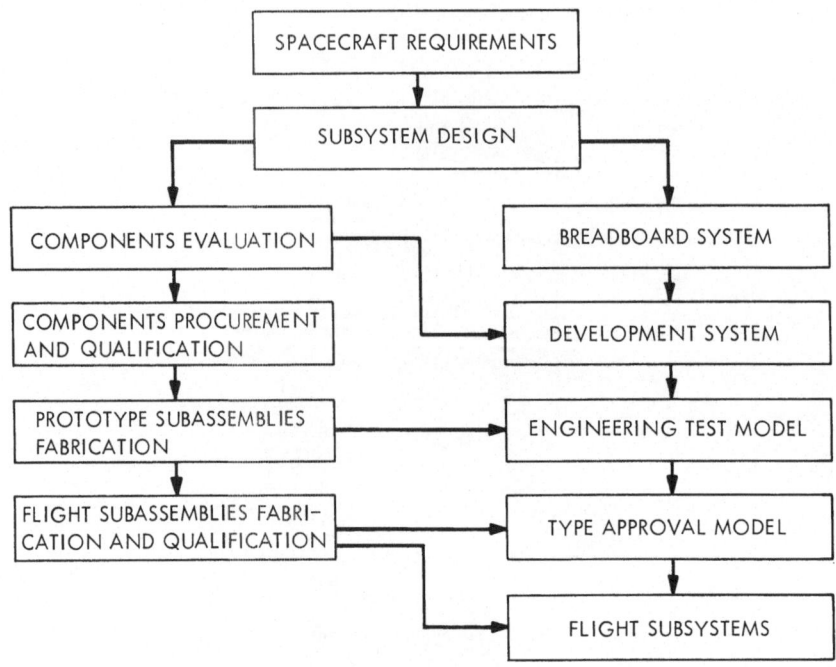

Fig. 6. Development sequence.

also served as a pathfinder for fabrication, assembly, checkout, and other operational aspects.

The ETM was loaded with solvents and subjected to flight-acceptance and type-approval vibration in a single axis. Five hot firing test series were conducted on the ETM in order to pathfind the conditions planned for the type approval (qualification) program and to evaluate performance after long-term (three-month) exposure to propellant. The results indicated that the ETM operated and performed satisfactorily.

Early in the design phase of the propulsion subsystem, the comparisons and trade-offs between welding and brazing of tube-to-tube and tube-to-component joints were resolved to a choice of the induction brazing process.

Aeroquip equipment (Aeroquip Corp., Aircraft Division, Jackson, Michigan) was utilized, consisting of (1) a 15 kV water-cooled induction generator/voltage regulator combination, (2) a remote console, which was connected via RF cable, water cooling, and argon gas lines to the induction generator, and (3) the water-cooled braze tools. During the course of the braze development and early stages of the assembly buildup, special considerations relative to cleanliness and preparation of material and techniques for maintaining inert environments in the braze joint zone were found necessary to consistently accomplish good brazes.

4. Flight Acceptance and Qualification Testing

At the completion of fabrication, each flight propulsion subsystem was subjected to the following test sequences:

(1) Proof and leak.

(2) Functional.

(3) Vibration.

(4) Vacuum chamber leakage.

(5) Postvibration functional.

In addition, the flight subsystems underwent thermal vacuum and vibration testing while installed on the spacecraft. Isopropyl alcohol and Freon were used as fuel and oxidizer simulants during vibration tests.

4.1. PROOF AND LEAK TEST

A proof pressure test was performed to demonstrate integrity of the subsystem at pressure levels of 1.5 times the normal working pressure for various components of the subsystem. The levels of pressure for parts of the subsystem varied from 41.4×10^6 N m^{-2} (6000 lbf in.$^{-2}$) for the pressurant bottles to 172×10^6 N m^{-2} (250 lbf in.$^{-2}$) for the rocket engine.

The purpose of the leak test was to verify that zero leakage was obtained at the many braze joints of the subsystem which had been added to interconnect subassemblies. Helium gas was used as the leak detection medium, with a portable helium mass spectrometer as the detector. In addition to the braze joints, various other areas of the propulsion subsystem such as the service valves and the rocket engine assembly flex hoses were leak-checked at working pressure.

4.2. FUNCTIONAL TESTS

The propulsion subsystem functional test had three objectives:

(1) To verify that all subassembly components meet their flight performance criteria.

(2) To observe any possible adverse interaction between components when they are operating under normal conditions.

(3) To provide assurance that the functional operation of a component has not been compromised as a result of other subsystem tests such as vibration.

The key functional tests were:

(1) Regulator lockup test.

(2) Relief valve assembly functional test.

(3) Check valves cracking pressure and leak test.

(4) Rocket engine valve and flow tests.

(5) Gimbal actuator functional test.

4.3. VACUUM CHAMBER LEAK TEST

The purpose of the vacuum chamber tests was to verify that the propulsion subsystem total external leakage was within specification when pressurized with helium at working pressure. A secondary purpose of the test was to verify that outgassing of various components on the subsystem, such as cabling, was within acceptable limits.

Although the two propellant tanks were pressurized, no attempt was made to pressurize the feed lines to the REA since these lines normally contain only liquid propellant. Furthermore, helium gas would soon have permeated the Teflon lining of the flex hoses and obscured the test results.

4.4. QUALIFICATION (TYPE APPROVAL) TEST PROGRAM

The broad objectives of the type approval (TA) program were, as nearly as practicable, to simulate the processes, interfaces, tests, environments, and duty cycle that an actual flight subsystem would experience. In addition, it was intended to expose the subsystem to limits or environments, where appropriate, beyond expected conditions so as to demonstrate a level of margin. Most notably, the extended conditions were (1) higher level and increased duration for mechanical vibration, (2) operation at extreme temperature limits, (3) two mission duty cycles, (4) extra handling and servicing, (5) additional functional and component checks, and (6) other extended operating limits such as high tank pressures, extreme nonoperating temperatures, and extreme engine valve temperatures.

The TA subsystem was assembled, tested, handled, and, in general, exposed to conditions similar to those that flight units would experience. Following this, the TA subsystem underwent two simulated mission duty cycles. In the vibration testing, the TA was subjected to more severe conditions than expected on the flight units, and in the two mission duty cycles, the TA unit was exposed to specific extended environments.

The fact that the pyrotechnic valves were irreversible in operation precluded a complete simulation, so the sequences were folded as shown in Tables III and IV. The total engine firing time of each test series exceeded that expected for flight. Therefore, the engine, filters, bladders, and service valves demonstrated margin in capacity and cycle capability.

The engine was heated to 338.7 K (150 °F) prior to the first midcourse firing, approximately 11 K (20 °F) hotter than the temperature predicted from solar radiation; the subsystem was heated to 303.7 K (87 °F), near the maximum of its specified range (305.4 K, 90 °F) and 6.7 K (12 °F) hotter than the temperature predicted for the first midcourse maneuver. Cold propellants were then loaded for test series 2; the propellants were at 280.3..K (45 °F) at the time of the orbit insertion firing, compared to an expected temperature of 297 K (75 °F). These temperature extremes were intended to demonstrate margin for bladder collapsing, engine valve operation, check valve operation, and the operation of other temperature-sensitive components. Saturated propellants were not used for testing, since analysis had shown that the predicted level of saturation at the time of orbit insertion would be well below the threshold required to affect performance. The orbit insertion firing of test series 2 was performed with propellant tank pressures initially 17 N cm^{-1} (25 lbf in.$^{-2}$) higher than expected, simulating tank heating or regulator gas leakage.

Except for a check valve momentary sticking problem, which was later satisfactorily resolved, all components functioned as expected. All specification requirements were

TABLE III

Type approval test series 1

Test event	Simulated flight event
Propellant vibration (3 axes)	Launch
Installation in vacuum chamber	2-week coast
Moog valve open	Moog valve open
P1, O1, F1 open	P1, O1, F1 open
8-s burn	Midcourse burn
1-day coast	1-week coast
P2 close	P2 close
O2, F2 close	O2, F2 close
Check valve test	–
1-day coast	6-month coast
O3, F3 open	O3, F3 open
P3 open	P3 open
10-s burn	Midcourse burn
1-day coast	3-week coast
900-s burn	Orbit insertion burn
2-day coast	2- to 4-day coast
0.4-s burn	Orbit trim burn
3-day coast	2- to 4-day coast
40-s burn	Orbit trim burn
–	Coast
–	Close P4, O4, F4
–	Orbit planet

TABLE IV

Type approval test series 2

Test event	Simulated flight event
900-s burn	Orbit insertion burn
3-day coast	2 to 4-day coast
16-s burn	Orbit trim burn
1-day coast	2 to 4-day coast
–	Orbit trim burn
–	Coast
P4, O4, F4, close	Close P4, O4, F4
1-day coast	Orbit planet
O5, F5 open	–
P5 open	–
10-s burn	–
1-day coast	–

satisfied. Reliability of the propulsion subsystem was demonstrated and all interfacing equipement, such as pyrotechnics, thermal, structure, and support equipment, operated satisfactorily.

5. Launch Preparation

The refurbished engineering test model was shipped to AFETR and used as a pathfinder for a complete exercise of the prelaunch operations to be performed on the flight subsystems. A typical sequence of testing was conducted which included helium leak test, functional test, squib installation, propellant loading operations, propellant unloading, and vacuum drying of the subsystem. All the procedures and support equipment, as well as facilities to be used in flight operations, were successfully employed. As a result of performing the operations on the pathfinder subsystems and conducting the propellant loading operations, modifications were made in the formal procedures for use during operations with the actual flight subsystems.

The pathfinder subsystem was later used with the PTM spacecraft for launch vehicle interface testing. All the testing was successful and provided an excellent proving ground before prelaunch preparations on the flight systems.

The preparations of the propulsion subsystems for launch at AFETR can be divided into three main areas:

(1) Performance of a subsystem leak test similar to the proof and leak test conducted at JPL but without taking the subsystem to the proof pressure levels.

(2) Repeat of the propulsion subsystem functional test.

(3) Installation of pyrotechnics, fuel and oxidizer fill, and pressurization of the subsystem.

In addition to the two flight subsystems, the PTM subsystem was also fueled and pressurized and maintained in readiness as a spare. After preparation, the subsystem pressures were monitored prior to delivery to the spacecraft. In addition, a toxic vapor detector was used to detect any possible propellant leakage.

After installation onto the spacecraft, and prior to launch, pressure monitoring was accomplished through the spacecraft telemetry system when the spacecraft was undergoing electrical tests. Toxic vapor detector monitoring was also continued up through launch.

6. Mission Sequence

The Mariner 9 propulsion subsystem performed a midcourse correction five days after launch and three maneuvers at Mars – orbit insertion and two trims. Pyrotechnic valves shown in Figure 2 provided positive isolation of propellants and pressurant for the five-month coast period between the first two maneuvers. The specific mission events and maneuver magnitudes are listed in Table V.

The loss of the Mariner 8 spacecraft caused a change in maneuver strategy. The maximum allowable ΔV was committed to orbit insertion in order to achieve a 12-h orbital period and maximum rotation of the orbital line of apsides. Commitment of 40 m s^{-1} was made for a single orbit trim to correct orbital period and time of periapsis passage. Periapsis altitude was allowed to float in order to decrease spacecraft risk by reducing the number of maneuvers.

A second midcourse maneuver was not required because of the extreme accuracy of the first maneuver. Orbit insertion and the first trim went according to plan; the orbit insertion maneuver was so accurate that only one-third of the allocated ΔV was required to synchronize the orbit to the Goldstone Tracking Station with the first

TABLE V

Propulsion event sequence

Event	ΔV, ms	Date, 1971
1. Launch with propellant tanks at low pressure	–	May 30
2. Vent air from liquid lines	–	Jun 1
3. Pyro valves open (P1, O1, F1) to pressurize tanks and lines	–	Jun 3
4. Midcourse 1 firing (5.1 s)	6.7	Jun 4
5. Pyro valve closed (P2) to isolate gas supply from propellant tanks	–	Jun 6
6. Pyro valves closed (O2, F2) to isolate propellants from engine	–	Jun 14
7. Pyro valves open (P3, O3, F3) to repressurize system	–	Nov 1
8. Orbit insertion firing (915 s)	1600.5	Nov 13
9. Orbit trim 1 firing (6.4 s)	15.3	Nov 15
10. Pyro valve closed (P4) to isolate gas supply from propellant tanks	–	Nov 17
11. Orbit trim 2 firing (17.3 s) in blowdown mode	41.8	Dec 30

trim. The resulting excess ΔV capability was used for a second trim to raise periapsis altitude from 1387 to 1650 km on December 30, 1971. This increased periapsis altitude allowed greater picture overlap; the primary mapping mission could, therefore, be accomplished in a shorter period of time after the Mars dust storm had subsided.

Another sequence change which did not affect total ΔV capability, but rather the readiness of that capability, was a decision to leave the propellant line isolation valves open after orbit trim 1. The gasline valve was closed to protect against regulator failure. The propellant lines were left open because (1) there was no evidence of rocket engine valve leakage as determined by observed valve and injector temperatures and lack of trajectory or attitude disturbances, and (2) the line-open state would allow performance of any additional maneuvers required in a propellant-tank blowdown mode without committing to a permanent line-open mode by opening valves O5, F5. The wisdom of that decision was borne out by the subsequent requirement for a second orbit trim maneuver.

7. Analytical Modeling and Flight Performance Correlations

A substantial effort was necessary to predict the in-flight performance of the subsystem. Accurate prediction was necessary, since up to 96% of the propellant was expected to be consumed during the Mars orbit insertion maneuver, and one must be able to commit to later orbit trim maneuvers without endangering the spacecraft as a result of propellant starvation of the rocket engine. Proper prediction of in-flight performance required the development of modeling tools to predict and account for nitrogen saturation of the propellant during the six-month coast period and to predict and statistically analyze in-flight data.

7.1. NITROGEN SOLUBILITY EFFECTS AND MASS TRANSPORT MODEL

A portion of the nitrogen pressurant gas used in the Mariner 9 propulsion subsystem dissolved into the propellants during the interplanetary cruise phase. The actual amount in solution was of interest because full saturation of the oxidizer at a tank pressure of 1.72×10^6 N m^{-2} (250 lbf in.$^{-2}$) would decrease operating mixture ratio about 6% from the unsaturated operation point. Preflight testing with an oxidizer flow bench which simulated the propulsion subsystem hydraulic circuit showed that (1) the engine injector is the only component which exhibits significant resistance increase due to the two-phase flow caused by gas coming out of solution, and (2) such a resistance increase does not occur until the partial pressure of N_2 in solution exceeds the injector inlet pressure, less vapor pressure. It was thus necessary to predict the nitrogen gas absorption level at orbit insertion so that the proper mixture ratio could be loaded.

A mathematical model of the Mariner 1971 propellant tanks was developed to predict (1) the rate of pressurant gas absorption into liquid propellants, (2) the volume of gas bubble inside the Teflon bladders, and (3) the propellant tank pressures with an isolated gas supply. The model was programmed for use on the Univac 1108 computer to print and plot gas concentration, average gas concentration, bubble volume and tank pressure as functions of pressurant supply profile, radius, and time. Required inputs are propellant and initial bubble volumes, propellant diffusivity and solubility, bladder permeation coefficient, and the bladder thickness.

Nitrogen gas will permeate through the Teflon bladders and diffuse into the liquid propellants because of a concentration gradient that develops after propellant tank pressurization. These processes can be compared to the transfer of heat due to a temperature gradient. Thus an existing heat transfer program was used as a basis for the gas transport program.

In the gas diffusion model, the mass transport equations were rewritten and solved as finite difference equations. The liquid propellant volumes were assumed spherical and contained within bladders exposed on the outside to pressurant gas. The liquid volume is divided into 10 concentric spheres and represented as 10 liquid nodes (Figure 7). The gas ullage volume outside the bladder and the bubble volume inside the bladder are represented by two gas nodes. Positions and volumes of each of the nodes and the conductance of each of the conductors are calculated by the program. The program recalculates each conductance as the bubble volume and surface area decrease.

Preflight predictions of saturation level at orbit insertion were calculated using the spherical permeation/diffusion model just described. Predicted propellant tank pressures obtained from this analysis are plotted, along with flight data, in Figure 8. Note that a significant difference in curve shape and final pressure exists for the oxidizer tank, while only the rate of pressure decay is different for the fuel tank.

Since certain assumptions were necessary to accomplish the orbit-insertion saturation predictions, it was desirable to update the model to match the observed pressure decay curves. Pressure profiles which matched the flight data were obtained with the

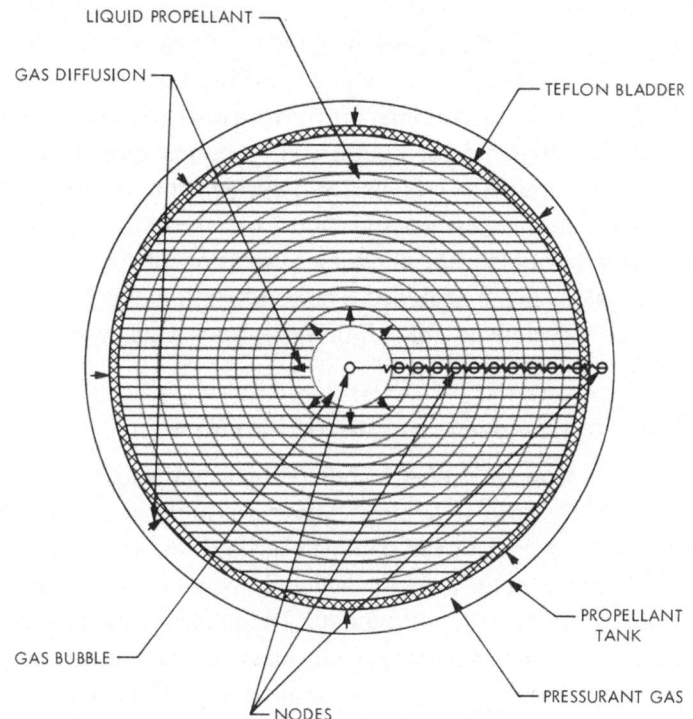

Fig. 7.   Propellant tank gas diffusion model.

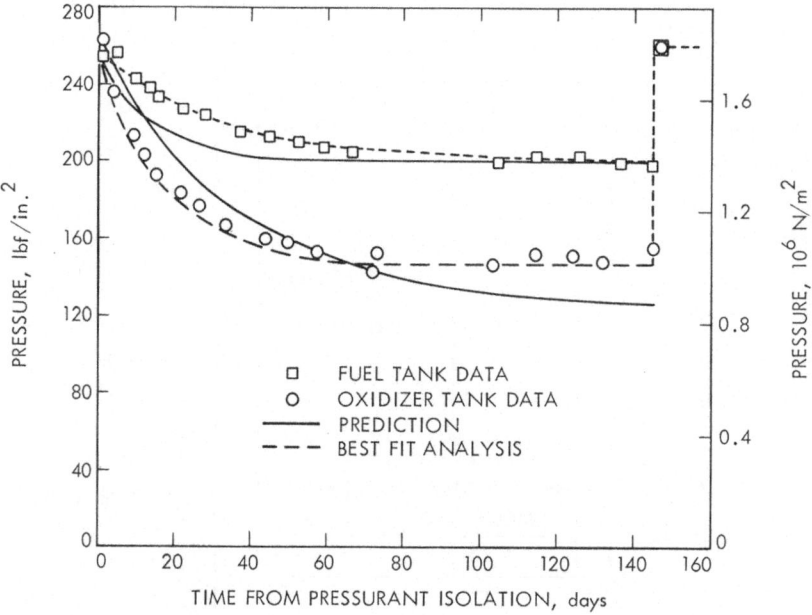

Fig. 8.   Propellant tank pressure compared with saturation model predictions.

permeation/diffusion computer model by assuming different values for bladder area available, diffusion rate, and solubility. The revised constants were used to calculate N_2 saturation pressures at orbit insertion of 1207×10^3 N m^{-2} (175 lbf in.$^{-2}$) for the oxidizer and 1469×10^3 N m^{-2} (213 lbf in.$^{-2}$) for the fuel. The oxidizer saturation level is very nearly equal to injector inlet pressure, so only a small amount of excess N_2 would be expected to come out of solution. No mixture ratio shift which could be attribute to excessive saturation was observed during orbit insertion. The flight data tend to support the ground test derived model of a saturation effects threshold of injector inlet pressure but do not allow an evaluation of saturation effects per se because the threshold was not exceeded. The combination of ground and flight data, however, should be especially useful to future programs which use N_2 as a pressurant gas.

7.2. OPERATION AND PERFORMANCE COMPUTER PROGRAM

A digital computer program called PSOP (Propulsion Subsystem Operation and Performance) was developed to support Mariner 9 flight analysis. PSOP is a low-frequency simulation model of the complete propulsion subsystem. Figure 9 is a simplified block diagram of the program which shows information flow from an input data list through the program to output data. The program was used to predict flight telemetry data, generate thrust and spacecraft-mass time functions for flight maneuver analyses, perform malfunction analyses, and investigate effects of variations in system initial conditions.

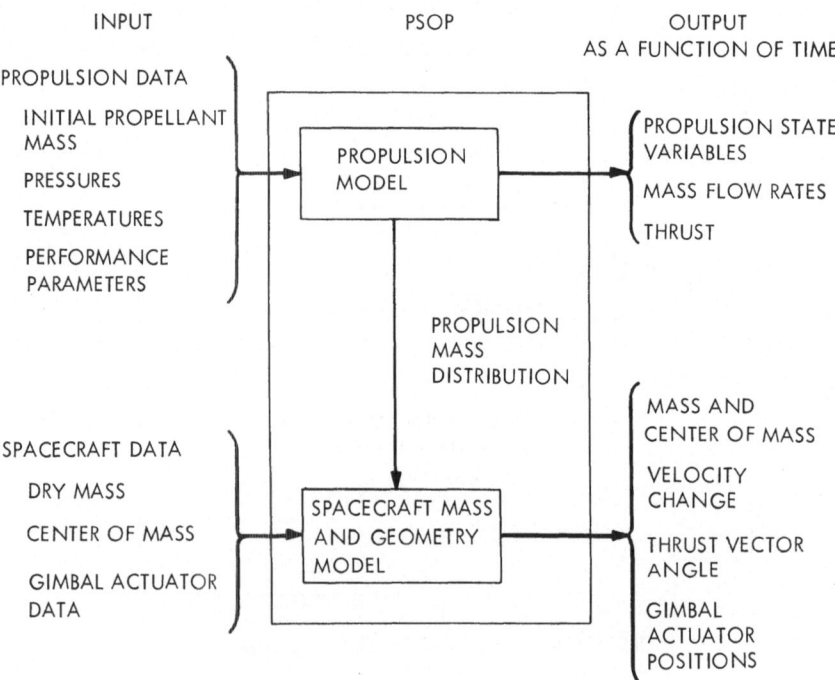

Fig. 9.   Subsystem performance prediction program.

On a given run, PSOP will simulate preburn, burn, and postburn behavior. A typical preburn event is tank pressurization by opening a pyrotechnic valve. Postburn activity includes regulator lockup and heat transfer between fluids and tanks. A burn simulation will continue until one of two conditions are met. Spacecraft velocity change can be input and total burn time will be determined or burn time can be specified to determine the spacecraft velocity increment.

The propulsion model was formulated by first describing the significant physical processes in each of the propulsion components and then organizing these descriptions into a large equation set. Like the physical hardware, component identity is retained and interactions between components (equations) are required to achieve a system solution. Figure 10 is a simplified block diagram of the propulsion model.

The propulsion performance portion of PSOP keeps track of the varying mass elements of the spacecraft, that is, the instantaneous mass of oxidizer, fuel, and gaseous nitrogen in the various containers. This information is coupled with the fixed spacecraft mass in an 8-element mass model to determine total mass and spacecraft center-of-mass location.

Since the spacecraft autopilot forces the thrust vector to point through the space-

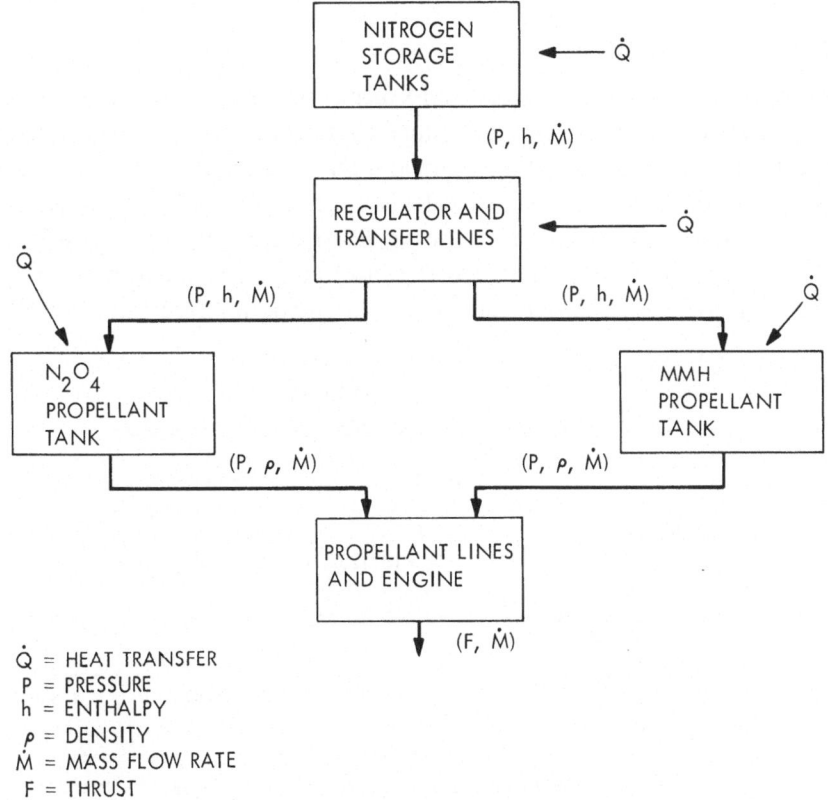

Fig. 10.   Propulsion model of PSOP block diagram.

craft center of mass, the thrust pointing angle can be determined from the location of the center of mass and the engine gimbal center. The gimbal actuator positions are computed from the thrust pointing angle and the results converted to telemetry output data number. Since the spacecraft center-of-mass movement during a burn is a measure of the integrated engine mixture ratio, and the gimbal actuator arm lengths indicate the direction of the spacecraft center of mass, average engine mixture ratio can be inferred from the gimbal actuator positions.

7.3. FLIGHT ANALYSIS COMPUTER PROGRAM

The PSOP program provided flight performance predictions, but a tool was also required to analyze the flight data and compare the data with predictions. A Propulsion Statistical Analysis Program (PSAP) was developed per the formulation of Alford [1] to perform this function. This program uses the statistical residual technique used by trajectory analysts and was readily available in computer subroutine libraries [2] treats the adaption for this application.

Some applications of the residual technique use time as a running variable and calculate a solution at every n s. One of the primary inputs for the problem at hand was total velocity change; this was not observable as a function of time from on-board sources, so a decision was made to keep the program simple and perform an average analysis for the entire burn.

7.4. PERFORMANCE RESULTS

Of the four propulsion maneuvers performed, only the orbit insertion maneuver was long enough to provide sufficient data for a thorough comparison with preflight predictions. PSOP was used with empirical input data obtained from the Mariner 9 and similar propulsion subsystems to calculate predictions. A weighted-least-squares fit of the flight data and predictions (using PSAP) resulted in the best-fit data list of Table VI. Also listed is the estimated 1-σ uncertainty of each parameter in the best-fit column. Burn time, chamber pressure, and engine mixture ratio are all within

TABLE VI

Orbit insertion propulsion performance summary

Parameter	Prediction	Best-fit data	1-σ uncertainty
Burn time[a], s	912.8	915.4	0.03
Regulator outlet pressure, N m^{-2} (lbf in.$^{-2}$)	1.76×10^6 (255.1)	1.75×10^6 (254.5)	6.89×10^3 (1)
Engine chamber pressure, N m^{-2} (lbf in.$^{-2}$)	7.99×10^5 (115.9)	7.97×10^5 (115.6)	6.89×10^3 (1)
Mixture ratio, kg$_O$/kg$_F$	1.575	1.582	0.011
Specific impulse, N-s kg^{-1} (lbf-s lbm^{-1})	2817 (287.3)	2816 (287.2)	6.8 (0.7)

[a] Burn was controlled by on-board accelerometer to produce a velocity change of 1600.5 m s^{-1}.

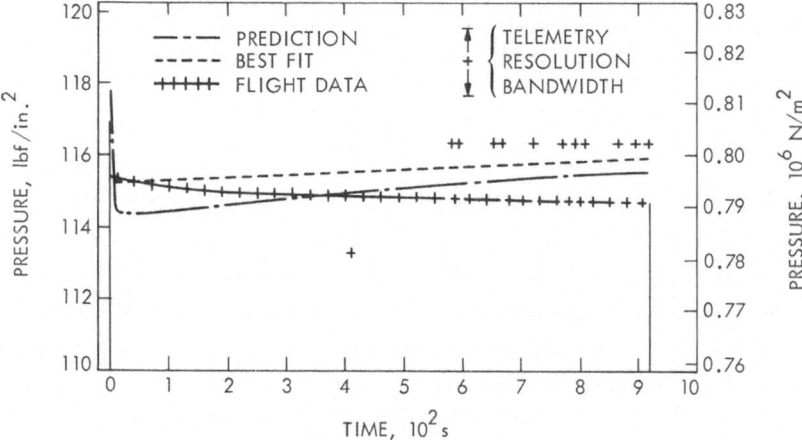

Fig. 11.   Thrust chamber pressure during orbit insertion.

TABLE VII

Mariner 9 propulsion maneuver summary

Parameter	Midcourse correction	Mars orbit insertion	Orbit trim 1	Orbit trim 2
Maneuver date (PST)	6/4/71	11/13/71	11/15/71	12/30/71
Time from launch (days)	4	167	169	214
Distance from Earth, 10^6 km	1.548	122.314	123.196	177.563
Commanded ΔV, m s^{-1}	6.73	1600.5	15.25	41.8
ΔV error, m s^{-1}	0.008	0.147	0.01	0.155[a]
Predicted burn time, s				
Preorbit insertion model	5.08	919.85	6.22	–
Best-fit post insertion	5.09	915.37	6.37	17.27
Measured burn time, s	5.1 ± 0.06	915.4 ± 0.06	6.36 ± 0.06	17.34 ± 0.06

[a] This value is in question.

0.5% of preburn predictions. The flight data were not sufficiently accurate as compared with engine acceptance tests to improve knowledge of specific impulse, so little change was noted there. The increase in mixture ratio and burn time compared with the corrected predictions was attributed to a 0.8% increase in fuel resistance. Note, however, that the fuel resistance change required to provide a data match is less than the 1-σ uncertainty of that parameter.

The revised propulsion model was used to calculate a set of best-fit performance predictions. An example of this, engine chamber pressure, is plotted in Figure 11 with preburn and best-fit PSOP output curves. One may note that the telemetry resolution available with the pressure measurements would have made detailed performance analysis difficult without the statistical program PSAP.

A performance summary for the four propulsion maneuvers is presented in Table VII. New performance calculations were performed after orbit insertion for the actual flight sequence. Burn-time predictions are listed for all burns in Table VII with the original and revised propulsion models. The excellent agreement between flight data and predictions provides a validation of the prediction tools used.

8. Conclusions

For the Mariner 9 orbiter mission, it was necessary to develop a large, bipropellant propulsion system to replace the small monopropellant system used on early Venus and Mars flyby Mariners. Cost effectiveness was achieved by the use of design simplicity rather than redundancy. Characterization of the subsystem through ground testing on a component, subassembly, and subsystem level provided confidence in the design and performance capability. Application of the propulsion module approach, coupled with the use of 'pathfinders' for all critical operations proved as successful on this larger sized subsystem as it had on previous Mariners.

In order to load the subsystem properly and to commit subsequently each propulsion maneuver, modeling tools were developed to predict in-flight performance. These tools led to a 0.5% agreement between observed performance and preflight predictions. These technologies are presently being applied to the Viking Orbiter 1975 spacecraft propulsion system.

References

[1] Alford, M. W., Jr., *Mathematical Aspects of Flight Analysis Computer Programs*, AIAA Paper 68-583, presented at the AIAA 4th Propulsion Joint Specialists Conference, Cleveland, Ohio, June 10–14, 1968, American Institute of Aeronautics and Astronautics, New York, N.Y.
[2] Cork, M. J., French, R. L., Leising, C. J. and Schmidt, D. D., *From Earth to Mars Orbit – Mariner 9 Propulsion Flight Performance with Analytical Correlations*, presented at the 8th Propulsion Joint Specialists Conference, New Orleans, La., November 29–December 1, 1972, American Institute of Aeronautics and Astronautics, New York, N.Y.

MARINER 9 AND THE EXPLORATION OF MARS

ROBERT J. PARKS

Jet Propulsion Laboratory, California Institute of Technology, Pasadena, Calif., U.S.A.

Abstract. The Mariner Mars 1971 Project, a follow-on to the earlier Mariners to Mars, utilized the very favorable 1971 Martian opportunity to put a spacecraft, Mariner 9, into orbit around Mars to provide high-resolution contiguous coverage (mapping) and time-variant (variable features) observations of the planet. The spacecraft was based upon those sent to Mars in 1969, with a special retropropulsion system added and some changes in the scientific instrumentation. The selected orbit was highly elliptical, was highly inclined, and had approximately a 12-h Earth-synchronous period to maximize the daily utility of the single 84-m tracking antenna located at Goldstone, California, and at the same time to provide, in conjunction with on-board data storage, contiguous, nearly complete, high-resolution mapping coverage of Mars and a particular ability to look for variations with time over a 17-Martian-day cycle.

Some of the unique, new design features of the spacecraft included:

(1) a selection and tailoring of the science instruments to match the orbiting mission characteristics and the science questions of maximum interest;

(2) a digital data storage and television instrument approach to provide improved-resolution, quantitative data;

(3) a ground-command reprogrammable, on-board control computer with significantly increased memory to provide considerable mission flexibility;

(4) a new pressure-fed, bipropellant retropropulsion system used for both orbit insertion and trajectory corrections.

Two spacecraft launches were attempted. The first did not succeed due to a difficulty in the Centaur stage of the launch vehicle. The second launch was successfully and accurately accomplished.

The ground system design for conducting the in-orbit flight operations also had some very unique features. An extensive adaptive mode concept was implemented that provided quick analysis and decision-making to reprogram the science collection plan for what was learned daily. It resulted in great flexibility, for example, to minimize the impact of the unexpected, extensive dust storm that was raging on Mars at the time of orbit insertion and that continued for some weeks thereafter.

After the dust storm cleared up, the full complement of scientific objectives of the mission were more than achieved. Complete high-resolution mapping of the planet was accomplished; all scientific experiments performed as intended and provided an extensive amount of valuable data, which has been referred to as a 'scientific bonanza', and changed completely the concepts about Mars. A much more dynamic, Earth-like planet is now recognized. Many large volcanoes and evidence of what appears to be water erosion and water ice were discovered. A large canyon, bigger and deeper than the Grand Canyon, was discovered, and relative altitude measurements were taken both of the canyon and other topographic features. The observation of the dust storm turned out, indeed, to be a bonus in that information on meteorology and dust particle size and material among other information was obtained.

In total, over 7000 pictures and an equivalent amount of the other scientific data were obtained, with over 40 000 total commands being transmitted to and acted upon by the spacecraft.

The Mariner Mars 1971 Project was the fifth in the Mariner Project series and the third with the objective of collecting scientific data about Mars. The first of the Martian series consisted of the launching of two early-vintage spacecraft in 1964. The first launch was not successful, however, the second launch worked well. The spacecraft, Mariner 4, proceeded successfully to fly by Mars in July 1965, and acquire the first closeup data ever received from that planet.

A photograph of this spacecraft is shown in Figure 1. Some of the spacecraft features which contributed significantly to its capabilities were:

(1) attitude stabilization and control;

L. G. Napolitano et al. (eds.), Astronautical Research 1972, 149–162. All Rights Reserved
Copyright © 1973 by D. Reidel Publishing Company, Dordrecht-Holland

Fig. 1.   Mariner 4 spacecraft.

(2) a trajectory correction maneuver capability;

(3) a single-degree-of-freedom scan platform on which were mounted the planetary scientific instruments;

(4) the first use of S-band radio.

In addition to other important scientific measurements, Mariner 4 returned radio occultation data which indicated that the Martian atmosphere is very thin, some 5 to 10 mbar at the surface, and provided 21 close-up pictures of the planet surface with a resolution of about 2.5 km. Figure 2 is the most noted of these first photographs and, with the others, gave the impression of a moonlike, heavily cratered surface.

In 1969 two new and improved spacecraft, Mariners 6 and 7, were launched to fly by Mars. Some of the more significant advancements included in these spacecraft were:

(1) a much more powerful set of planetary scientific instruments;

(2) a two-degree-of-freedom scan platform on which were mounted the science instruments;

(3) a control and sequence computer, reprogrammable from the ground, that permitted last-minute changes in the pointing of the scan platform and thus the science instruments;

MARINER IV PHOTO ENHANCED

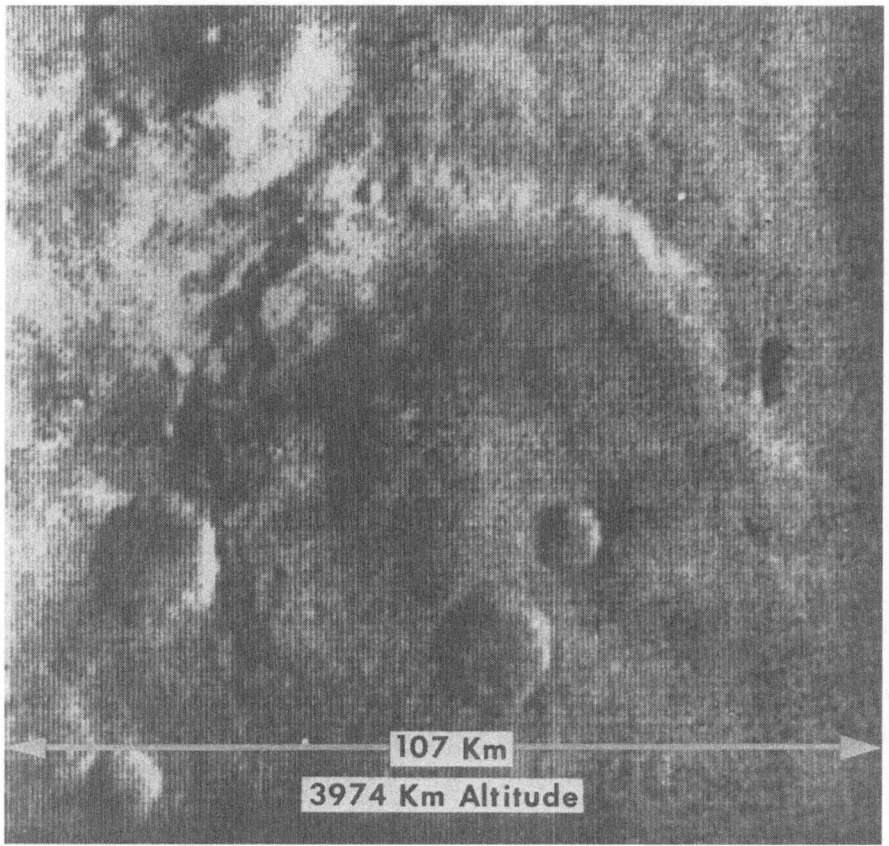

107 Km

3974 Km Altitude

Fig. 2. Mariner 4 photograph of Martian craters.

(4) an improved telecommunications capability which permitted receiving data at the rate of 16 000 bps from Mars.

In addition to many UV and IR interferograms and IR radiometer temperature measurements, over 200 photographs were returned from selected areas and sites on Mars. The South Pole was of particular interest, and the data indicated that the white cap is at least predominantly dry ice or frozen carbon dioxide. Furthermore, the thin atmosphere is predominantly carbon dioxide. The UV radiation largely penetrated to the surface, and there was no measurable amount of liquid water. Trace amounts of water vapor were detected.

This scientific information was highly significant relative to the critical question of whether life had evolved and was still present on Mars. The findings were not favorable to supporting the existence of life but did not completely rule out the possibility. Clearly, much more data was needed to give a fuller understanding of the questions of life on Mars, the origin and nature of the planet, and the insight that Mars could offer towards a better understanding of our own planet.

In the 1967/1968 period it became clear that a set of fortunate circumstances would exist in 1971 which would offer a unique opportunity to considerably increase our scientific knowledge of Mars. The 1971 Martian flight opportunity was the optimum from a launch vehicle energy requirement standpoint for orbiting the planet Mars. A spacecraft that had been built as a spare for the 1969 mission together with all of its associated ground support and test equipment would be available in 1971. That spacecraft, augmented by a special orbit insertion retropropulsion system could be launched to Mars in 1971 by the Atlas/Centaur launch vehicle.

In order to take full advantage of an orbiter, a high data return rate would be required. The 16000 bps data return rate developed for use in the 1969 missions would be adequate, but only one ground station, the 64-m station at Goldstone, California, would be available. A photograph of this antenna is shown in Figure 3. Studies indicated that the use of a special, highly elliptical, 12-h Earth-synchronous orbit such that the periapsis on every other orbit would occur at the same time as Goldstone zenith would permit the data to be played out at this high rate during the once-a-day view period of the Goldstone station. The point on the planet over which the periapsis would occur would slowly rotate with time around the planet, and complete longitudinal coverage of the planet every 18 Earth days would result. The use of a highly inclined orbit would permit full latitudinal coverage.

After detailed planning, it was decided to make some specific changes to the one available spacecraft, to build another one like it, and to launch both during the 1971 opportunity. A photograph of these spacecraft is shown in Figure 4. Each would be expected to operate in orbit for at least 90 days, the time designated as the standard mission period.

The significant changes in the spacecraft that were made for the 1971 opportunity were:

(1) the science instruments were selected and tailored to match the orbiting mission characteristics and provide data on the science questions of maximum interest;

(2) an all-digital tape recorder was substituted for the analog unit;

(3) the memory capacity of the ground reprogrammable on-board computer was increased;

(4) a large orbit insertion propulsion system was substituted for the small trajectory correction propulsion system.

The Mariner 1969 scientific instruments consisted of both a UV and IR spectrometer and an IR radiometer in addition to the TV cameras. Improvements were accomplished in all instruments for use in Mariner 1971 with the exception of the IR spectrometer, which was replaced completely. Figure 5 lists the experiments and the associated science groups. The UV spectrometer could furnish information on the upper atmosphere, particularly its constituents, and could provide a relative altitude measurement by, in effect, measuring the total amount of carbon dioxide

Fig. 3. Goldstone 64-m antenna.

through which the radiation traveled. The IR radiometer could measure surface temperatures, and the IR spectrometer could provide information on the composition of the surface and the lower atmosphere, particularly with respect to water vapor content. The TV could photograph the surface and any visible phenomena in the atmosphere, such as clouds, with a resolution of 1 km for one camera and 100 m for the other camera. In addition, information on the thickness and nature

Fig. 4.   Mariner 8 and 9 spacecraft.

of the atmosphere could be obtained from the occultation effect on the spacecraft's radio signal of the atmosphere of the planet.

The scientific data from the spacecraft instruments needs to be collected at a faster rate than it can be transmitted to Earth; therefore, a large-capacity data storage device must be utilized. All of the Mariners sent to Mars have included a tape recorder for this purpose. In 1969 a combination analog and digital tape recorder technique was used, with most information being recorded in an analog format.

It was recognized, however, that for Mariner 1971 a recently-developed all-digital unit would give better quality and more quantitative data and would simplify the ground processing and analysis of the data. The capacity of the tape recorder was such two loads of stored data could be read out during the single daily Goldstone view period, and the 12-h orbit was selected to permit two full loads to be collected and read out each day.

An orbiting mission requires, of course, a retropropulsion system of sufficient total impulse to permit the spacecraft to be placed into the desired orbit. The pressure-fed, bipropellant system that was selected is shown in Figure 6. Hypergolic propellants are carried in separate tanks; each is contained within a Teflon expulsion bladder. The thrust of the engine is approximately 1330 N (300 lb), and this thrust level re-

MARINER MARS 1971

EXPERIMENTS

EXPERIMENT	WEIGHT kg	COGNIZANT PRINCIPAL INVESTIGATOR
INFRARED INTERFEROMETER SPECTROMETER	24.2	R. HANEL, GODDARD SPACE FLIGHT CENTER
INFRARED RADIOMETER	3.6	G. NEUGEBAUR, CALIF. INST. OF TECH.
TELEVISION	26.4	Δ H. MASURSKY, USGS, FLAGSTAFF ARIZONA G. de VAUCOULEURS, UNIVERSITY OF TEXAS J. LEDERBERG, STANFORD UNIVERSITY B. SMITH, NEW MEXICO STATE UNIVERSITY G. BRIGGS, JPL.
ULTRAVIOLET SPECTROMETER	15.7	C. BARTH, UNIVERSITY OF COLORADO
S-BAND OCCULTATION	---	A. KLIORE, JPL
CELESTIAL MECHANICS	---	Δ J. LORELL, JPL I. SHAPIRO, MASS. INST. OF TECH.
		Δ = Team Leader
TOTAL	**70.0**	

11600–02

Fig. 5.   Mariner 9 scientific experiments.

quired a burn time of about 15 min to provide the total required orbit insertion deceleration.

The launch of Mariner 8, the first spacecraft, was not successful owing to a failure in the autopilot of the Centaur stage of the launch vehicle. This meant that the job that two spacecraft were planned to do now had to be done as best as possible by one. Twenty-two days after the unsuccessful launch of Mariner 8, and after adequate steps were taken to eliminate the cause of the first launch vehicle failure, the second spacecraft, Mariner 9, was successfully launched on 30 May 1971.

A compromise orbit was selected that emphasized the area coverages or mapping objectives while still permitting some variation with time studies to be carried out. This orbit had a period of 11.98 h, a periapsis of 1500 km, and an inclination of 65°.

The mapping objectives were planned to be accomplished as shown in Figure 7. As indicated earlier, a latitude band starting in the southern hemisphere could be covered for all longitudes, and then a new, more northerly latitude band could be covered until essentially the entire planet would be photographed.

Several weeks prior to arriving at the planet, it became apparent from ground-based observations that a major surface obscuration phenomenon, probably a dust storm, had developed on Mars. Such occurrences had occasionally been witnessed in the past, but this particular storm appeared to be much larger and more engulfing than any seen before. As the spacecraft approached the planet, photographs were

Fig. 6.   Mariner 9 retropulsion motor system.

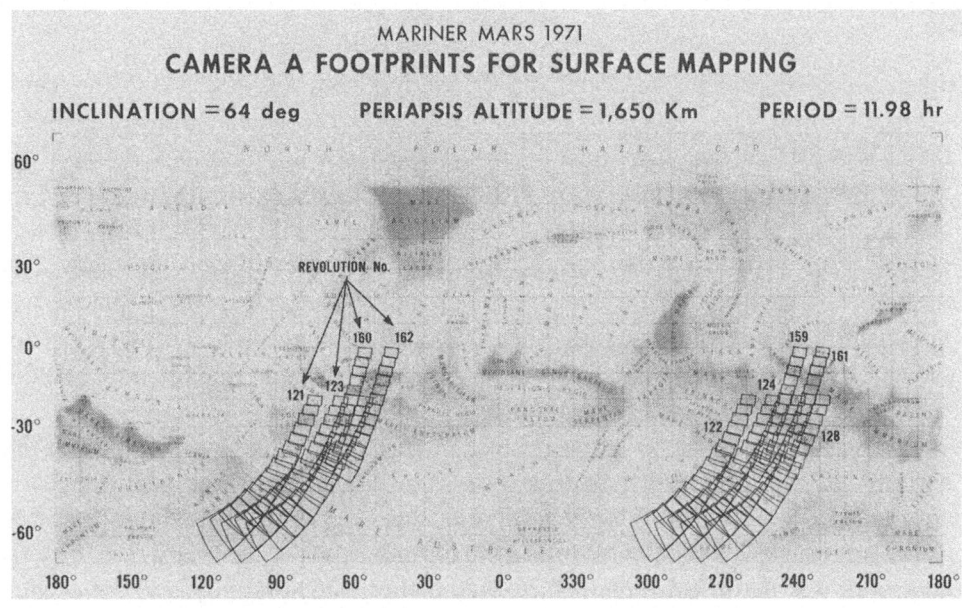

Fig. 7. Mariner 9 wide-angle camera footprints.

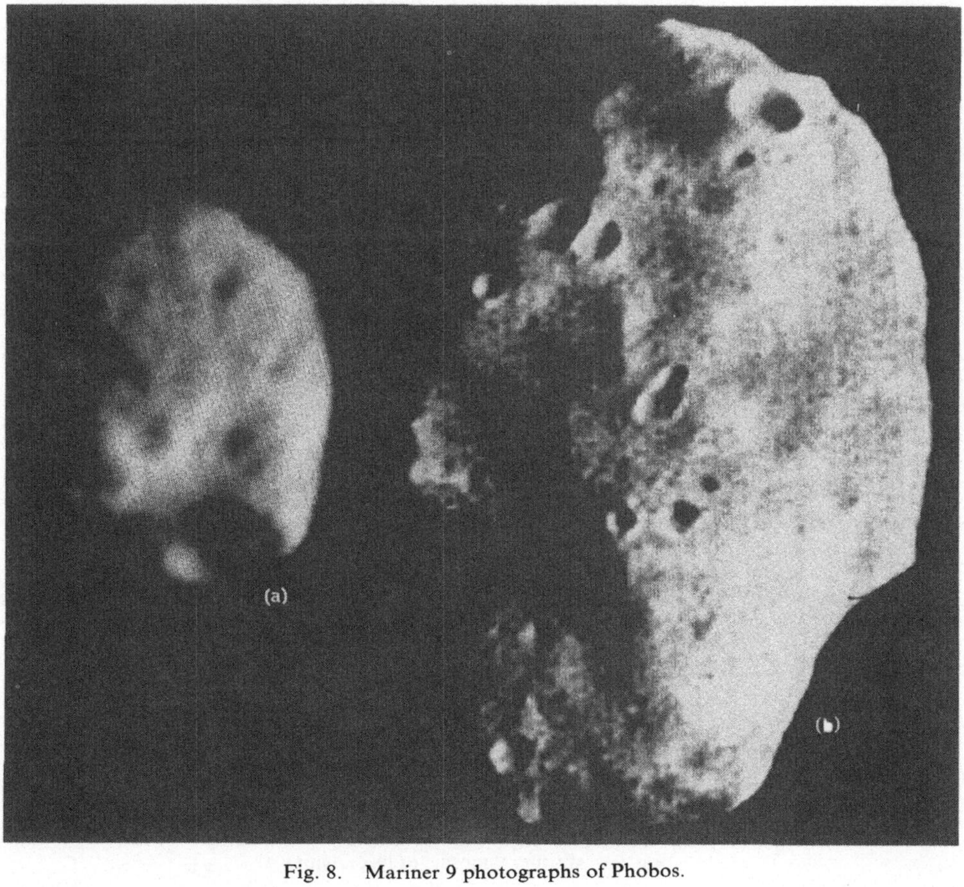

Fig. 8. Mariner 9 photographs of Phobos.

taken which confirmed that essentially the entire planet was engulfed in the storm.

Once in orbit, the original data collection plan had to be abandoned because of the lack of surface visibility. Instead, effort was concentrated on looking for atmospheric phenomena as evidenced by the action of the storm, trying to measure the characteristics of the obscuration material, and searching for signs of local settling of the storm. Data from the infrared spectrometer indicated that the material in the atmosphere comprised particle sizes up to 10 μ and contained silicate-type material, thus confirming the dust storm hypothesis. In addition, during this period the first closeup observations of the two Martian moons, Phobos and Diemos, were made. Figure 8 shows Phobos, the inner moon.

Clearing of the storm was first noticed in the south polar region, and the first surface data was obtained from that area. In addition, some high peaks appeared above the level of the dust. These peaks were soon determined to be mountain peaks with characteristics very similar to Earth volcanic calderas. Several weeks after the spacecraft was put in orbit the planet cleared surprisingly rapidly, and the original data collection plan was initiated on the 50th day in orbit.

In order to provide flexibility in the science data collection activities, a so-called 'adaptive mode concept' was implemented. An extensive ground-based system was

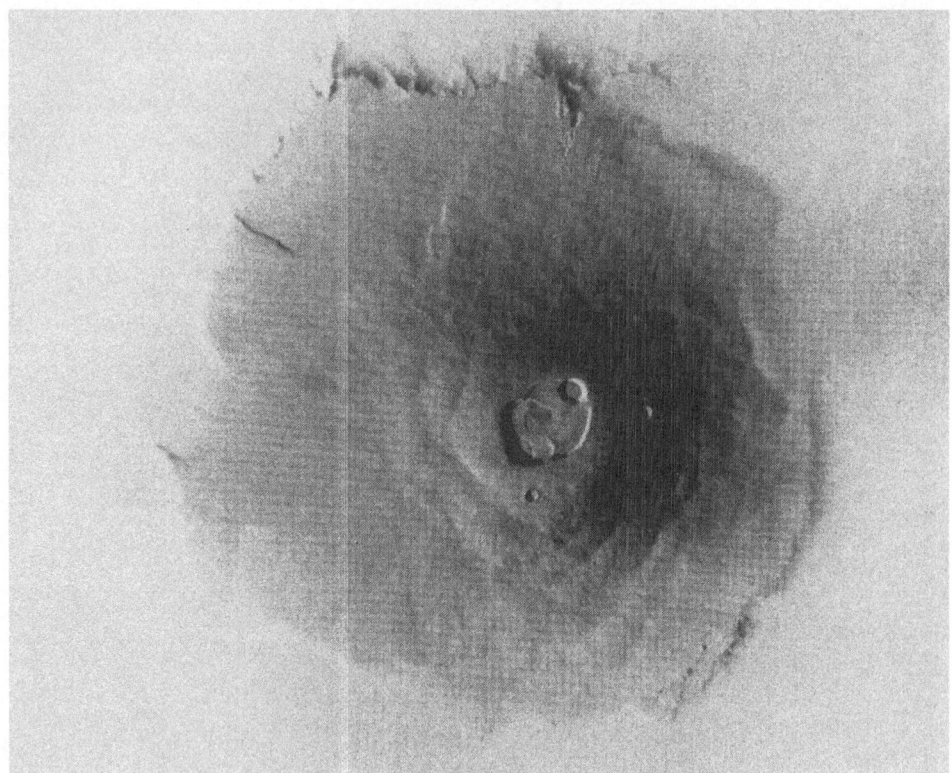

Fig. 9.   Mariner 9 mosaic of Nix Olympica.

designed and implemented which complemented the high flexibility inherent in the spacecraft, and which permitted rapid data display, analysis, and decision-making. This ground-based system utilized multipurpose computers with special software programs.

Within the first 24 days after the initial orbit, two equipment failures occurred in the spacecraft. On the 12th day, the filter wheel on a TV camera failed to change position on command and, as a consequence, no polarized or color photographs could be obtained after the dust storm abated. On the 24th day, the S-band RF traveling wave tube power amplifier failed rather suddenly and without warning. Fortunately, this was one of the relatively few places where redundancy had been included in the spacecraft design. The alternate traveling wave tube amplifier was switched on by ground command, and normal performance was restored.

At the end of the standard mission, which covered 90 days in orbit, an extended mission was authorized and implemented. The spacecraft was still operating normally, and the initially planned science objectives had not been fully achieved because of the time lost when the dust storm was highly active. In addition, an extended mission offered the possibility of obtaining even more data than had been originally planned.

Fig. 10.   Mariner 9 mosaic of Martian equatorial region.

As was expected, a high science return rate was maintained for the first month and a half of the extended mission. There then followed a two-month period when solar occultations of the spacecraft occurred twice daily; i.e., the spacecraft went into the shadow of Mars on each orbit for up to two hours.

The spacecraft survived the occultation period without damage, and full science data collection was reinitiated. In order to recover the data after this period it was necessary to store it in the tape recorder, as was normally done, and then command the spacecraft to maneuver so that during the Goldstone viewing period the spacecraft antenna would be pointed at Earth and the data stored in the recorder could be transmitted.

The spacecraft is now in a quiescent state and is in the process of completing a general relativity experiment. This experiment, involving the measurement of the effect of the Sun's gravity on the velocity of the radio waves passing between the spacecraft and Earth, is made during the period when the spacecraft is nearly on the opposite side of the Sun from Earth, i.e., in superior conjunction. A similar experiment was conducted in 1969 and validated one part of the general relativity theory

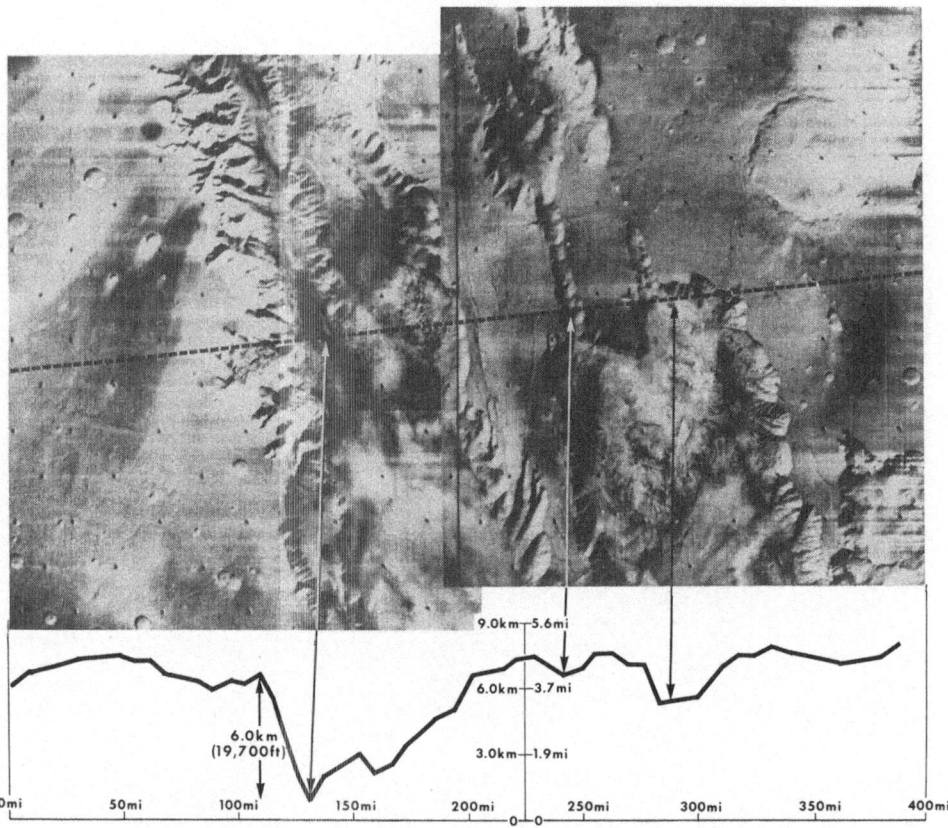

Fig. 11.   Mariner 9 UVS elevation measurements of Martian 'Grand Canyon'.

to about 4% accuracy – an improvement of several factors over any prior experiment. In the case of Mariner 9, since it is in effect locked to Mars, opportunities for more accurate measurements are possible, and it is hoped that the accuracy of the experiment can be improved to about 1%.

In the next several weeks, the remaining attitude control gas will be used in making the final two or three maneuvers of the spacecraft to complete the scientific data collection. The spacecraft has been operating in orbit for about 11 months, has transmitted over 7000 photographs and a large amount of other scientific data on Mars, and has responded to over 40000 commands from Earth.

The next several figures illustrate some of the science data returned. Figure 9 shows the 500-km-diam volcanic mountain Nix Olympica. The main crater at the top is 60 km in diameter. The data indicates that this peak is over 16 km high. Figure 10 is a mosaic showing the Grand Canyon of Mars with an overlay showing that this canyon would stretch from coast to coast in the United States. Figure 11 shows a closeup of part of this canyon and illustrates the capability of the UVS instrument to measure relative altitudes. Altitude differences of 6 km and greater are shown. Figure 12 is one of several pictures giving evidence that water may have flowed on the surface

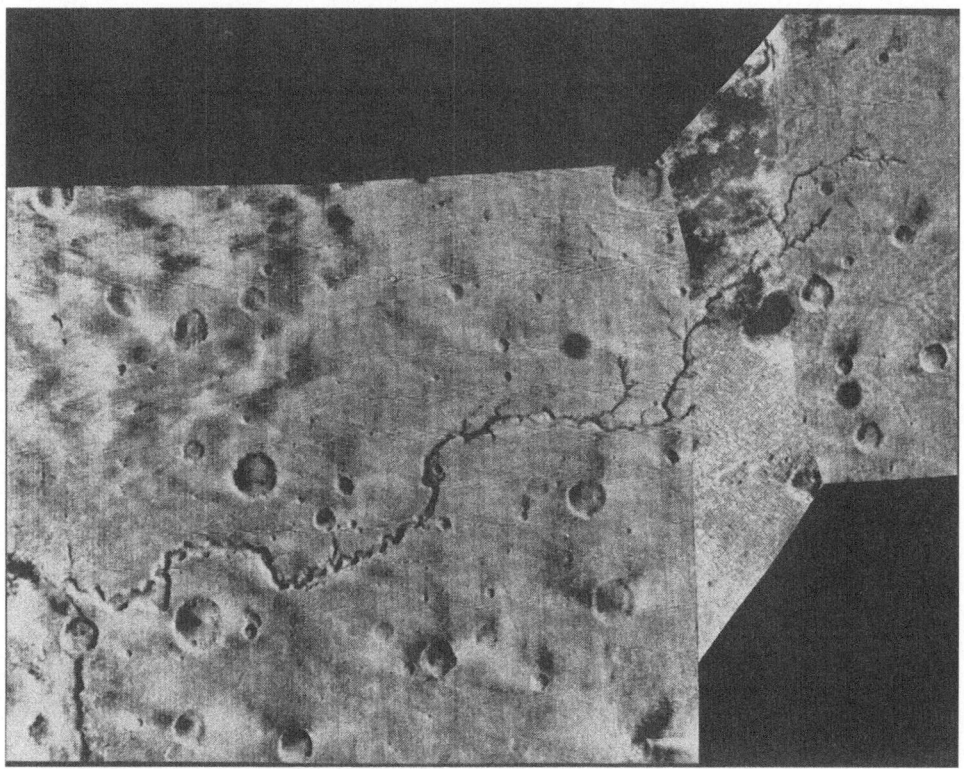

Fig. 12.   Mariner 9 photograph of Martian 'river'.

of Mars at one time. This feature resembles a dry desert river bed. There is also some evidence that water ice may exist on or below the surface.

These findings from the Mariner 9 mission are indicative of the wealth of scientific information that has now been added to our body of knowledge of Mars. Time does not permit covering more; indeed, much of the data recovered is still in the process of being analyzed by the scientific community. It is sufficient to say, however, that the findings to data have already changed completely many of the prevailing concepts about Mars. It is evident that this neighbor of ours is a much more interesting, dynamic, and Earthalike planet than had been realized.

DESIGN PRINCIPLES FOR CONTAMINATION ABATEMENT IN SCIENTIFIC SATELLITES

ROBERT J. NAUMANN

*Space Sciences Laboratory, Marshall Space Flight Center,
National Aeronautics and Space Administrati9n, U.S.A.*

1. Introduction

As space experiments develop in complexity and sophistication, the problems of interference from the self-induced local environment become more severe. Extreme care in design and operational procedure is demanded to insure that the experiments measure the intended phenomena rather than extraneous self-induced environmental effects. Optical experiments, particularly those operating in the far ultraviolet, are highly susceptible to contamination from absorption and scattering; hence, the study of such interference is referred to as 'Optical Contamination' or, in a more general sense, 'Contamination'.

There have been a number of experiments on both manned and unmanned spacecraft that have failed or have been severely degraded because the effect of contamination were not considered. Scattered light from ice crystals or other debris has prevented astronomical observations in the sunlit portion of the orbit. Windows and other optical surfaces have become coated with contaminating films and globules which produce scattering and absorption. Since organic molecules are particularly good absorbers in the far ultraviolet, a film of only a few monolayers can destroy the usefulness of a mirror or grating. Cooled infrared detectors have become coated with layers of ice from condensing water vapor. Mass spectrometers have become swamped by water vapor and other outgassing products. High voltage power supplies have been destroyed by arc-over because the ambient pressure was not yet below the corona region when the high voltage was activated. Such difficulties are not restricted to the operation of experiments in space. Many optical surfaces have become contaminated in vacuum chambers during tests.

This type of problem was recognized as being extremely critical to the Apollo Telescope Mount (ATM) experiments and other Skylab experiments. Since ATM must operate in the sunlight portion of the orbit, the problem of scattering from particulate debris cannot be avoided. Therefore, it is mandatory that extreme care be taken to prevent the production of particulates. Also, since many of the measurements are made in the extreme ultraviolet where obtaining good reflectivity is a difficult problem, the deposition of even a few monolayers of an organic contaminant cannot be tolerated. For this reason, extreme care must be exercised in the selection of nonmetallic material to be used in the vicinity of optics. Also, elaborate precautions must be taken in storage and during vacuum testing of the optics to prevent dust particles, outgassing products, and vacuum pump oil from depositing. All structures must be as dust free as possible to prevent such particles from forming a debris cloud

L. G. Napolitano et al. (eds), Astronautical Research 1972, 163–176. All Rights Reserved

in orbit or deposition on the optics during boost. Care must be taken to vent all possible sources of outgassing material and regions that contain trapped atmospheric gasses so that pressures in the vicinity of high voltage systems will rapidly fall well below the corona regime, and measurements for verifying that the pressure is in fact low enough are required before activation of high voltage systems. There was even some question as to whether the life support system with its associated effluents could be made compatible with the mission requirements.

This paper develops a physical basis for estimating and controlling the contamination problems that may be encountered in scientific satellites, both manned and unmanned, and shows how the contaminant environment associated with a spacecraft as complex as Skylab can be kept to an acceptable level.

2. Deposition of Contaminants

The possibility of vapor deposition of high molecular weight polymeric fragments on critical surfaces must be considered. Typical damage produced by thin contaminant films on grazing incidence X-ray optics [1] is shown in Figure 1. Since organic molecules absorb strongly in the ultraviolet, thin films can also produce severe degradation in ultraviolet optics. Furthermore, since deposited films, because of the surface mobility of the molecules, tend to cluster about active sites such as dislocations or other surface imperfections, even visible optical surfaces can be degraded by unwanted scattering. Water vapor, which is usually the most abundant species of vapor, does not cause problems with optics at ambient temperatures, but can cause severe degradation of cooled infrared detectors.

For a canister containing internal sources, the time required to deposit a monolayer on surface 1 is found from elementary theory to be

$$t_1 = \frac{\tau_1 \sum\limits_{i=1}^{N} A_i}{\sum\limits_{i=2}^{N} A_i \left(\dfrac{\tau_1}{\tau_i} - 1 \right)}, \tag{1}$$

where τ_i is the stay time on the ith surface which has area A_i. The stay times are a function of temperature and may be obtained from the Frenkel relation,

$$\tau = \tau_0 e^{H_v/RT}. \tag{2}$$

The constant τ_0 is on the order of 10^{-13} s for most materials. The heat of evaporation H_v may be obtained from the vapor pressure or measured weight loss of the material in question.

If the time to form a monolayer in Equation (1) is negative, this signifies that material is evaporating from surface 1 faster than it arrives. If this is the case, the residual deposition after equilibrium is reached will be given by the BET equation [2].

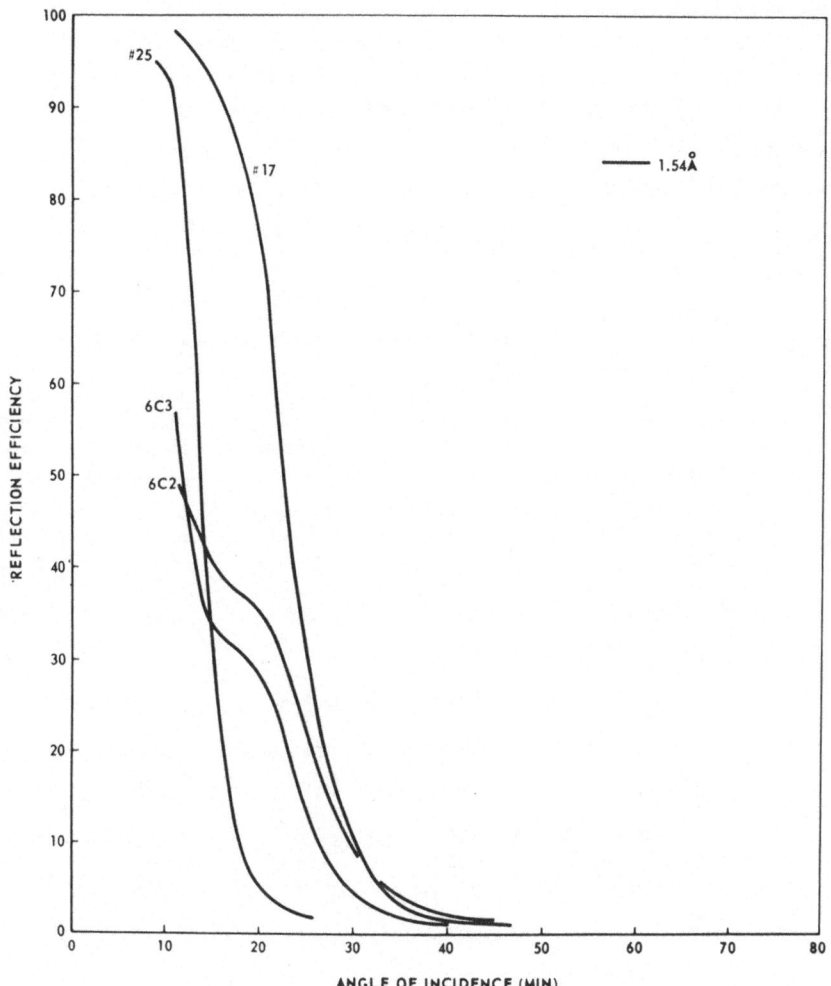

Fig. 1. Reflection efficiency for grazing incidence 1.54-Å X-rays. Sample #25 is a quartz flat, #17 is a quartz flat with 1000 Å Ni overcoat, #6C2 is a quartz flat contaminated by indirect exposure to an RCS engine plume, and #6C3 is a Ni coated quartz flat similarly contaminated.

It is not uncommon for the heat of adsorption between the substrate and the first monolayer to be much larger than for subsequent monolayers. In this limit, the BET equation reduces to

$$\Theta \rightarrow \frac{1}{1 - P/P_v} = \frac{-t_1}{\tau_1}, \tag{3}$$

where Θ is the number of monolayers (or fraction thereof), P is the partial pressure, and P_v is the vapor pressure of the contaminant in question.

As an illustrative example, consider a cylindrical container 30 cm diam by 100 cm long. Let one end be an open aperture and the other end contain an optical surface, say a telescope mirror. The source of contaminant in this case is assumed to be an insulated wire in the canister, 1 mm in diameter and 2 m long. The wire is assumed to be operating above ambient temperature because of the current it is carrying. Good insulations have an outgassing rate of $\sim 3 \times 10^{-11}$ gm cm^{-2} s^{-1} at 50 °C. A wire with 'bad' insulation that has 10 times this outgassing rate is also considered for comparison. The times required for deposition of a monolayer are given in Table I for a variety

TABLE I

Times required for decomposition of a monolayer

	$T_s = 323$ K 'Good' insulation	$T_s = 333$ K 'Good' insulation	$T_s = 323$ K 'Bad' insulation	$T = 333$ K 'Bad' insulation
Walls at 300 K, mirror at 300 K, aperture closed	3.9 days (0.256 t)	1.2 days (0.833 t)	8.8 h (2,73 t)	13.34 min (10.8 t)
Walls at 300 K, mirror at 300 K aperture open	12.04 days (0.083 t)	1.54 days (0.650 t)	2.1 days (0.476 t)	4.3 h (5.58 t)
Walls at 300 K, mirror at 299 K, aperture open	2.51 days (0.398 t)	23.3 h (1.03 t)	6.0 h (4.00 t)	2.6 h (9.23 t)
Walls at 299 K, mirror at 300 K, aperture open	−4.8 days (12.8 t)	1.04 days (0.961 t)	−8.5 h (11.3)	10.52 h (2.28 t)
Walls at 300 K, mirror at 300 K, 5 eV atom^{-1} binding, aperture open	9.3 h (2.58 t)	7.6 h (3.16 t)	47.5 min (30.3 t)	40.7 min (35.4 t)

$A_{wall} = 9424$ cm^2
$A_{mirror} = 707$ cm^2
$A_{aperture} = 707$ cm^2
$A_{source} = 62.8$ cm^2
'Good' insulation has $m = 3 \times 10^{-11}$ gm cm^{-2} s^{-1}, $M = 100$, $\varrho = 1$ gm cm^{-3}
'Bad' insulation has $\dot{m} = 3 \times 10^{-10}$ gm cm^{-2} s^{-1}, $M = 100$, $\varrho = 1$ gm cm^{-3}

Note: The numbers in parenthesis represent the number of monolayers grown in t days. For negative values of t_1, the value in the parenthesis represents the maximum number of monolayers predicted from the BET equation with C large.

of mirror and wall temperatures. Also shown is the case where it is assumed that the binding energy of the deposit on the mirror surface is increased approximately 5-fold to 5 eV atom^{-1} or 155 kcal mol^{-1}. Such values are typical for chemical bonds that could be formed by photolytic polymerization or other chemisorption on the surface. The stay time for this binding energy at ambient temperature is 10^{71} s, which is for all purposes infinite.

Several observations are in order. The effect of increasing the mirror temperature above the surrounding walls results in an evaporation of the contaminant film in all but the worst cases. Use of a 'bad' insulation greatly increases the rate of contamiant deposit in all cases, but the optic will clean itself far more rapidly. If chemisorption takes place on the surface of interest, very rapid buildup will result and the optic will never clean itself up.

This treatment assumed a randomized velocity and is invariant with geometry. For sources situated such that the surface of interest subtends a large solid angle, the geometrical treatment described in the following section should be used.

2.1. EXTERNAL SOURCES

For an external source such as a solar cell panel, the flux at some point in its line of sight is given by

$$n = \frac{n_{evap}}{\pi} \int_{\Omega} \cos\theta_1 \, d\omega, \tag{4}$$

where θ_1 is the angle between the surface in question and the line of sight to the source, and Ω is the solid angle subtended by the source.

Since the incident flux on a surface from Equation (4) will always be less than n_{evap}, the evaporation rate from the source, it is necessary that the reevaporation rate from the surface in question be less than the source rate for contamination to accumulate. This implies that contamination can only build up if the surface in question is cooler than the source.

McKeown [3] using a quartz crystal microbalance on OGO-6 measured a flux of 9.2×10^{-11} gm cm^{-2} s^{-1} from solar cells that subtended an area of 2.23 m^2 at a distance of 3.05 m. Later, when the sensors looked toward space, he observed an evaporation rate of 1.2×10^{-13} gm cm^{-2} s^{-1} for a temperature of 7 °C. From this, he estimated a heat of adsorption of 26 kcal mol^{-1}, which is typical of polymers found in solar cell assemblies. From these data, it can be shown that this flux rate corresponds to a solar cell temperature of 70 °C, which is very close to the measured temperature, and that the surface in question must be cooler than 60 °C for any deposition to occur.

2.2. RETURN MECHANISMS

There has been much concern over the induced atmosphere surrounding a spacecraft. The gaseous component for the most part is directed radially and expanding freely. There are only two mechanisms that have any reasonable chance of redirecting the molecules back to the spacecraft in any significant flux: scattering from the ambient atmosphere and self-scattering between out-going molecules.

The probability of return of a molecule by atmospheric scattering at the stagnation point is [4]

$$\frac{n_{\text{ret}}}{n_{\text{out}}} = \frac{R_0 N_a v_a \sigma_a}{v_r} \left(\frac{\pi^2 - 4}{8}\right),$$

(5)

where

v_r is the radial velocity ~ 400 m s^{-1}

v_a is the satellite velocity ~ 8000 m s^{-1}

σ_a is the scattering cross section $\sim 1.04 \times 10^{-19}$ m^2

N is the ambient density at 420 km $\sim 1.5 \times 10^{14}$ m^{-3}

R_0 is the spacecraft radius ~ 10 m.

For these values,

$n_{\text{ret}}/n_{\text{out}} = 0.0023$ at the stagnation point.

Typical outgassing rates for Skylab are:

0.208 gm s^{-1} H_2O

0.133 gm s^{-1} $O_2/N_2/CO_2$

0.008 gm s^{-1} outgassing products.

For the outgassing products, assume $M = 200$. This represents 2.4×10^{19} mol s^{-1}. Taking the spacecraft to be a sphere 20 dm in diameter, the surface area is 1.25×10^7 cm^2. The average efflux is, therefore, 1.9×10^{12} mol cm^{-2} s^{-1}. The return flux at apex is $n_{\text{ret}} = 4.4 \times 10^9$ mol cm^{-2} s^{-1}. Since the surface density $\sigma_s = 2.08 \times 10^{14}$ mol cm^{-2}, the time to form a monolayer would be 13.25 h, provided every incident molecule was permanently chemisorbed on the surface. If the molecules are to strike the surface faster than they leave, they will require stay times equal to this amount. For heats of adsorption on the order of 26 kcal mol^{-1}, this corresponds to a surface temperature of 312 °K. Even assuming a reduction by a factor of 4 from averaging over an orbit, the temperature above which contaminants will not accumulate is 310°. Therefore, some additional protection is required, such as covering the MDA window when it is not being used. The ATM optics are protected by the fact that they have very restricted viewing angles and, therefore, receive only a small fraction of the return flux.

For H_2O, the return flux at the stagnation point is 1.28×10^{12} mol cm^{-2} s^{-1}. Since σ_s is 1.038×10^{15} mol cm^{-2}, the required stay time to cause buildup is 810 s. The stay time for H_2O is given empirically by

$$\tau = 1.05 \times 10^{-16} e^{6049/T}.$$

(6)

Therefore, T would have to be lower than 139 °K to allow H_2O buildup at this flux. The need for additional protection from cryogenically cooled detectors is obvious.

The returning molecules from atmospheric scatter will be fairly energetic, having collided with relative velocities of 8 km s^{-1}. This is equivalent to temperatures in excess of 50 000 °K or more than 5 eV. Chemical changes will certainly take place, and the returning molecules may be lighter fragments, free radicals, ions, or atomic species. It is possible that some of the species may chemically combine with the surfaces, but the impact energy is so high that they will probably be more effective in cleaning the surface rather than contaminating it. In addition, the atmospheric flux of 2.10×10^{14}

mol $cm^{-2} s^{-1}$ at 420 km will also cause sputtering. McKeown [5] observed an additional mass loss rate of 2.3×10^{-3} gm $cm^{-2} s^{-1}$ when his detector was oriented toward the velocity vector. From this he calculates a sputtering yield of 7×10^{-5} molecules of contaminant per atmospheric colision.

It is also possible to get a return component from collisions between molecules leaving the spacecraft. Because of the net positive radial velocity of the outgoing molecules, it is very unlikely that a scattered molecule will acquire a negative radial velocity necessary for it to return. Even if this is ignored and all collisions are considered isotropic in the lab system, which greatly overestimates the return fraction, this fraction will be less than the atmospheric scatter component.

Some speculation has been advanced concerning the possibility of the spacecraft 'dragging' its contamination cloud along with it through charge or dipole interactions. This does not seem feasible for the following reasons. First, the molecules would either have to be charged or have dipole moments. Second, charges on spacecraft are never more than a few volts because of the large discharge currents produced by the ionosphere or the solar plasma. The solar plasma alone can provide currents as high as 10^{-7} A m^{-2}, which would discharge a 2-m diameter sphere at the rate of 20 000 V s^{-1} [6]. Therefore, even if the spacecraft takes on a high potential by firing a thruster or dumping a liquid, it will soon return to equilibrium. Finally, the Deybe length is only a few centimeters in the ionosphere and a few meters in cislunar space. Therefore, the particle would not be influenced by any spacecraft charge at any great distance. Since ion production rates are generally less than 10^2 $cm^{-3} s^{-1}$ at 350 km where the density is 4×10^8 mole cm^{-3}, the lifetime before ionization must be 4×10^6 s. By this time, the molecule should be far enough away to be completely screened from the spacecraft potential.

Hoffman [7] with his Apollo 15 and 16 mass spectrometers which looked away from the spacecraft detected neutral H_2O, CO_2, and other gases. These molecules are believed to be sublimation products of the residual ice crystals produced by liquid waste dumps.

Having established in the previous discussion the general behavior of contamination deposition, methods of controlling may be considered. Perhaps the most obvious step is to eliminate or carefully control the use of nonmetallic materials. Most agencies have their own specification for acceptance or rejection of materials. These are generrally based on steady state weight loss at elevated temperatures, amount of condensing material on a cold surface near the heated sample, no mass fragments above a certain molecular weight, etc. These tests represent screening methods to eliminate the worst offenders, which is necessary as seen from the previous examples. However, there is a tendency to lose sight of the fact that the criteria are somewhat arbitrary and that 'acceptable' materials still cause contamination.

Careful consideration must be given to the intended use, the amount used, the location relative to critical surfaces, and, above all, the maximum expected temperature of the material. Wires and electronic components that dissipate heat demand special care. They will be the primary internal sources. They should never be allowed to be in direct line of sight with a sensitive surface. The very stable polymers used in insulation and paints with low outgassing properties will still come off at elevated temperatures in sufficient quantities to cause problems to nearby surfaces in their direct line of sight. Because of their stability, they will be extremely persistent if they do condense on a cold surface. Keeping the surface in question at a higher temperature than its surroundings is an excellent way to protect it from contamination. As given in Table I, only a $\pm 1°$ difference in temperature relative to its surroundings determines whether a surface contaminates rapidly or clean itself.

It is even better practice to isolate optical surfaces from sources of contamination by shrouding them directly to the outside by means of an unpainted metal tube. The OAO is designed this way and operates with its mirrors at $-45°C$. It has seen no degradation in the optical system in its several years of operation. This concept was incorporated into HEAO for the X-ray telescope. Since ATM was not designed this way, much more care was necessary in controlling the choice, placement, handling, and cleanliness of all material used in the canister. All nonmetallic components go through a thermal vacuum bake before they are installed to eliminate some of the initial outgassing. All handling and assembly of the internal portions of ATM is done in a class 10 K clean room, while the external portions are assembled in a class 100 K clean facility.

Particular care must be exercised in venting regions around high voltage components. Potting voids and poorly vented regions or sheets of multilayer insulation opening near high voltage components have damaged or destroyed a number of experiments from high voltage arcing. Because of the open window phototubes and large amounts of multilayer insulation in ATM, pressure gages have been installed to assure the pressure is below the corona region before activating the high voltage system.

Cooled detectors required for infrared experiments are especially susceptible to contamination, particularly from H_2O vapor. Again, the multilayer insulation usually used to insulate the detector is the source of a large quantity of trapped H_2O vapor. Care must be taken to vent this away from the detector surface.

Finally, care must be taken to locate vents, RCS thrusters, or other sources of expelled material such that the plume does not contact the surface. Whether suitably located RCS thrusters can be operated without contaminating extremely sensitive optics is still open to question. A decision was made early in the Skylab program to avoid this problem by using cold gas thrusters to desaturate the CMG's. The J series Apollo vehicles use RCS engines without apparent damage to the SIM Bay experiments; however, the thrusters that fire over the SIM Bay are inhibited when the SIM Bay is open. There is a certain amount of splatter from liquid dump plumes. Astronauts on Apollo flights report seeing ice crystals from liquid dumps strike the CM windows. Whether material from the RCS engines does this is not yet known. It is known that a thin liquid film of very low vapor pressure is ejected from the perimeter of the nozzle. It would appear best at this time to avoid firing of RCS thrusters when critical surfaces are exposed and to provide some means of covering these surfaces during RCS maneuvering.

3. Induced Atmosphere

The second concern in conducting optical experiments from a spacecraft is the effect of the atmosphere surrounding the spacecraft on the intended observations. This atmosphere will consist of mostly molecular species such as O_2, CO_2, N_2, and H_2O which originate from outgassing, leakage, purges, waste dumps, and thruster firings. A smaller amount of higher molecular weight material is present from outgassing of nonmetallic material in the spacecraft. Also present will be particulate matter which may come from dust from the surface or interior of the spacecraft, flakes of paints and insulation material, and ice crystals from H_2O jettisoned into vacuum. The effects of concern are absorption and scattering of radiation which may interfere with optical experiments. Scattering from particulates is particularly troublesome. On several occasions, star trackers have been known to lose their reference star and track a small bright particle instead. Dim light astronomical observations from spacecraft during sunlight have been hampered by unwanted scattering from particulates and/or struc-ture.

3.1. COLUMN DENSITIES

The most significant parameter of the induced atmosphere is the column density N_c. For an isotropic source of material with radius R,

$$N_c = \frac{\dot{N}_s}{4\pi R v_r},$$

where $\dot{N}_s$ is the source rate (number/unit time), R is the radius, and v_r is the average radial velocity.

For a point source with a Lambertian distribution, such as a vent or leak, a more general expression is

$$N_c = \frac{\dot{N}_s \cos\gamma}{\pi v_r} \left\{ \frac{1}{R_r} - \frac{1}{R_f} - \left(\sqrt{\frac{R_1^2}{R_0^2} - 1} \right) \left(\sqrt{1 - \frac{R_0^2}{R_f^2}} - \sqrt{1 - \frac{R_0^2}{R_r^2}} \right) \frac{1}{R_0} \right\},$$

where

$R_f = v_r t_1$; R_1; R_1' (whichever greater)

$R_r = v_r t_2$; R_1; R_1' (whichever greater)

γ is the angle between the plume axis and the line of sight, R_0 is a vector from the nozzle to the sight vector making a right angle with the sight vector, R_1 is a vector from the sight vector to the nozzle making a right angle with the nozzle axis, R'_1 is the distance from the nozzle to viewing port if the viewing port is in front of the nozzle plane (otherwise $R_1 = R'_1$), t_1 is the time the vent was initiated, and t_2 is the time it was terminated ($t_2 = 0$ if not terminated).

This expression allows the computation of the column density as a function of time for each vent location and can predict clearing times or steady state values.

For molecules, the velocities are of the order of 400 m s^{-1}. However, predictions of velocities for particulates are difficult. Photographs of Apollo water dumps indicate an approximate velocity of 6 m s^{-1} for the ice crystals [8]. Dust particles leaving the spacecraft probably have much less velocity. If it is necessary to include a distribution of velocities, this equation must be integrated over the velocity distribution.

For Skylab, the column densities resulting from the various vents seen from the ATM are summarized in Table II. It may be seen that the molecular column densities

TABLE II

Scattering from induced atmosphere at $\theta = 0$

Source	Source rate (gm s^{-1})	N_c (No. cm^{-2})	$(d\sigma/d\Omega)_0$ (cm^2)	$B/B_\odot$
All condensate water in 1-μ sphere	0.05	1.26×10^4	2.47×10^{-8}	1.7×10^{-8}
All condensate water in 10-μ spheres	0.05	1.26×10^{-1}	2.47×10^{-4}	1.7×10^{-7}
All condensate water in 100-μ spheres	0.05	1.26×10^{-2}	2.47×10	1.7×10^{-6}
Actual particles from vent	10^{-11}	1.5×10^{-7}	2.47×10^{-8}	2.2×10^{-19}
All gasses (visible)	0.349	1.90×10^{13}	7.7×10^{-28}	1.9×10^{-19}

For reference

Source:	$B/B_\odot$
Near solar corona (K + F):	4.9×10^{-6}
Zodiacal light (10°):	1.3×10^{-11}
Gegenschein:	9.1×10^{-14}
Perfect sky from ground (zenith):	1.5×10^{-14}
Unresolved stars (fainter than $M_v = 6$):	3.4×10^{-14}
Galactic and zodiacal background:	2.0×10^{-14}

from induced atmosphere are substantially less than the 6.3×10^{14} mol cm^{-2} from residual atmosphere at 420 km. Therefore, there seems to be no detrimental effects from spacecraft effluents in the gaseous phase.

The primary threat from the induced environment is from forward scattering from particulates in the vicinity of the spacecraft. As may be seen in Table II, the water from the condensate vent if allowed to vent continuously in the form of ice crystals will provide a scattering background which would preclude the observation of dim-light phenomena.

It was originally thought that water from the Environmental Control System

(ECS) condensate could be disposed of by overboard dumps at high velocity to assure rapid clearing. It was later realized that large ice deposits could build up in the vicinity of the nozzle and instead of a sharp, well defined cutoff, particle production could continue for very long times after the dump. Figure 2 shows the actual versus predicted

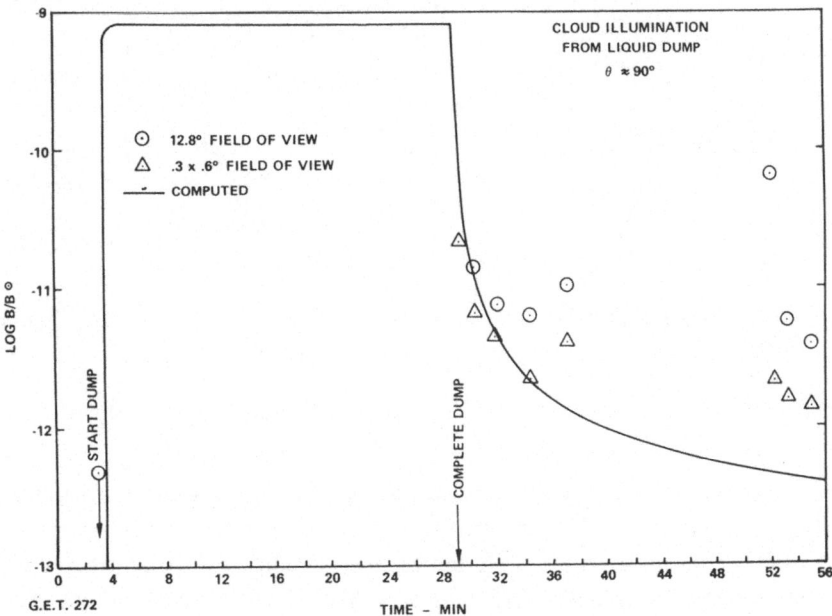

Fig. 2.   Observed and computed clearing time for approximately 12 kg of liquid dumped throughout a 20-min interval during the Apollo 15 mission. The decay at the dump termination is reasonably well predicted by the theory, but the fact that material continues to leave the spacecraft after the dump makes the clearing time much longer than anticipated.

clearing times for an Apollo liquid dump. This could severely interfere with ATM observing time; therefore, it was decided to eliminate the overboard dumps. Instead, all water is now dumped into the OWS storage tank. This presented several problems that required solution. It is necessary to vent the OWS tank to prevent possible build-up of pressure from bacterial action in the waste material that could result in leakage into the crew compartment. To avoid the possibility of dumping liquid water or saturated vapor through the external vents, the flow rate of water into the tank must be controlled to keep the partial pressure below the triple point. Large quantities of ice, which will be formed when the water is dumped into the tank, must be kept from finding their way through the vents. This is accomplished by separating the region near the dump lines from the main tank volume and the main tank volume from the vent ports by extremely fine mesh screens. A diagram of the tank geometry is shown in Figure 3. Figure 4 shows an electron micrograph of the screen itself. The mesh has nominal size of 2μ and an absolute size of 9μ.

This concept was successfully demonstrated in the Skylab Contamination Ground

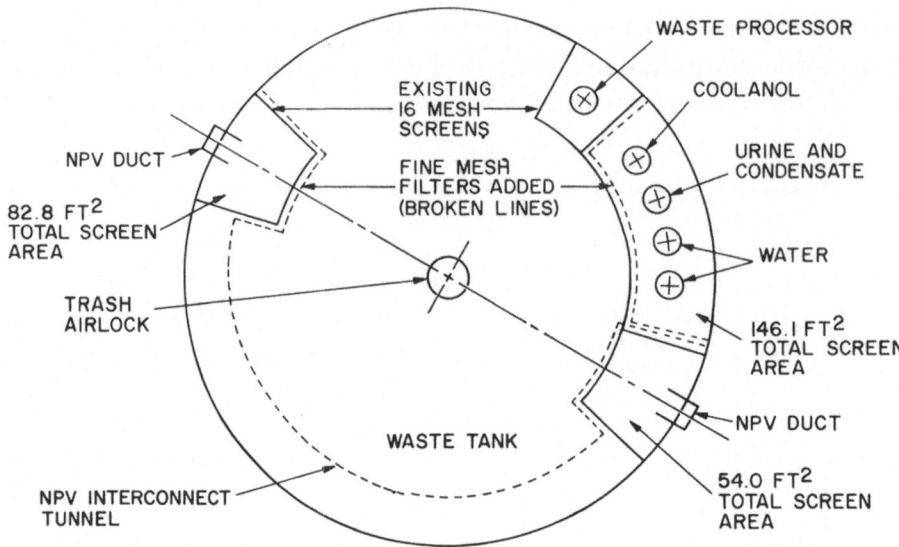

Fig. 3.   Skylab waste tank configuartion showing location of nozzles and filter screens.

Test Program conducted at Martin Marietta Corporation. During normal as well as contingency operation of the Skylab OWS waste tank, the screens were extremely successful in preventing particulate matter from emerging through the vent [9]. The maximum particle production rate was only 40 s^{-1}, with a nominal rate of 20 s^{-1}. Of the particles, 95% were between 0.1 and 1.6 μ. This corresponds to a mass flux of 10^{-11} gm s^{-1}. Velocities ranged from 0.5 to 10 m s^{-1}. The estimated scattering shown in Table II is completely negligible.

4. Summary

Details of the mechanisms by which scientific experiments can become contaminated have been developed. Deposition of contamination film on optics can be controlled by: (1) isolating critical optical surfaces from the rest of the spacecraft; (2) using as little nonmetallic material as possible, particularly near or in line of sight of optical surfaces; (3) choosing nonmetallics carefully to avoid materials with high vapor pressures; (4) vacuum baking materials before use to drive off the more volatile outgassing products; (5) keeping critical surfaces warmer than surroundings; (6) avoiding having nonmetallic material run at elevated temperatures; (7) paying special attention to optics exposed to intense ultraviolet, X-ray, or particulate radiation; (8) permitting no source of water vapor, such as multilayer insulation vents, near cooled detectors; and (9) directing RCS plumes away from critical surfaces and providing suitable covers that can be closed during RCS maneuvering.

For control of particulate contaminants: (1) maintain stringent cleanliness requirements on all spacecraft surfaces; (2) provide protection such as dust covers for critical

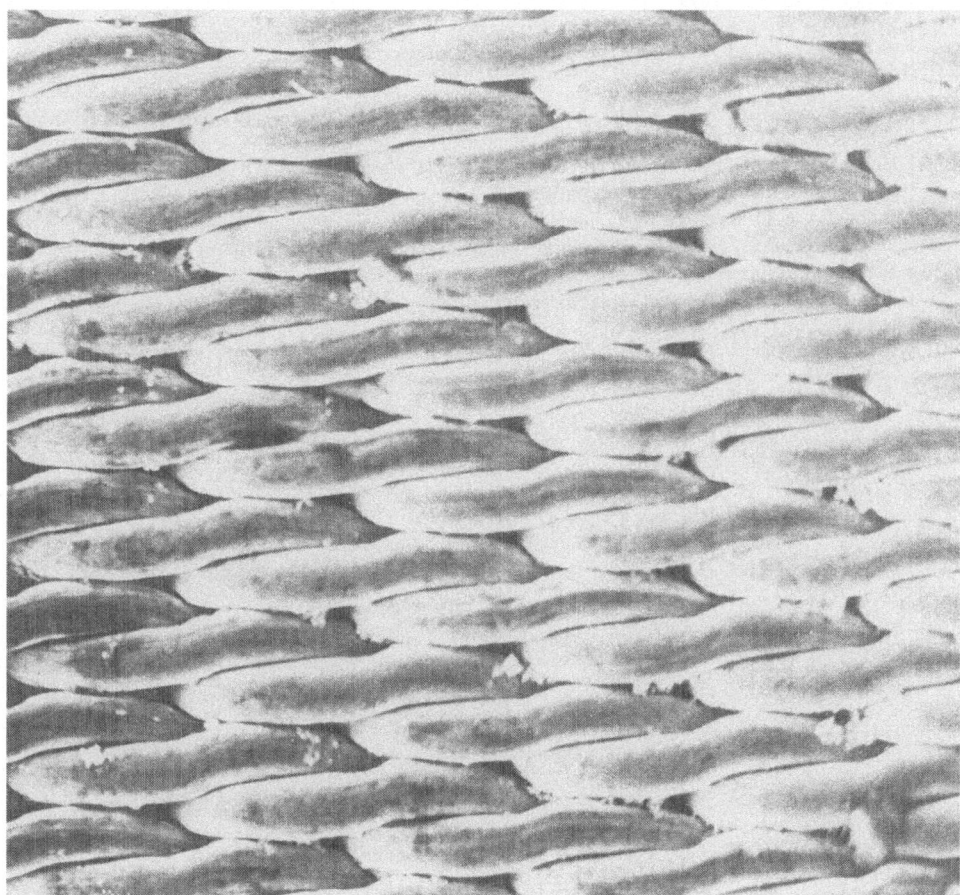

Fig. 4.   Scanning electron micrograph (300 ×) of the stainless steel 'Dutch Twill' screens used as filters in the Skylab waste tank. The small particles trapped in the screen resulted from a urine flow test.

surfaces during launch; (3) protect critical surfaces during storage in clean rooms; (4) design critical surfaces for easy access so that final inspection and cleaning may be made just prior to flight; (5) avoid dumping liquid waste overboard – convert it to vapor phase first; and (6) make sure there are no loose edges of multilayer insulation or fabric material that may create particles by abrading.

The most important thing is to develop an awareness of contamination problems and always be alert to uncover unsuspected sources and effects.

References

[1] Reynolds, J. M., Fields, S. A., and Wilson, R. M., 'X-Ray Reflection Efficiency of Ni-Coated Quartz Flats', IN-SSL-T-69-9, Marshall Space Flight Center, NASA, November 1969.
[2] Brunauer, S., *The Absorption of Gases and Vapors*, Oxford Univ. Press, London, 1943.
[3] McKeown D., Corbin Jr., W. E., 'Space Measurements of the Contamination of Surfaces by OGO-6', presented to the Space Simulation Conference, NBS, Gaithersburg, Md., September 1970.

[4] Robertson, S. J., 'Backflow of Outgas Contamination onto Orbiting Spacecraft as a Result of Inter-molecular Collisions', LSMC-HREC D 306000, Lockheed Missiles and Space Co., Contract NAS8-26554, June 1972.

[5] McKeown D., and Corbin, Jr., W. E., 'Removal of Surface Contamination by Plasma Sputtering', Proceedings of AIAA 6th Thermophysics Conference, U. of Tenn. Space Institute, Tullahoma, Tenn., April 1971.

[6] West, W. S., Gore, J. V., Kasha, M. A., and Bilsky, H. W., 'Spacecraft Charge Buildup Analysis', NASA SP-276, 1971.

[7] Hodges, Jr., R. R., Hoffman, J. H., Yeh, T. T. J., and Chung, G. K., 'Orbital Search for Lunar Volcanism', *J. Geophys. Res.* (in publication).

[8] Buffalano, A. C., Kratage, M. L., and Sharma, R. D., 'Interpretation of Visual Observations of Apollo Water Dumps – Case 340', Bellcomm Memorandum B7107014, July 1971.

[9] *Skylab Orbital Assembly Systems*, Design Certification Review Report, Contamination, August 1972.

[10] Austin, P. R., 'Personel and Their Contribution to Product Contamination', *Contamination Control*, October 1965, p. 45.

UTILIZATION AND APPLICATIONS
OF SPACE TECHNOLOGY

THEORIES AND APPLICATIONS

ASPECTS OF THE FINITE ELEMENT METHOD AS APPLIED TO AERO-SPACE STRUCTURES

J. H. ARGYRIS

Imperial College of Science and Technology, University of London, England

and

Institut für Statik und Dynamik der Luft- und Raumfahrtkonstruktionen, Universität Stuttgart, F.R.G.

J. ST. DOLTSINIS

Institut für Statik und Dynamik der Luft- und Raumfahrtkonstruktionen, Universität Stuttgart, F.R.G.

J. F. GLOUDEMAN

Space Division of North American Rockwell Corporation, Downey, Calif., U.S.A.

and

K. STRAUB and K. J. WILLAM

Institut für Statik und Dynamik der Luft- und Raumfahrtkonstruktionen, Universität Stuttgart, F.R.G.

1. Introduction

A unique combination of aero-spacecraft technology is necessary for the success of the 'Space Shuttle Program' which forms the next major manned space flight program in the Western World. The primary design objectives involve analytical problems of so far unseen complexity and magnitude. The parallel burn at lift off involves liquid and solid rocket engines which results in accelerations up to 1.5 g's. Moreover, during the early atmospheric flight the vehicle will experience severe aerodynamic forces and induced aeroelastic effects due to its geometric characteristics. Having achieved Earth orbit, the Shuttle Orbiter will serve a number of functions, involving both low and high power thrusts for different maneuvers. The return flight to Earth is likely to be in the 8000 ms^{-1} range forming a severe challenge to the analysis since the large scale structure is exposed to extreme environmental conditions. Both the success of a given mission, and the system reliability for an envisaged 100-flight vehicle, depend on the solution of these problems.

Following an appraisal of the shuttle in the first part some aspects of the finite element method will be reviewed when applied to the static and dynamic analysis of aero-spacecraft components. These techniques combined with interactive graphic capabilities in conjunction with the development of a central data bank form the prerequisite for the successful design of such a complex structure.

First the static formulation is reviewed and illustrated on pertinent examples. A selection of higher-order elements is described and applied to demonstrate the power of the finite element approach in conjunction with recent advances in computing technology. Then a mathematical model is constructed to characterize physically nonlinear phenomena within the scope of thermo-elasto-plasticity and creep. For the solution of these problems incremental and iterative procedures are reviewed and

L. G. Napolitano et al. (eds.), Astronautical Research 1972, 179–238. All Rights Reserved

discussed with regard to accuracy as well as convergence. These techniques are illustrated with two examples, the thermo-elasto-plastic analysis of a conical re-entry vehicle and a similar analysis of a pressure vessel nozzle considering also creep.

The dynamic analysis of structures is then briefly reviewed with regard to both spectral decomposition and dynamic response methods. Solution techniques such as mode superposition and direct integration procedures of the equations of motion (finite elements in time) are discussed both as far as accuracy and efficiency are concerned. A comparison of analytical and experimental mode shapes and frequencies on the FS 28 power glider illustrates the spectral decomposition technique on the example of a free-free system. Dynamic response analyses of a rocket show certain features of time integration procedures to deal with propellant cut off, stage separation and the build up of thrust. In conclusion, the example of a spherical container vessel subjected to a sudden pressure pulse illustrates the dynamic response analysis of a complex structural component. Various time marching algorithms are employed for the solution of this problem using mode superposition and direct integration procedures.

2. The Space Shuttle System

The National Aeronautics and Space Administration in the United States recently awarded the Space Shuttle contract to the Space Division of the North American Rockwell Corporation which is responsible for the design, development and production of the Shuttle's payload carrying orbiter vehicle and for the integration of all elements of the system. A close contact on software development is kept with the Institut für Statik und Dynamik, ISD, of the University of Stuttgart, which led to this cooperative effort.

The primary objective of the Space Shuttle Program is to provide an economic space transportation system that will support a wide range of scientific, defense, and commercial applications in Earth orbit. Major milestones are first manned orbital flight by 1978 and initial operational capability before the end of this decade. Since a number of aspects of the Space Shuttle Program are covered elsewhere we restrict ourselves to associated computer software developments with special emphasis on the finite element method.

2.1. THE SPACE SHUTTLE VEHICLE

The vehicle shown in Figure 1 consists of a relatively compact airplane-like orbiter with a single external liquid propellant tank, dual solid rocket motors, and support equipment balanced with the onboard capabilities of the vehicle. The payload bay is about 5 m in diameter and 20 m long. It is environmentally controlled and can handle a payload of approximately 30 tons.

The orbiter resembles a large-scale jet airplane and is designed to carry a crew of at least four men including two flight crew-men in the upper part of the double-deck compartment and two mission crew-men below. To be constructed primarily of aluminium, the orbiter is 33.5 m long, 17 m high and has a wing span of 24 m.

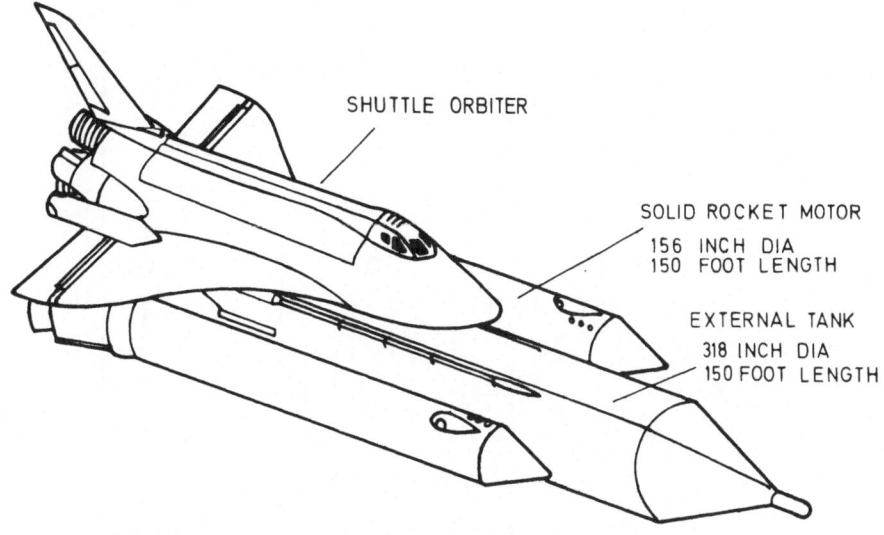

ORBITER CHARACTERISTICS			
● CONFIGURATION			● STRUCTURE
LENGTH	FT	110.7	ALUMINIUM FUSELAGE AND AERO SURFACES
SPAN	FT	79.6	MULTISPAR WING. TRUSS CARRY THROUGH
WING ASPECT RATIO		2.10	WELDED ALUMINIUM FLOATING CREW CABIN
PAYLOAD BAY	FT	15 × 60	CERAMIC REUSABLE SURFACE INSULATION
CARGO LOAD	Lbs	65 000	CARBON CARBON WING LEADING EDGES AND BODY NOSE
CREW COMPARTEMENT	MEN	4	TRICYCLE GEAR AND DECELERATION CHUTE

The external tank provides the orbiter's three liquid rocket engines with its liquid oxygen and liquid hydrogen propellants. Essentially of aluminium monocoque construction with an external insulation of urethane foam, the external tank is approximately 47 m long and 8 m in diameter. In the vertical position the Space Shuttle Vehicle stands 64 m high as shown in Figure 2 for readily identifiable comparison with other aero-space vehicles.*

2.2. THE SHUTTLE MISSION

The Shuttle Vehicle is launched by a parallel burn of its three liquid propulsion engines and the two mated solid rocket motors. The pressure-fed liquid propulsion engines each generate a thrust of 470 000 pounds which combine to provide 1 410 000 pounds of thrust. The two reusable solid rocket motors, 46 m long and 4 m in diameter, each generate a maximum thrust of 3.5 million pounds giving the combined propulsion system a thrust of nearly 8.5 million pounds. The total weight of the Shuttle Vehicle at lift-off is approximately 4.7 million pounds which yields an initial thrust to weight ratio of 1.7.

A combination of main engine thrust vector control and aileron deflection provides guidance throughout the boost phase. The parallel burn of the liquid and solid

---

* Author's note: The physical descriptions correspond to the design configuration of the time the paper was presented. They are of course subject to change.

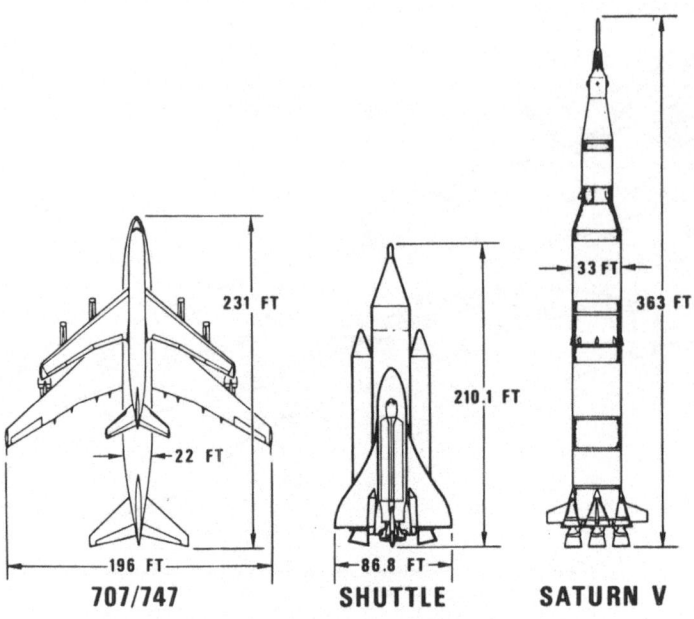

Fig. 2.   Comparison of aero-space vehicles.

systems continues until a staging velocity of 1300 m s^{-1} is achieved at an altitude of about 43 km, approximately 109 s after lift-off. A maximum dynamic pressure $p_{max} = 0.32$ kp cm^{-2} is reached at an altitude of approximately 12 km fifty-four seconds after lift-off. The separation of the Solid Rocket motors (SRM) requires auxiliary rockets having a total thrust of 160 000 pounds, the first of which are re-covered by barge 270 km down-range following parachute descent. The SRM's are initially decelerated using drogue parachutes which open at an altitude of 8.0 km, 307 s after lift-off. 43 s later the main parachute opens at an altitude of 4.9 km fol-lowed by the splashdown 463 s after lift-off.

On a normal mission, the three liquid engines continue to burn until the depletion of the external tank propellants at an altitude of about 80 km 551 s after lift-off. The external tank is subsequently separated and sinks in a safe ocean area. The Orbit Maneuver System (OMS) of the Orbiter is then employed to complete the required mission.

Once orbit has been achieved, the Orbiter can be used to perform a number of prescribed tasks including satellite emplacement, repair, servicing and retrieval. In addition, propulsive stages can be delivered to designated orbits or short-duration science and application missions can be conducted. Designed to fly over 100 missions, the Shuttle Orbiter will travel an estimated four million miles during a typical seven-day mission. A modern jetliner of comparable size travels approximately 31 000 miles in week's time. In-orbit maneuvers will be accomplished using the Reaction Control System (RCS) which consists of two sets of twelve 1000 pound thrusters in the OMS pads in the aft fuselage and two 8-thruster sets located forward on the nose.

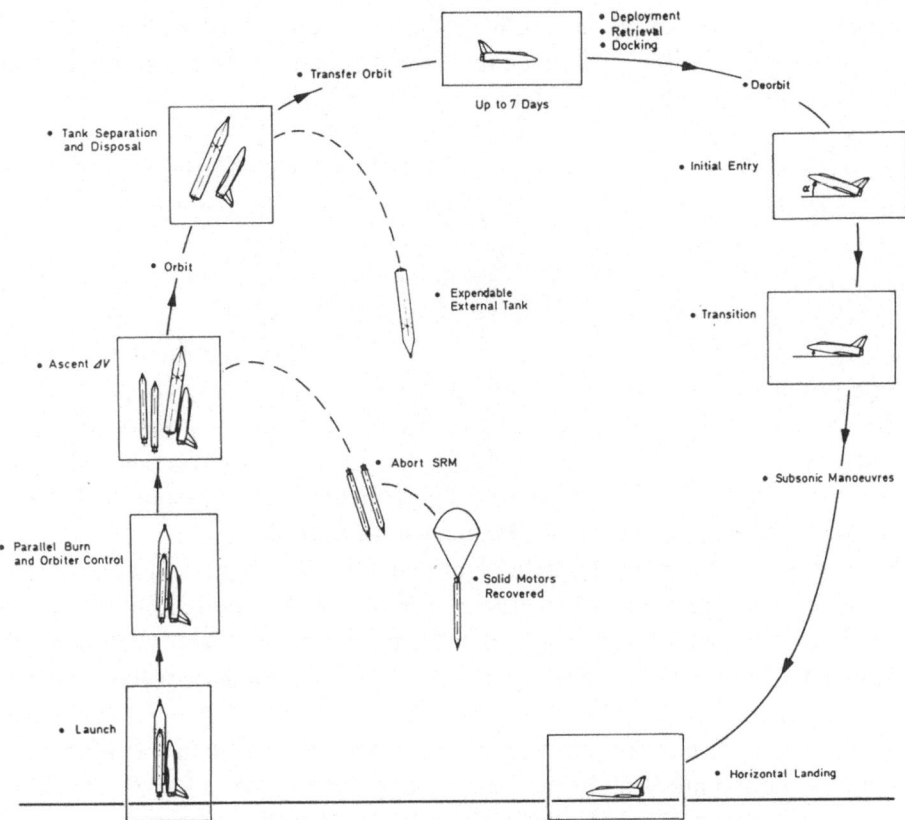

Fig. 3.   Mission profile of space shuttle system.

Following mission completion, the Orbiter returns to Earth much like an ordinary aircraft, the primary difference being the loading conditions resulting from the high speed of $8\,000$ m s^{-1} following de-orbit and re-entry during which the most severe thermal impact will be encountered.

In the hypersonic regime, flight control is via aileron and yaw reaction control meters. Designed for stability in the supersonic and subsonic flight regimes, control is achieved by aileron deflection for pitch and roll and by rudder deflection for yaw. The Orbiter lands on a runway of at least 3300 m length. It is then subjected to required maintenance and is to be ready for flight 14 days after landing.

Figure 3 summarizes the main stages of a typical shuttle mission profile.

3. The Finite Element Method

From the previous condensed account of the space shuttle we are very much aware that effectively none of the complex analyses can be performed without the active participation of the digital computer and a highly sophisticated software. It has been demonstrated conclusively over the past 10 to 15 years or so that, for example, no meaningful static and dynamic analysis of complex systems is feasible unless the classical concepts of structural mechanics are reshaped and extended with a computer oriented philosophy. In this area a true revolution has taken place with the impact of the 'Matrix or Finite Element' Methods as they are predominantly called now.

What are finite elements? Perhaps the simplest definition for the present occasion is that they represent a piecewise application of the Rayleigh-Ritz procedure to discrete domains of the complete structure or continuum. In fact, the assembly of these domains forms the so-called idealised representation of the physical system. We should emphasize that this is only a very limited specification of the finite element method whose general philosophy can be clothed in a much broader framework. Let us only add that this technique is in no way restricted to structural or solid mechanic problems but is also applicable to a wide range of physical phenomena described by a set of field equations. For example, this holds for hydro-, thermo-, and structural dynamics as well as coupled problems such as pogo. The extensive surveys of the ISSC 1969 Conference and the 12th Lanchester Memorial Lecture of 1969 [5], [6] demonstrate some applications in these fields.

In the following we aim to spotlight some significant concepts and applications of the finite element method as they arise in the analysis and design of aero-space-crafts. For computer-based methods two approaches may be taken starting either with differential or integral formulations of the initial boundary value problem. Of the latter the virtual work or energy methods are suited better for numerical solution as they involve differential expressions of lower order than the differential equation formulation and lead normally to a symmetric system of algebraic equations. Moreover, boundary conditions are subdivided into essential and natural ones, the latter being satisfied in a global sense only. Some of the most common techniques for the computer analysis of structural problems are summarized in Figure 4.

Collocation and finite difference techniques are normally applied to discretize the differential formulation either by approximating directly the field variables or by replacing the differentials by suitable difference operators. In general, the boundary conditions pose difficulties particularly in the case of complex geometric configuration. In addition the algebraic problem often requires the solution of an unsymmetric

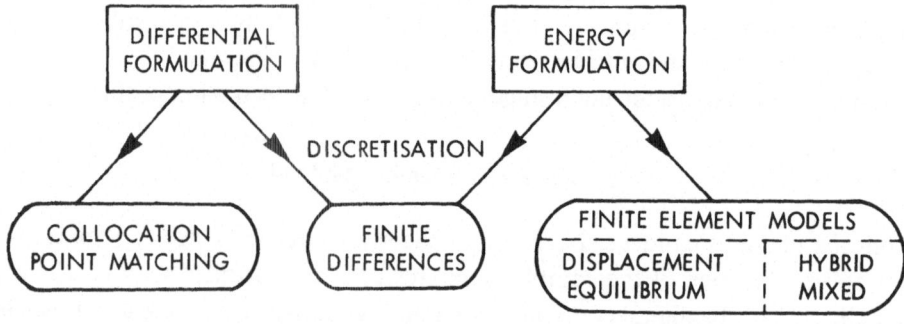

Fig. 4.   Computer-based methods of structural analysis.

system of equations. In contrast, the discretization schemes based on virtual work or energy principles yield symmetric algebraic equations which are also positive definite for problems of stable equilibrium. Furthermore, the approximation of the field variables within subdomains strongly decouples interaction within the structure providing a high degree of sparsity in the ensuing system of equations. We remark that finite difference schemes have also been proposed recently for discretisation of the energy formulation in order to obtain algebraic systems with these convenient properties [7].

Originally, two finite element formulations were proposed and denoted as the matrix force and displacement methods [1], [2]. Since both methods are derivable from extremum variational principles or equivalent arguments, energy bounds are obtained as long as the governing continuity requirements are satisfied. Subsequently, hybrid and mixed models have been suggested in an attempt to improve the accuracy by forfeiting boundedness of the solution and/or positive definiteness of the system of equations. In spite of these alternative formulations to date by far the greatest number of successful applications have been achieved with displacement models. These prove best for numerical computations and are easily extended to dynamic and geometrically non-linear problems. In the following, only this finite element formulation will be discussed.

3.1. LINEAR ELASTO-STATICS

Let us briefly review now some of the underlying concepts of the finite element (matrix) displacement method as applied to the solution of linear elastic problems for which inertia effects remain negligible.

3.1.1. Basic Concepts

Recalling the principle of virtual work

$$\delta W_i = \delta W_e \tag{1}$$

we arrive – in the special case of the virtual displacement principle – at an integral formulation of static compatibility if we equate the internal and the external virtual work of the static field σ, $\mathbf{p}_v$ and $\mathbf{p}_s$ with the virtual kinematic field $\delta\gamma$ and $\delta\mathbf{u}$

$$\int_v \delta\gamma^t \sigma \, dv = \int_v \delta\mathbf{u}^t \mathbf{p}_v \, dv + \int_{S_\sigma} \delta\mathbf{u}^t \mathbf{p}_s \, ds.^* \tag{2}$$

We can easily verify that this statement is equivalent to the differential form of static equilibrium

$$\mathbf{D}\sigma - \mathbf{p}_v = 0. \tag{3}$$

---

* Bold founts denote vectors and matrices; the superscript t stands for the transpose of a vector or matrix.

Together with the stress boundary conditions on the surface S_σ

$$[\mathbf{D}n]\,\boldsymbol{\sigma}=\mathbf{p}_s. \tag{4}$$

Implicitly we only permit compatible virtual strain- and displacement fields $\delta\boldsymbol{\gamma}$ and $\delta\mathbf{u}$

$$\boldsymbol{\gamma}=\mathbf{D}^t\mathbf{u} \quad \text{and} \quad \delta\boldsymbol{\gamma}=\mathbf{D}^t\delta\mathbf{u}. \tag{5}$$

At the same time displacement boundary conditions on the surface S_u must also be satisfied

$$\mathbf{u}=\mathbf{u}_s. \tag{6}$$

In the above relations $\mathbf{D}$ denotes the well-known differential operator (see [3], [18]). The finite element displacement method utilizes the principle of virtual displacements to construct via an approximate displacement field a discrete set of equilibrium equations. Let us now consider the prescribed kinematic definition of a single finite element:

3.1.1.1. *Displacement field*

$$\mathbf{u}=\boldsymbol{\omega}\boldsymbol{\varrho}. \tag{7}$$

The interpolation functions $\boldsymbol{\omega}$ define the spatial variation of displacements $\mathbf{u}$ in terms of the nodal degrees of freedom $\boldsymbol{\varrho}$.

3.1.1.2. *Total strain field*

$$\boldsymbol{\gamma}=\mathbf{D}^t\mathbf{u}=(\mathbf{D}^t\boldsymbol{\omega})\,\boldsymbol{\varrho}=\boldsymbol{\alpha}\boldsymbol{\varrho}. \tag{8}$$

The total strain field is defined by linear contributions of the displacement gradient field as derived from Equation (7).

3.1.1.3. *Stress field.* For a linearly elastic material

$$\boldsymbol{\sigma}=\mathbf{E}\boldsymbol{\varepsilon}=\mathbf{E}(\boldsymbol{\gamma}-\boldsymbol{\eta})=\mathbf{E}\boldsymbol{\gamma}+\boldsymbol{\tau}. \tag{9}$$

We note that the concepts of initial strains and initial stresses are equivalent since

$$\boldsymbol{\tau}=-\mathbf{E}\boldsymbol{\eta}. \tag{9a}$$

They provide a mechanism to account for non-elastic deformations which may arise either from environmental conditions, such as temperature, moisture, irradiation or from mechanical conditions, such as plasticity and creep.

We emphasize that the virtual work formulation does not depend on the stress-strain law. In other words, no assumption has to be made in regard to the existence of energy potential from which stresses may be derived.

Invoking the principle of virtual work we obtain the following discrete form of element equilibrium

$$\mathbf{k}\boldsymbol{\varrho}=\mathbf{Q}+\mathbf{J}, \tag{10}$$

where the individual element quantities are defined as

ELEMENT STIFFNESS

$$k = \int_v \alpha' E \alpha \, dv \qquad (11a)$$

NODAL LOADS (kinematically equivalent to distributed body forces and surface tractions)

$$Q = \int_v \omega' p_v \, dv + \int_{S_\sigma} \omega' p_s \, ds. \qquad (11b)$$

INITIAL LOADS (kinematically equivalent to initial strains or stresses)

$$J = \int_v \alpha' E \eta \, dv = - \int_v \alpha' \tau \, dv. \qquad (11c)$$

The individual element contributions can now be assembled using the Boolean connectivity matrix a_I which transforms element into structural degrees of freedom

$$\varrho_I = a_I r.$$

Hence the overal structural stiffness

$$K = \sum_I a_I^t k_I a_I \qquad (12)$$

and the structural loading

$$R = \sum_I a_I^t (Q_I + J_I) + R_G.$$

The structural load vector R contains contributions from distributed body and surface tractions, Q_I, from initial stresses or strains, J_I, and from concentrated nodal forces R_G. Structural equilibrium is then described by

$$Kr = R. \qquad (13)$$

This system of simultaneous equations can be solved for the nodal displacements r which in turn define the state of deformation and stress within each element of the total structure ϱ, γ, σ. Details of different solution techniques are not discussed here, it is only noted that symmetry, sparsity and positive-definiteness (after removal of rigid body modes) facilitates a direct solution method [20], [76].

3.2.1. *Finite Element Models*

The versatility of the finite element displacement method requires no commentary and is demonstrated best by the all too numerous publications in this field. Here we merely attempt to summarize some of the latest developments in element modelling procedures.

We recall that the choice of modal functions is restricted by two aspects, completeness requirements, inter-element continuity requirements. Completeness is satisfied if the chosen polynomial expansions, which are suited best for numerical work, contain all terms of the order m where $2m$ denotes the order of the elliptic boundary value problem. This is an equivalent statement to the constant energy criterion which requires that the modal functions satisfy the following condition at each point of the element domain defined by the homogeneous element coordinates ζ

$$\sum_{i=1}^{n} \omega_i(\zeta) = 1. \tag{14}$$

In addition, the geometric approximation must be a subset of the displacement expansion to represent all constant energy states within the element

$$\omega_x(\zeta) \subseteq \omega_u(\zeta). \tag{15}$$

Continuity is satisfied if no overlapping or disruption takes place between contiguous elements before and after deformation. This implies that both, geometric and displacement interpolation including derivatives up to the order $(m-1)$ have the same variation along the common inter-face of two neighbouring elements a, b

$$\omega_x^a(\zeta) = \omega_x^b(\zeta)$$
$$\omega_u^a(\zeta) = \omega_u^b(\zeta) \tag{16}$$
$$\vdots \qquad \vdots$$

The old question of the higher order versus lower order elements has been answered, at least partially, in favour of the higher order expansions [8], [62]. The discretisation error depends mainly on the element subdivision h, the order of polynomial expansion p and the order of differentiation m. The energy norm is then a measure for the discretisation error

$$\|u - \hat{u}\|_{\varepsilon} = c_1 h^{2(p+1-m)}, \tag{17}$$

where c_1 denotes a proportionality factor. In contrast, the numerical error due to round off and initial truncation is hardly affected by the order of polynomial expansion p and depends mainly on the spectral conditioning number for the inversion [76]

$$\operatorname{cond} \mathbf{K} = \lambda_{\max}/\lambda_{\min} \sim c_2 h^{-2m}. \tag{18}$$

Here $\lambda_{\max}, \lambda_{\min}$ denote the extreme eigenvalues of $\mathbf{K}$, which increase mainly with the number of degrees of freedom. This clearly indicates that the accuracy is improved much more effectively with the order of expansion rather than by the refinement of the mesh size. A word of caution should be appended to this discussion. It is straightforward to construct examples for which a TRIM 3 idealisation would be more effective than a TRIM 6 one. However, the main theme of our above assertion remains, in general, correct.

In what follows we consider polynomial expansions of the Lagrangean type for which only function values, but not their derivatives are used as nodal quantities. The discussion is restricted to triangular and quadrilateral elements and uses for convenience cartesian coordinates instead of the more effective and elegant parametric representation in homogeneous coordinates.

Triangular elements of the TRIM-type have the marked advantage that the modal functions are expressible by *complete* polynomials [3], [14]. This implies that the functional variation in the element domain is independent of linear coordinate-transformations ensuring 'isotropic' properties of the expansion.

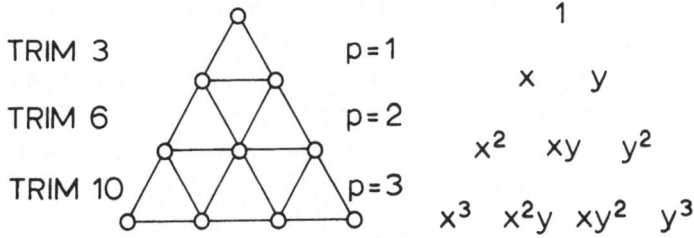

$$
\begin{array}{c}
1 \\
x \quad y \\
x^2 \quad xy \quad y^2 \\
x^3 \quad x^2y \quad xy^2 \quad y^3
\end{array}
$$

TRIM 3   p=1
TRIM 6   p=2
TRIM 10   p=3

Quadrilateral elements of the QUAM-type are in contrast based on modal functions which are products of one-dimensional Lagrangean interpolation schemes [11], [13]. These expansions contain complete polynomials of the order p but also include higher order terms up to the order $2p$. This implies that the functional variation in the element domain depends strongly on linear coordinate transformations loosing the 'isotropic' properties of the expansion.

QUAM 4   p=1
QUAM 9   p=2
QUAM 16   p=3

$$
\begin{array}{c}
1 \\
x \quad y \\
x^2 \quad xy \quad y^2 \\
x^3 \quad x^2y \quad xy^2 \quad y^3 \\
x^3y \quad x^2y^2 \quad xy^3 \\
x^3y^2 \quad x^2y^3 \\
x^3y^3
\end{array}
$$

In this context a word of caution is appropriate with regard to geometric distortions. For simplicity consider a one-dimensional line-element whose geometry and displacements are defined in the homogeneous coordinates ζ by polynomial expansions of the order q and p

$$
x = x(\zeta^q)
$$
$$
u = u(\zeta^p). \tag{19}
$$

Inversion of the mapping between the global space x and the parameter space ζ yields the following approximation for the displacements

$$\zeta \Rightarrow \zeta\left(x^{1/q}\right)$$
$$u \Rightarrow u\left(x^{p/q}\right). \tag{20}$$

This clearly indicates that the geometric representation is primarily improved at the expense of the displacement approximation in the global space. This fact suggests the use of subparametric elements in which the geometry is represented by linear interpolation $q = 1$. It also follows that the case of $q = p$ should be applied in a manner ensuring the best possible centering of the midside and interior nodes in order to reduce the effects of geometric distortions.

The superior convergence properties of higher order expansions are illustrated below on two examples. Figures 5 and 6 show a truncated cone subjected to a torque which is idealised by triangular axisymmetric solid elements using linear, quadratic and cubic expansions, TRIAX 3, TRIAX 6 and TRIAX 10 [15], [66]. Figure 7 indicates the relative error in tangential displacements as a function of the total number of degrees of freedom. Note that the semi-logarithmic scale indicates different orders of magnitude of the relative error. The convergence characteristics of stresses are shown in Figure 8 illustrating similar behaviour.

The results of an analogous investigation for a square plate subjected to a concentrated midpoint load are shown in Figure 9. The relative error in energy indicates again the superiority in accuracy of the quintic expansion of the TUBAC 6 element, in comparison with the cubic TRIB 3 element, shown in Figures 12 and 13.

In conclusion some of the recent element developments are discussed which are currently available in the integrated finite element software package ASKA [19], [20], [76]. Figure 10 shows some quadratic rod and membrane elements which

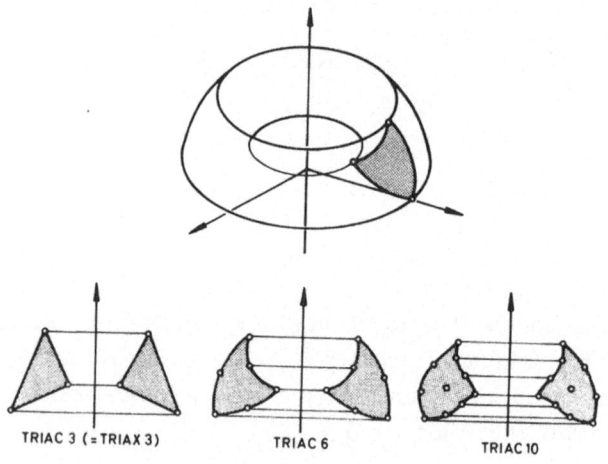

TRIAC 3 (= TRIAX 3)   TRIAC 6   TRIAC 10

Fig. 5.   The TRIAX family of ring elements with curved edges for Axisymmetric and Harmonically distributed loading.

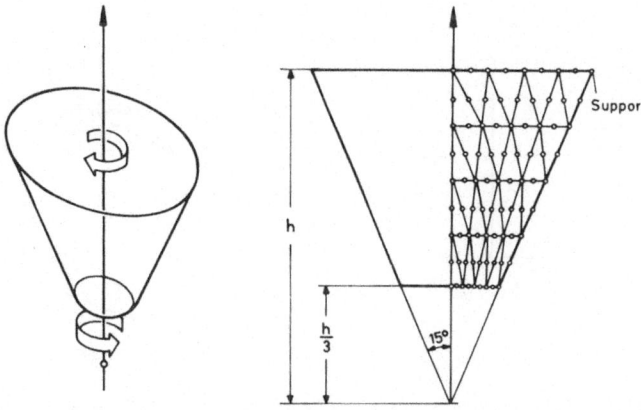

Fig. 6.   Truncated cone under torsion. Geometry and typical idealisation.

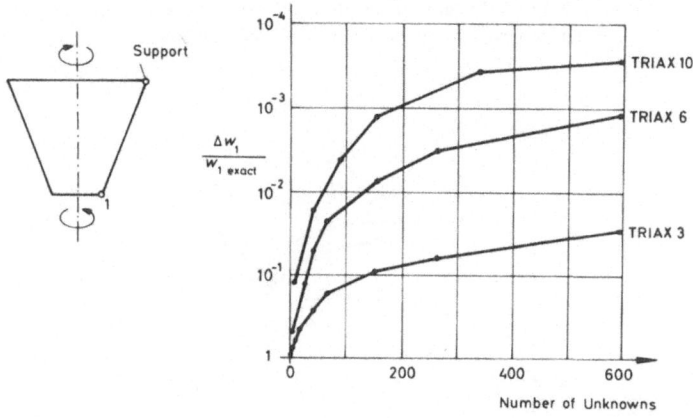

Fig. 7.   Truncated cone under torsion. Accuracy of the calculated tangential
displacement w for different idealisations.

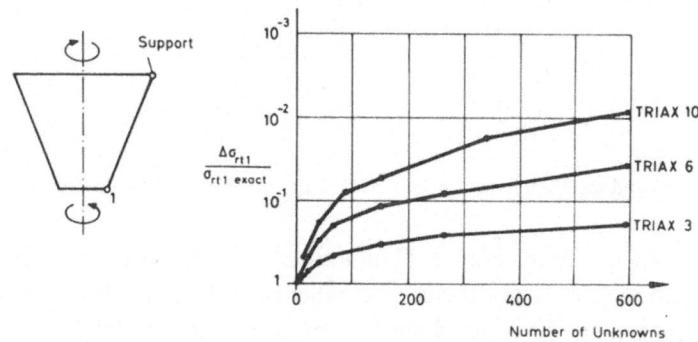

Fig. 8.   Truncated cone under torsion. Accuracy of the calculated shear stress σ_{rt} for different idealisations.

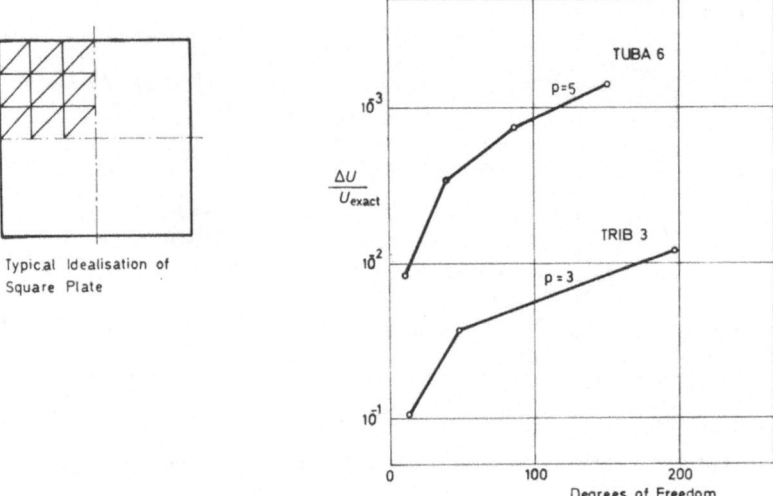

Fig. 9.   Simply supported square plate subject to concentrated center load.
Relative error of energy for different idealisations.

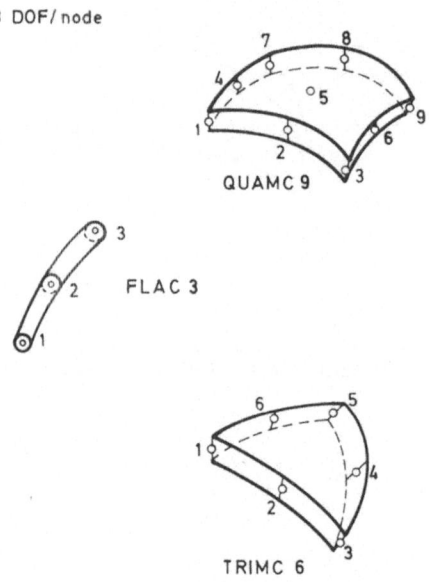

Fig. 10.   Rod and membrane elements. Quadratic approximation of geometry and displacements.

may be curved in space. Figure 11 illustrates the extension to three-dimensional
solids in form of curved tetrahedra, pentahedra and hexahedra. Again geometry as
well as displacement field are described by quadratic modal functions. Figure 12
demonstrates the aforementioned very accurate plate bending element TUBAC 6.
The displacement expansion is based on a complete quintic polynomial involving

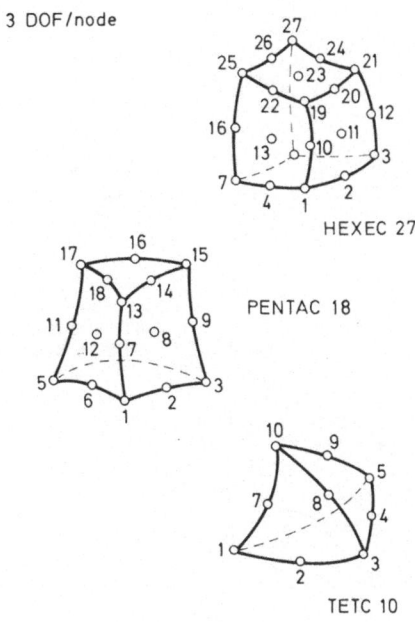

3 DOF/node

HEXEC 27

PENTAC 18

TETC 10

Fig. 11. Three dimensional solid elements. Quadratic approximation of geometry and displacements.

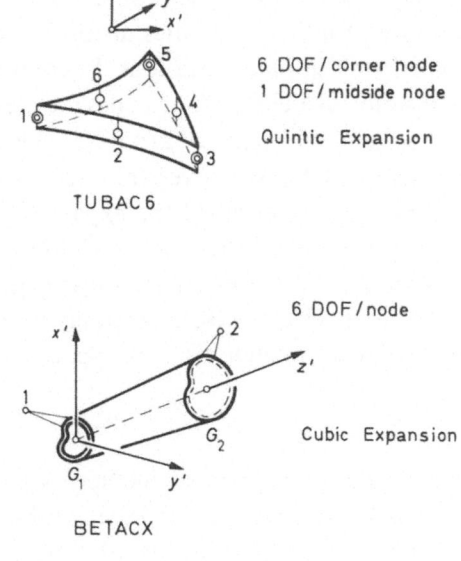

6 DOF / corner node
1 DOF / midside node

Quintic Expansion

TUBAC 6

6 DOF / node

Cubic Expansion

BETACX

Fig. 12. Beam and plate bending elements.

six degrees of freedom at each corner node (w', w'_x, w'_y, w'_{xx}, w'_{yy}, w'_{xy}) and one at each midside node w'_n. The same figure shows also a typical tapered beam element having a closed section and eccentric connections. The next figure, Figure 13, illustrates

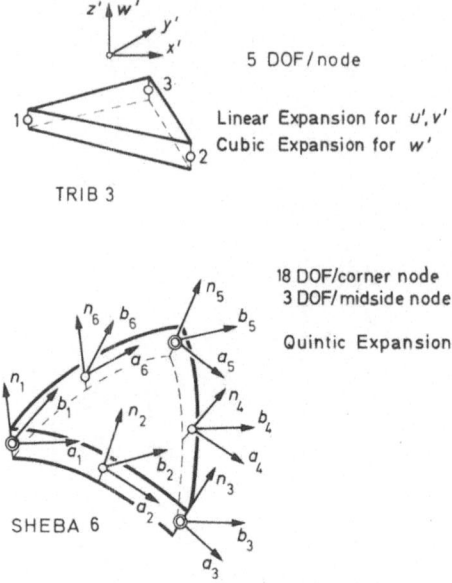

Fig. 13.   Flat and curved shell elements.

a flat and a curved shell element. The difficulties with the flat 'shell' element are well understood by now; here we just recall the basic problem in approximating curved geometries [9], the appearance of incompatibilities at folds after deformation and the lack of stiffness associated with in-plane rotations. In contrast, the curved shell element SHEBA 6, a generalisation of TUBAC 6, provides the means for an extremely accurate description of the actual shell geometry avoiding folds along inter-element boundaries before and after deformation. Moreover, this highly sophisticated element features rigid body modes and completeness of the expansion (quintic), all at the expense of 63 deg of freedom per element [16], [72], [73], [74].

The wide spectrum of elements and the extensive substructure capability of the integrated finite element software package ASKA inevitably brought about its selection as main program for the structural analysis of the space shuttle orbiter.

3.1.3. *Examples*

In the following the power of the finite element method is demonstrated on two aero-spacecraft structures. The first problem involves the static analysis of a supersonic aircraft while the second one illustrates the capabilities of the method on the thermal protective system of the shuttle orbiter. Both problems have been solved at the Los Angeles Division of North American Rockwell using ASKA. Due to the classified nature of the activity, it is only possible to outline some of the results; all of them have been substantiated either by other analyses or by tests.

3.1.3.1. B-1 supersonic aircraft. Figure 14 gives a general view of the B-1 which is

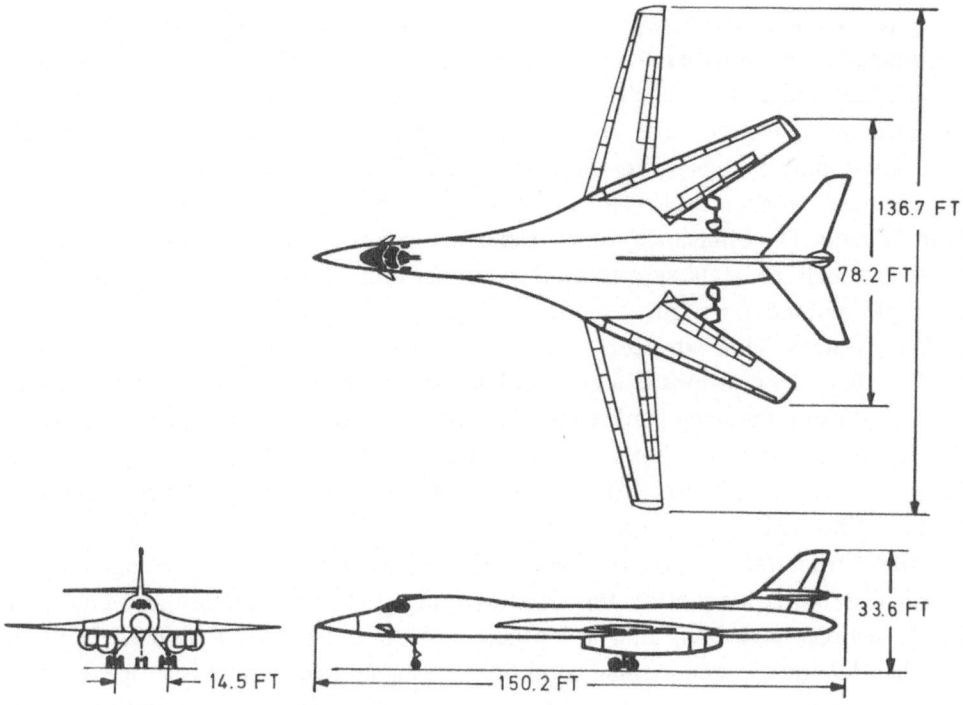

Fig. 14.   Lay-out of B-1 supersonic aircraft.

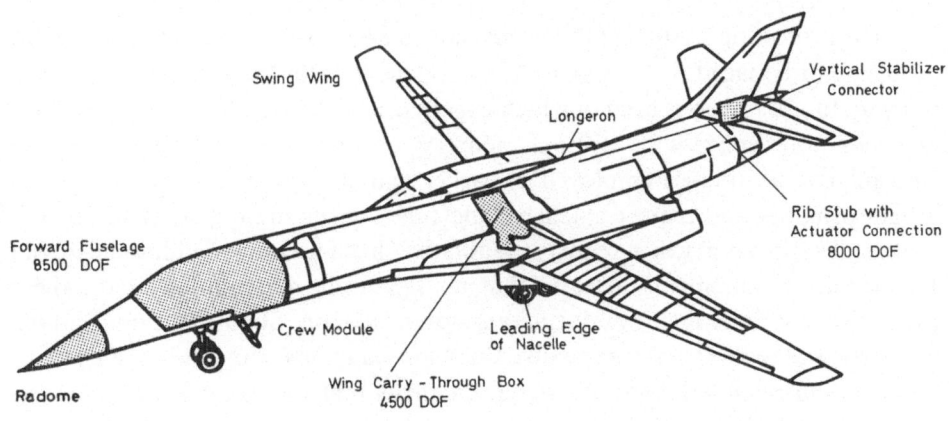

Fig. 15.   Substructure analysis of different components.

currently under development by the Los Angeles Division of North American Rock-well. Several major portions of the system have been analysed using ASKA including the forward fuselage, the wing carry-through box, the nacelle leading edge, the composite materials and the vertical stabilizer connector as indicated in Figure 15 [21]. Forward Fuselage:

The forward fuselage includes the radome and the crew compartment, the first of which was analysed using flat shell elements (TRIB 3C). The crew compartment including the pressurized bulkhead was modeled with 2200 membrane, 600 rod, 1950 beam and 250 flat shell elements involving more than 8500 deg of freedom. Substructuring proved most useful in the internal load analysis of the crew module. The environment considered included elevated temperature, air pressure loads, inertia loads, rocket ejection forces, parachute forces, internal pressure, and bird impact loads. The windshield, which is of complex form and constraints was analysed with flat shell elements as well as three dimensional solid elements to model the multiple lamination design.

Wing Carry – Through Box:

The swing wing proposed for the B-1 led to the design of a wing carry-through box to transmit the wing loads to the fuselage. In addition to the fuselage and wing loads, the carry-through box is also subjected to loads from the landing gear. The analysis involved 650 rod, 70 membrane, 900 flat shell and 310 beam elements resulting in a 4500 deg of freedom model.

Other structural components such as the longeron, the nacelle leading edge, the vertical stabilized connector, the rib stub with actuator connection and finally the wing sections have been analysed using shell, plate, beam and solid elements for mechanical and thermal loading conditions taking into account anisotropic material properties and composites. The substructuring capability made it possible to perform detailed stress analyses while still retaining cognizance of the overall structure.

3.1.3.2. *Thermal protection system of shuttle orbiter.* The design and analysis of Thermal Protection System (TPS) is envisioned as one of the most critical technical problems to be faced in the design of the Space Shuttle Orbiter. The requirement to fly up to 100 missions each of which experiences entry temperatures up to 3000 °F demands an effective TPS without incurring any risk for the safety of the crew. The ablative techniques employed in earlier manned spacecrafts impose excessive weight penalties and require extensive and time consuming replacement due to the in-flight deterioration which renders the ablative shield useless for additional missions. Re-usability requirements demand that no significant degradation of the thermal protective system is incurred during any given mission. Should a minor failure be experienced, the design concept must allow for ready detection and repair in a cost-effective and reliable manner. The requirements that the Orbiter land like a conventional aircraft on a relatively standard runway and launch-related minimum weight requirements call for a light-weight thermal protection system.

Currently it is proposed to use three types of materials on different parts of the Orbiter. A low-weight elastomer, a rubber-like material, will be bonded to those surfaces which will experience temperatures up to 650 °F. These surfaces represent approximately one-third of the total external surface area of the Orbiter and are found primarily on the upper portions. The wing leading edges and nose cap are to be protected from temperatures expected to reach 3000 °F by a newly-developed oxidation inhibited, reinforced carbon-carbon material. This represents approximately three percent of the total Orbiter external surface area. The remainder of the Orbiter will be protected against temperatures in the range from 650 °F to 2500 °F by a newly-developed ceramic insulation. The areas protected by the ceramic insulation include the nose, underside, aileron and forward wing portions.

This ceramic insulation consists of panels ranging in size from 8 to 12 in. square and is built-up in four layers. The first layer is a silicone elastomer adhesive used as the bonding material. The second layer is a chemically foamed methyl-phenyl silicone elastomer pad ranging in thickness from 0.3 in. to $1\frac{1}{2}$ in. This pad fills in surface irregularities, isolates the basic insulation from structural strain and serves as a back-up insulation. The third layer ranges in thickness from 0.5 to 2 in. and is the basic insulation of mullite fibers stiffened with a binder of aluminium-boria silica refractory glass. The fourth layer is a waterproof ceramic coating on the top and sides of the mullite panels for protection against weather and ground handling. Small gaps will be left between panels to allow for heat expansion during entry. These gaps will be partly filled with a low-density quartz gasket to protect the structure.

Analysis of Re-usable Surface Insulation (RSI):

Two nearly identical configurations were analysed, the hot-windward and the cold windward panel insulation. As shown in Figure 16 the hot-windward specimen is a 4 in. by 8 in. mullite tile with a depth of 1.9 in. The tiles are bonded to the aluminium substrate with 1.1. in. thick PD 200 foam pads faced on upper and lower surfaces with RTV (Room Temperature Vulcanizing) 560 adhesive. The expansion gap (0.125 in.) filler for this study was an omni-weave gasket which has since been eliminated in favour of a woven quartz fabric. The 0.8 in. deep by 0.03 in. grooves are intended to provide strain relief for the coating to minimize the possibility of cracking during thermal cycling.

The specimen for the cold windward region of the Shuttle Orbiter, has 1.70 in. thick mullite tiles with a 0.30 in. thick PD 200 foam pad. Except for these thickness differences, the cold windward and hot windward specimen are identical.

Both specimen were analysed with the three-dimensional finite element idealization shown in Figure 17; 385 membrane, plate bending and solid elements were used to model each structural component of the 'composite' structure. A total of 2624 deg of freedom were utilized to assemble the individual elements.

Table I summarizes the in-plane and transverse stresses in each of the different structural components. The results are given for the cold and hot windward specimen both of which are subjected to cold soak conditions corresponding to a temperature gradient of -250 °F on the outside and -50 °F on the inside. The cold windward specimen was also exposed to a temperature gradient of 1250 °F on the outside and 350 °F on the inside surface. Results indicated with an asterisk (*) denote stresses for which the coupon tests yield considerably lower values than the analysis. In this context we should remark that similar two-dimensional computations appreciably underestimate the values of the three-dimensional investigation.

TABLE I
Summary of maximum stresses

Structural Element	Finite Element Type	Maximum Stress (psi)	Cold Windward Specimen				Hot Windward Specimen	
			Temperature–Gradient		Cold Soak – 250 F		Cold Soak – 250 F	
			In–Plane	Transverse	In–Plane	Transverse	In–Plane	Transverse
Aluminium Flanges	QUAM 8	Tension	27		383		3898	
		Compression	6		2027		6273	
		Shear	5		401		2108	
Aluminium Plate	QUAM 9	Tension	40		4087		9885	
		Compression	40		6500		15000	
		Shear	9		1157		8097	
RTV 560 Adhesive	QUAM 9	Tension	–		2172		2168 *	
		Compression	5		–		–	
		Shear	–		1519		2334	
PD 200 Foam Pad	HEXE 27	Tension	–	–	612	238	577	310
		Compression	2	1	–	401	–	479
		Shear	1	–	631 *	–	584 *	–
RSI Mullite	HEXE 27	Tension	21	113 *	10	490 *	12	943 *
		Compression	81 *	90 *	489 *	107 *	473 *	100 *
		Shear	85 *	–	111 *	–	170	–
SR-2 Ceramic Coating	TRIB 3	Tension	–		18030 *		24000 *	
		Compression	14390 *		22610 *		31060 *	
		Shear	3311 *		12520 *		18050 *	
SR-2 Ceramic Coating	QUAM 9	Tension	5215 *		–		–	
		Compression	–		1813		1528	
		Shear	3948 *		1299		1190	
SR-2 Ceramic Coating	QUAM 9	Tension	1983 *		1090		3545 *	
		Compression	1226		9102 *		15830 *	
		Shear	417		2509		4256 *	
RSI Densified Mullite	QUAM 9	Tension	12870 *		367		404	
		Compression	1092		735		1444	
		Shear	4339 *		543		252	
RSI Densified Mullite	QUAM 9	Tension	3992 *		679		1181	
		Compression	84		489		556	
		Shear	585		136		252	
RSI Densified Mullite	QUAM 9	Tension	1512		608		1306	
		Compression	–		1810		2970	
		Shear	205		459		778	

3.2. THERMO-ELASTO-PLASTIC PROBLEMS INCLUDING CREEP

The severe environmental conditions to be experienced by the Space Shuttle System described earlier make it clear that linear elasto-static methods are inadequate towards obtaining accurate and reliable analytical results in support of design assurances. Most materials exhibit pronounced non-linearity, rate sensitivity, irreversibility and memory effects, phenomena which fall into the scope of so-called physical nonlinearities. Moreover, large deformations, i.e. large displacements and particularly large displacement gradients are responsible for the so-called geometric nonlinearities. Both sources of nonlinear phenomena call for powerful non-linear computer techniques as extensions of the linear finite element method.

The following discussion is concerned with two aspects: First, different concepts

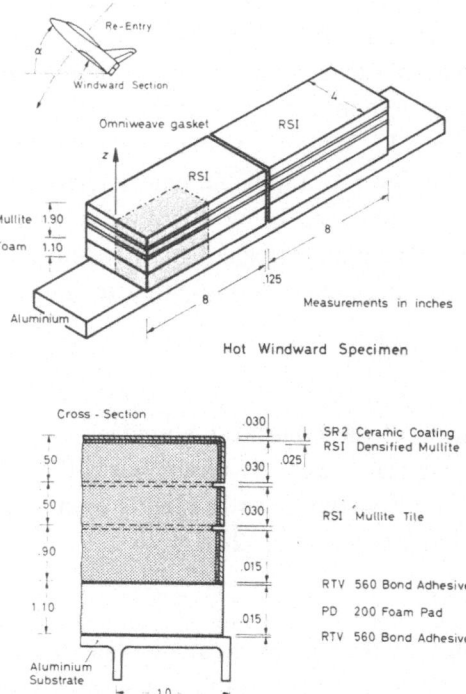

Fig. 16. Thermal protective system. Reusable surface insulation.

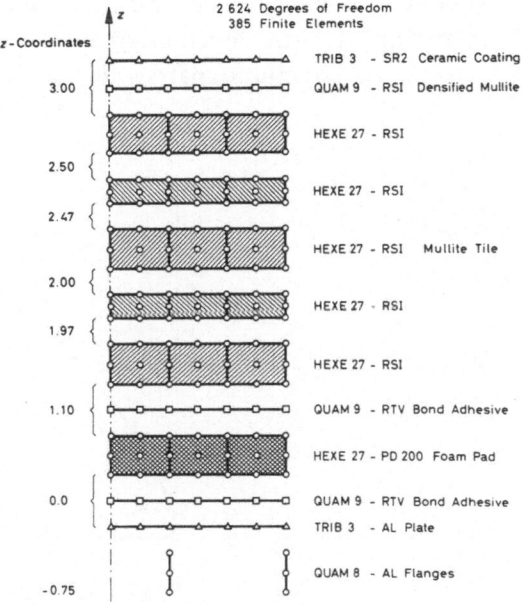

Fig. 17. Section through 3-dimensional reusable surface insulation. Finite element idealisation.

of numerical analysis are reviewed in regard to the solution of the non-linear mechanical field problem. Subsequently, a material constitutive law is formalized characterizing thermo-elasto-plasticity and creep. To simplify our exposition, the discussion is limited to physical non-linearities under small deformations for which no distinction need to be made between the deformed and the undeformed configuration and for which the linear strain-displacement relationships form a proper measure of deformation. For the time being attention is restricted to quasistatic motions for which inertia effects remain negligible. In conclusion, different thermo-elasto-plastic examples involving creep are presented to illustrate the concepts of non-linear analysis.

3.2.1. *Solution Methods*

The incremental formulation of the non-linear boundary value problem has been the subject of numerous publications in the recent past [17], [23]. Here, we only remark that the principle of virtual work can be readily used to discretize the quasi-static motion of a solid in differential form [22]

$$\mathbf{K}(r, t)\,\dot{\mathbf{r}}(t) = \dot{\mathbf{R}}(t), \tag{21}$$

where the dots denote time derivatives. A great number of numerical integration schemes have been proposed to solve these coupled first order differential equations, all of which are based on piecewise linearization of the non-linear response using incrementation and/or iteration.

3.2.1.1. *Tangential stiffness method.* Euler's algorithm is the most common technique for constructing a step by step procedure in which the solution is advanced by incrementing the load and updating the function gradient according to the value at the beginning of each step. In structural analysis this method is known under the

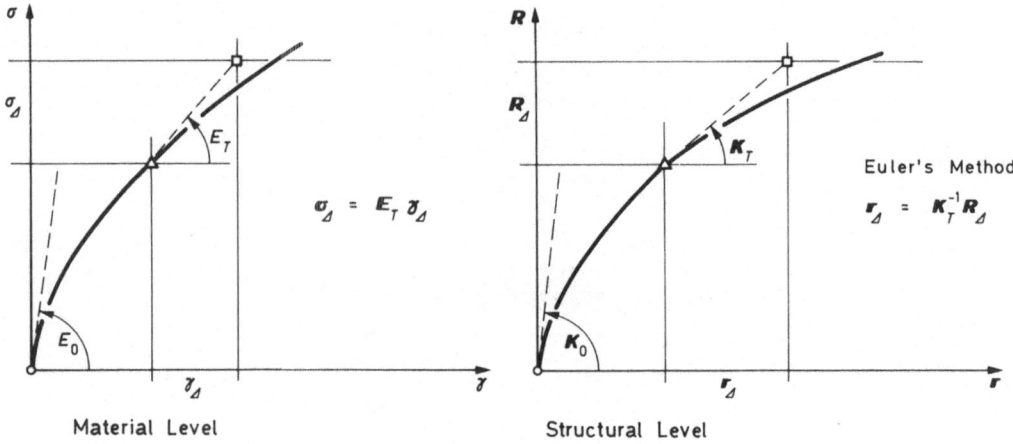

Fig. 18.   Tangential stiffness method.

name of incremental tangential stiffness. Figure 18 illustrates this technique in its simplest form on the constitutive and on the structural level. The non-linear material and structural response is linearized within each load step using a 'tangential' material law which is projected onto the structural level in the form of a 'tangential' stiffness matrix. We remark here that an out of balance load should be added to each load increment in order to reduce the accumulation of errors arising from the linearization. Since the new solution satisfies only equilibrium and compatibility but not the constitutive law this corresponding error could be corrected in a more refined analysis by iteration on the residual forces.

Alternative methods have been proposed involving the multiple evaluation of out of balance loads and tangential stiffnesses within each load increment. The improved and the modified Euler's method are very efficient schemes, but both require two evaluations of the tangential stiffness to obtain an improved estimate for the governing secant stiffness within a load step. More sophisticated multistep techniques, predictor-corrector algorithms and Runge-Kutta schemes have also been proposed in analogy to the refined numerical integration methods for the solution of ordinary differential equations.

In this context partial modification techniques of the structural stiffness provide an efficient tool for treating localized non-linearities avoiding the new formation and triangular factorization of the tangential stiffness matrix at each load step [24]. Substructuring further reduces the computational effort if non-linear regions of the structure can be identified in advance, e.g. in the case of plasticity near crack tips, etc.

3.2.1.2. *Initial load method.* Iterative root finding techniques have been proposed for a long time within the context of numerical analysis. Probably the most popular ones are the Newton-Raphson method and its modification in which the iteration is carried out without updating the function gradient. In structural analysis the latter is known under the name of initial load iteration in which case the nonlinearity of the stiffness matrix is simulated by 'equivalent' perturbations of the load vector [1]. Figure 19 illustrates the method on the constitutive and on the structural level. The non-linear material and structural response is obtained by iteration of residual stresses or equivalent initial loads within each load step. In the case of the modified Newton-Raphson method the constitutive law and the equivalent structural stiffness remain unaltered during the process requiring only a single triangular factorisation. This method is analogous to the Gauss-Seidel iterative scheme using successive substitution in contrast to the Newton-Raphson technique which involves updating of the stiffness and a new factorisation at each iteration step.

Two problems arise with iterative methods, the question of range and rate of convergence. The solution strongly depends on the initial guess as most iteration schemes are locally convergent only. These topics are discussed in detail in reference [30] in which different iterative initial load methods are presented for the solution of elasto-plastic problems.

Here we note that for path independent problems the load may be applied in a single step as long as the range of convergence is not exceeded. In this case the iteration history is of no influence because the solution is unique as long as the strain energy density remains positive definite. In contrast, dissipative processes require incremental procedures to minimise the deviation from the true path during the evolution of the problem. In this case particular care has to be exercised in regard to the convergence criterion and the choice of the step size. Strictly, for determining a solution it is not sufficient to reduce the residual loads or the change of displacements to a preset accuracy. It is rather necessary to seek sufficient convergence of the initial stress or strain history during the sequel of iterations [75]. On the other hand, there is no guarantee that the solution is the correct one if the evolution of the process has not been traced properly during the step by step linearizations; hence in practice less stringent convergence criteria are commonly used such as vector norms on differences of subsequent iteration values.

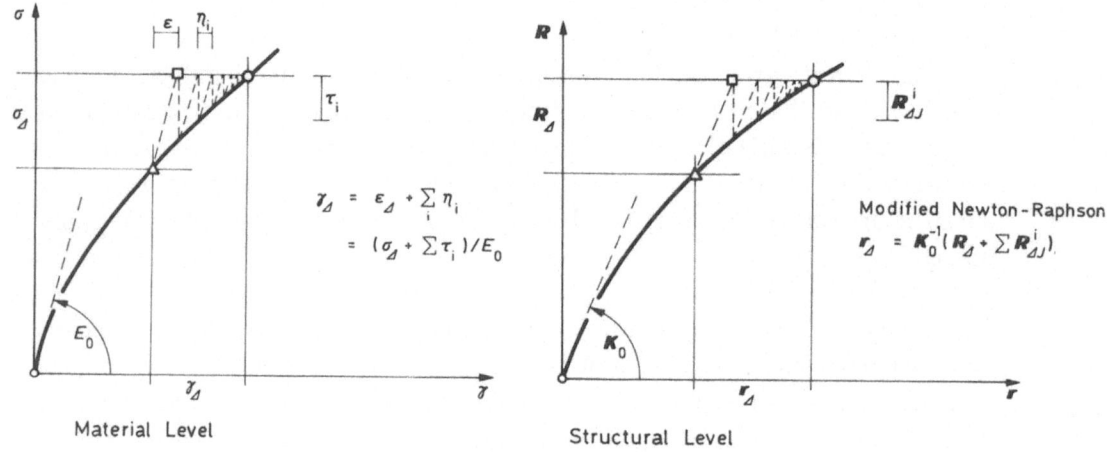

Fig. 19.   Initial load method.

We may conclude that a flexible non-linear solution algorithm should combine the advantages of the incremental step by step methods with those of the initial load iteration techniques in order to provide the means for reducing the residual loads to a preset accuracy. At one end of the spectrum this procedure would degenerate to the tangential stiffness method in its simplest form and on the other end it would be the initial load iteration technique without incrementation.

3.2.2. Material Characterisation

Based on the flow theory of plasticity [25], [26] elasto-plastic stress-strain relations are formulated below for non-isothermal conditions taking into account temperature dependent material behaviour. This relationship is then generalised to account for creep phenomena without memory effects. The formulation is restricted to isotropic conditions and the yield function is assumed to depend only on the second deviatoric stress invariant. For cyclic loadings, Prager's method of kinematic hardening or Ziegler's modification should be incorporated to account for the Bauschinger effect. In the case of non-metallic materials the yield condition and flow rule could be adjusted to include the effects of hydrostatic pressure similar to Drucker's formulation for materials with internal friction and cohesion. For dual transformation consider the following definition of stress and strain vectors for three-dimensional continua

$$\boldsymbol{\sigma} = \{\sigma_{xx}\ \sigma_{yy}\ \sigma_{zz}\ \sqrt{2}\sigma_{xy}\ \sqrt{2}\sigma_{yz}\ \sqrt{2}\sigma_{zx}\}$$

$$\boldsymbol{\varepsilon} = \left\{\varepsilon_{xx}\ \varepsilon_{yy}\ \varepsilon_{zz}\ \frac{1}{\sqrt{2}}\varepsilon_{xy}\ \frac{1}{\sqrt{2}}\varepsilon_{yz}\ \frac{1}{\sqrt{2}}\varepsilon_{zx}\right\}. \tag{22}$$

Using the fundamental law of decomposition – valid for small strains – we can express the increment of total deformation in terms of an elastic, stress producing strain increment $\boldsymbol{\varepsilon}_A$, and initial strain increments $\boldsymbol{\eta}_A$ due to plastic, thermal and creep effects

$$\boldsymbol{\gamma}_A = \boldsymbol{\varepsilon}_A + \boldsymbol{\eta}_{PA} + \boldsymbol{\eta}_{TA} + \boldsymbol{\eta}_{cA}. \tag{23}$$

The thermo-elastic behaviour is described by

$$\boldsymbol{\sigma}_A = \mathbf{E}\boldsymbol{\varepsilon}_A = \mathbf{E}(\boldsymbol{\gamma}_A - \boldsymbol{\eta}_{TA}) \tag{24}$$

$$\boldsymbol{\eta}_{TA} = \mathbf{e}_{3,3}\alpha T_A, \tag{25}$$

where

$$\mathbf{E} = 2G\left(\mathbf{I}_6 + \frac{v}{1-2v}\,\mathbf{e}_{3,3}\mathbf{e}_{3,3}^t\right) \tag{26}$$

and

$$\mathbf{E}^{-1} = \frac{1}{2G}\left(\mathbf{I}_6 - \frac{v}{1+v}\,\mathbf{e}_{3,3}\mathbf{e}_{3,3}^t\right). \tag{27}$$

$\mathbf{I}_6$ denotes the unit matrix and the vector $\mathbf{e}_{3,3}$ is defined by

$$\mathbf{e}_{3,3} = \{1\ 1\ 1\ 0\ 0\ 0\}. \tag{28}$$

G is the shear modulus, v Poisson's ratio and α the coefficient of thermal expansion. We remark that the hypo-elastic formulation of Equation (24) describes a path dependent process which coincides with the hyper-elastic description only if the material properties remain constant. Above formulation presumes an isotropic material.

Using the von Mises yielding criterion and the Prandtl-Reuss flow rule we obtain the following expression for the plastic strain increment

$$\boldsymbol{\eta}_{PA} = \bar{\eta}_{PA}\mathbf{s} \qquad \text{where} \qquad \mathbf{s} = \frac{3}{2\bar{\sigma}}\,\boldsymbol{\sigma}_D \tag{29}$$

Equivalent Stress $\qquad\qquad\qquad\qquad\qquad$ $\bar{\sigma}^2 = \frac{3}{2}\sigma_D^t\sigma_D$

Equivalent Plastic Strain Increment $\quad$ $\bar{\eta}_{PA}^2 = \frac{2}{3}\eta_{PA}^t\eta_{PA}$ $\qquad$ (30)

Note that the orientation of the plastic incremental vector is determined by the direction of the deviatoric stress σ_D, while the equivalent plastic strain increment yields a measure for its length. It is clear that the expression for the plastic strain increment is valid only in a differential sense. In view of the necessarily finite step length the change of direction of the deviatoric stress σ_D and the length $\bar{\eta}_{PA}$ are adjusted iteratively within each load increment to account for the non-linear character of the process. The following technique is used to relate the equivalent plastic strain increment to experimental evidence obtained from uniaxial tension tests under different temperatures. Assuming that the incipient yielding stress σ_y is described by the associated rule

$$\sigma_y = f(\eta_P, T) \qquad \bar{\sigma} = f(\bar{\eta}_P, T) \tag{31}$$

the change of the thermo-mechanical state causes a variation in the plastic strains

$$\delta\bar{\eta}_P = \frac{\partial\bar{\eta}_P}{\partial\bar{\sigma}}\,\delta\bar{\sigma} + \frac{\partial\bar{\eta}_P}{\partial T}\,\delta T. \tag{32}$$

In an incremental form this becomes

$$\bar{\eta}_{PA} = \frac{1}{\zeta}\,\bar{\sigma}_A + \varphi T_A. \tag{32a}$$

Figure 20 illustrates the meaning of the plastic loading parameters ζ and φ. Using Equation (32) in the flow rule we obtain for the plastic strain increment

$$\eta_{PA} = \bar{\eta}_{PA}\mathbf{s} = \frac{1}{\zeta}\,\mathbf{ss}^t\sigma_A + \varphi\mathbf{s}T_A. \tag{33}$$

A schematic interpretation of this method known under the name of 'initial strain' is shown in Figure 21. We remark that for perfect plasticity $\zeta = 0$ in which case the above formulation breaks down.

If we rewrite the elastic stress increment of Equation (33) as

$$\sigma_A = \mathbf{E}(\gamma_A - \eta_{PA} - \mathbf{e}_{3,3}\alpha T_A) \tag{34}$$

we arrive at an alternative formulation for the flow rule in terms of total strain and temperature increments

$$\eta_{PA} = \frac{2G}{\zeta + 3G}\,\mathbf{ss}^t\gamma_A + \frac{\zeta\varphi}{\zeta + 3G}\,\mathbf{s}T_A. \tag{35}$$

A schematic interpretation of this method known under the name of 'initial stress' is shown in Figure 22. Note that this formulation is restricted to $\zeta + 3G > 0$, hence perfect plasticity can be accomodated.

In analogy to the theory of plasticity the flow rule can be used to describe the energy

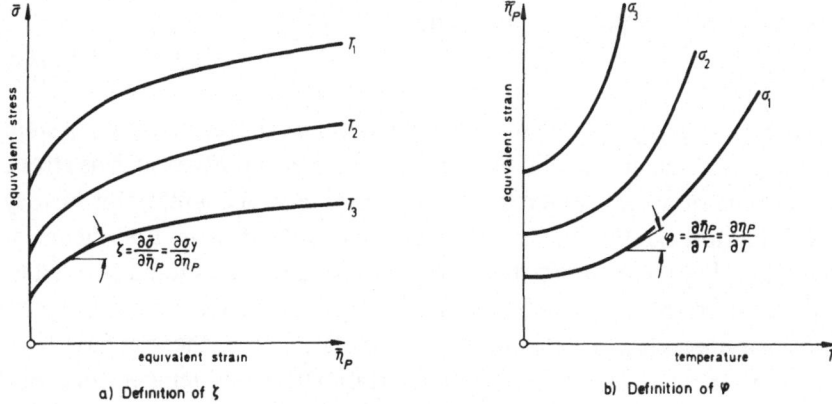

a) Definition of ζ b) Definition of φ

Fig. 20.   Thermo-plasticity under transient conditions.

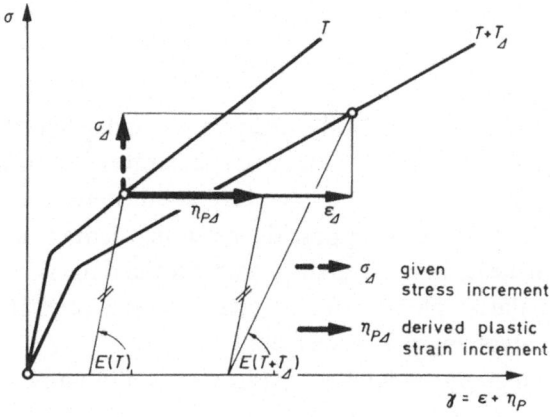

Fig. 21.   Plastic strain increment from stress increment.

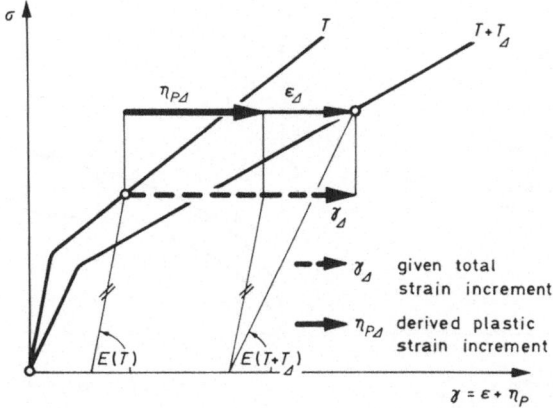

Fig. 22.   Plastic strain increment from total strain increment.

dissipation due to creep by equivalent uniaxial quantities

$$\boldsymbol{\eta}_{c\Delta} = \bar{\eta}_{c\Delta}\mathbf{s}. \tag{36}$$

Within the engineering theory of creep we assume that there exists an equation of state which defines the creep rate of the solid as a function of state variables such as stress, temperature and time or an equivalent hardening parameter accounting for history effects. This formulation exhibits shortcomings similar to the deformation theory of plasticity suggesting that the memory theory of viscoplastic solids should be used to characterise the creep response of metals. For the solution of engineering problems a simple approach is necessary, hence a number of cumulative creep theories have been proposed extending creep laws obtained from constant stress and temperature tests to time-varying processes.

Time Hardening

$$\bar{\eta}_{c\Delta} = f_1(\bar{\sigma}, T, t) \tag{37}$$

Strain Hardening

$$\bar{\eta}_{c\Delta} = f_2(\bar{\sigma}, T, \eta_c). \tag{38}$$

Note that for secondary creep, in which the creep rate is independent of time, both formulations yield identical expressions. Figure 23 illustrates the difference of both hardening rules for varying stress and temperature conditions at a specified time. Although experimental data tend to support the strain-hardening rule, this method may lead to difficulties in describing creep during unloading if the accumulated creep strains are large. Hence, the simpler time-hardening method is generally favoured for numerical calculations involving primary creep.

Now we have all the ingredients for the characterisation of an isotropic solid ex-

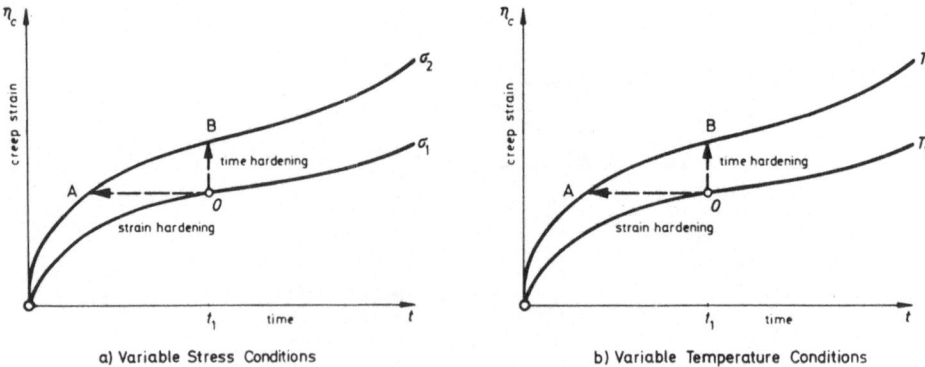

a) Variable Stress Conditions   b) Variable Temperature Conditions

Fig. 23.   Thermal creep under transient conditions.

hibiting elastic, plastic, thermal and creep effects. Equation (34) can be cast into

$$\gamma_\Delta = E^{-1}\sigma_\Delta + s\bar{\eta}_{P\Delta} + e_{3,3}\alpha T_\Delta + s\bar{\eta}_{c\Delta}. \tag{39}$$

Substitution and partial inversion yields the incremental stress-strain temperature relation

$$\sigma_\Delta = F\gamma_\Delta + GT_\Delta - s\frac{2G\zeta}{\zeta+3G}\bar{\eta}_{c\Delta}. \tag{40}$$

The instantaneous behaviour is characterised by the 'tangential' material moduli

$$F = E\left(I_6 - \frac{2G}{\zeta+3G}ss^t\right) \tag{41}$$

$$G = -E\left(\alpha e_{3,3} + \frac{\zeta\varphi}{\zeta+3G}s\right). \tag{42}$$

Formulation (40) furnishes a linear relationship between stress-, total strain- and temperature increments. However, it is strictly valid in the differential sense only as F, G and s vary with the loading. It should be emphasized that the creep rate components have to be considered prescribed since they are determined from the state of stress, temperature and time but do not involve their increments. For a detailed exposition of the material formulation see references [17] and [32], the latter of which provides detailed information on its numerical implementation in the finite element software package ASKA.

3.2.3. Initial Load Iteration

In the following, three methods are discussed for the numerical application of the initial load formulation. They are denoted in short by DIM, the direct incremental method, NIM, the normal iterative method, and VIM, the improved iterative method. A schematic illustration of these techniques is shown in Figure 24. For a detailed discussion the reader is referred to [30], [70].

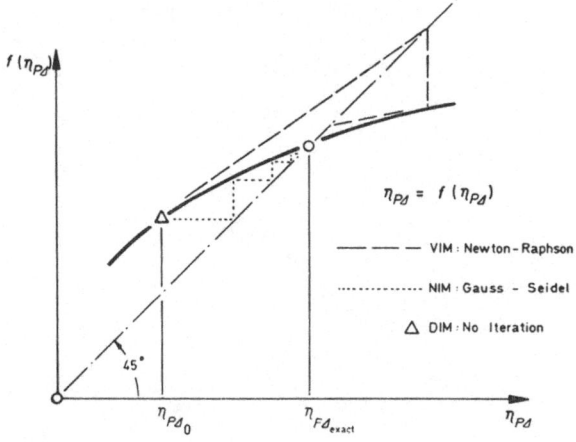

Fig. 24.   Initial load methods. Elasto-plastic solution schemes.

DIM is based on an incrementation of the load without change of stiffness. The non-linearity is accounted for solely by 'equivalent' initial loads which correspond to the plastic strain increments from the preceding loading step without subsequent iteration.

Alternatively, NIM incorporates an iterative scheme using successive substitution (Gauss-Seidel) for tracing the nonlinear behaviour of the equivalent plastic strains within each load increment. Again, the initial stiffness remains unaltered requiring only a single triangular factorisation during the whole process.

Finally, VIM is a Newton-Raphson iterative procedure involving the repeated evaluation of a function gradient matrix for the plastic variables. This method has been developed to improve the convergence properties of NIM.

In the following, an academic example illustrates some common characteristics of an elastoplastic analysis using initial loads. Figure 25 shows the lay-out of a statically indeterminate truss. The material properties refer to the aluminium alloy 2024-T3 with

$$\begin{array}{lll} \text{Elastic Modulus} & E = 11.4 \times 10^6 \text{ lbf/in}^2 & \\ \text{Poissons Ratio} & v = 0.30 & \\ \text{Yield stress} & \sigma_y = 34\,500 \text{ lbf/in}^2 & (43) \\ \text{Ramberg-Osgood Law } \eta_P = \dfrac{1.1\,\sigma_y}{mE}\left[\left(\dfrac{\sigma_y}{1.1\,\sigma_y}\right)^m - \left(\dfrac{1}{1.1}\right)^m\right]; & m = 10. \end{array}$$

A linear solution is first used to determine the load level at incipient yielding. Thereafter, the load is incremented by 10% of this value. Figure 25 shows the results obtained with DIM and NIM using the initial stress as well as the initial strain formulation for computing the plastic strain increment. We observe that the latter causes divergence of either DIM or NIM illustrating the small range of convergence of the

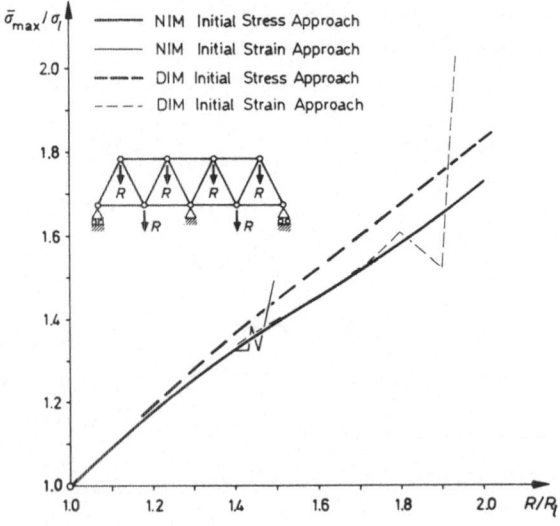

Fig. 25.   Indeterminate truss. Elasto-plastic solution using initial load method.

initial strain method. On the other hand, the *rate* of convergence of the initial strain approach – when converging – is greater than for the initial stress method. Hence, the initial stress formulation is best suited for automatic computation because the new equivalent plastic strain increments are determined from total instead of elastic strain increments. It should be mentioned that VIM has also been applied extending considerably the range of convergence of the initial strain formulation and improving the rate of convergence for the present structure with few kinematic constraints. These results are not shown in the figure as they coincide with those of NIM using the initial stress formulation, which may be considered exact. The same problem has also been solved using the tangential stiffness method in its simplest form yielding very close results to those of NIM using the initial stress formulation. However, it should not be forgotten that a new triangular factorisation of the stiffness is required in each load step.

The initial load technique is also very well suited for the solution of creep problems. Recalling that the creep strain increments must be considered prescribed, the DIM procedure readily applies to this time-variable phenomenon if we increment time instead of loading. In this case it is tacitly assumed that the stresses remain constant (no relaxation) within each time step imposing a restriction on the choice of the time interval.

In conclusion we note that the different initial load techniques represent perturbations on the load vector which are 'equivalent' in some sense to the change of stiffness due to nonlinearities. It is quite clear that the success, meaning the stability and rate of convergence, depend strongly on the degree of nonlinearity.

3.2.4. *Examples*

Two realistic engineering problems are now discussed to illustrate application of the initial load technique.

3.2.4.1. *Spherical shell with nozzle.* The first example deals with the analysis of a spherical vessel with a radial nozzle subjected to temperature loading and internal pressure. The geometry and idealisation of the vessel are shown in Figure 26, 230 axisymmetric TRIAXC 6 elements with 551 nodal points being used for the spatial discretisation. The variation of temperature-, plastic- and creep strains is approximated within each element by linear interpolation. The steady state temperature distribution is determined first, assuming that the inside surface of the vessel is exposed to a prescribed temperature of 250 °C while the ambient air temperature on the outside remains at 20 °C. The upper flange ring and the bottom rim are insulated against heat flow. The thermal material properties are taken to be

$$
\begin{aligned}
&\text{Thermal conductivity} && k = 0.398 \text{ kcal/cm } h\,°C \\
&\text{Specific heat} && c = 0.114 \text{ kcal/kg } °C \\
&\text{Heat transfer coefficient} && \lambda = 0.0426 \text{ kcal/cm}^2 \; h\,°C \\
&\text{Density} && \varphi = 7.9 \times 10^{-3} \text{ kg/cm}^3.
\end{aligned}
\tag{44}
$$

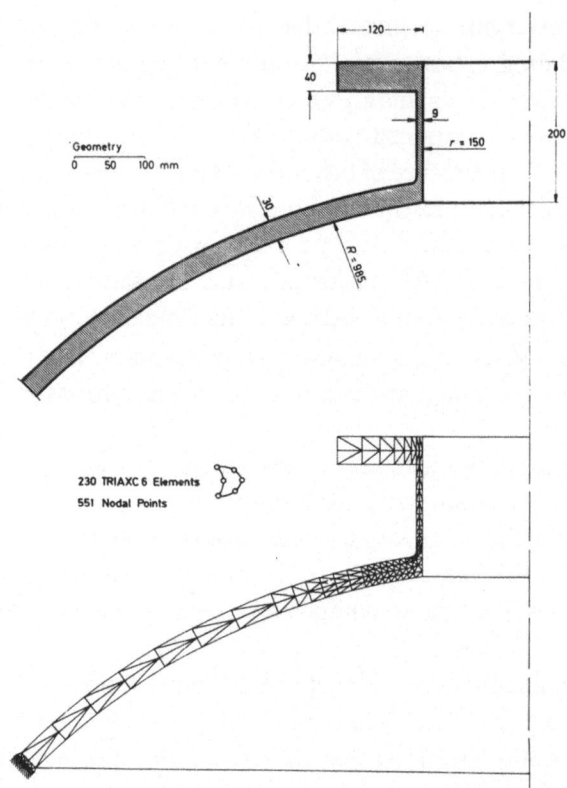

Fig. 26.   Spherical shell with nozzle. Geometry and idealisation.

The resulting temperature distribution is reproduced in Figure 27 in form of contour lines.

For the mechanical analysis the properties of mild steel are

Elastic Properties $E = 21\,000 \text{ kp/mm}^2; \nu = 0.30$

Thermal Expansion $\alpha = 1.7 - 0.7\left(1 - \dfrac{T}{720\,^\circ\text{C}}\right) \times 10^{-5}$

Plastic Properties   Ramberg-Osgood stress-strain law with $m = 20$ and

$$\sigma_y = 26\left[1 - \left(\frac{T}{850\,^\circ\text{C}}\right)^2\right] \text{ kp/mm}^2$$

Creep Properties $\bar{\eta}_{cA} = \left(\dfrac{\bar{\sigma}}{23.9 \text{ kp/mm}^2}\right)^{6.5} \dfrac{t_A}{10^9 \text{ s}}.$

Starting from a stress free reference state a fraction of the steady state temperature distribution is applied at each loading step. The final thermo-elasto-plastic condition is reached after 13 increments. Subsequently, the incremental analysis is continued to account for secondary creep effects. The time integration is performed in such a manner that the maximum equivalent stress in the vessel never varies by more than 5%

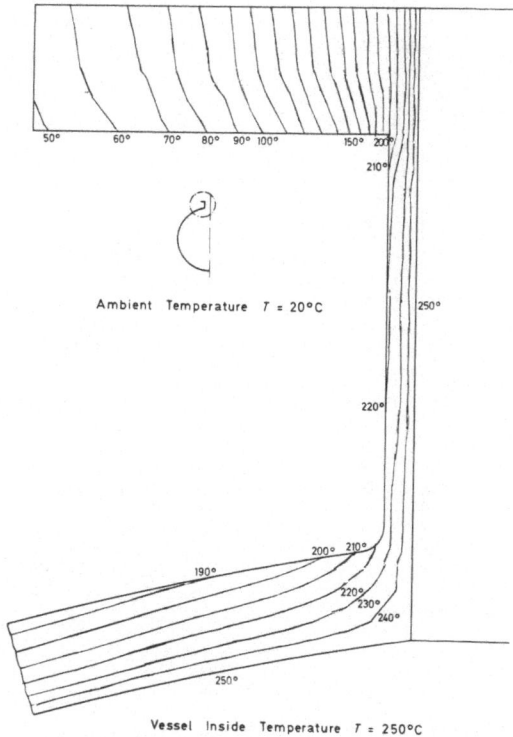

Fig. 27.   Spherical shell with nozzle. Steady state temperature distribution.

of the current maximum value within each time step. The procedure was terminated after 21 increments which corresponds to a time lapse of 181 days. Figure 28 illustrates the equivalent stress distributions at different instants of time. We observe that the plastic zones (shaded areas) develop first at the inside of the flange. This occurs due to the relatively low temperature at the outside of the flange which prevents the expansion of the hot inner portion, thus inducing large compressive hoop stresses. Due to the temperature dependent yield stress plastic flow is achieved both by stress and temperature. Figure 29 shows the stress relaxation due to creep causing a significant redistribution of stresses.

In conclusion, the spherical shell with nozzle is also subjected to a monotonically increasing pressure in which case temperature and creep effects are not considered. Figure 30 shows the equivalent stress distribution for a pressure $p \times 33.4$ kp cm^{-2} at which the transition zones between sphere and nozzle plastify. The internal pressure is then incremented up to 90.2 kp/cm^2, for which the theory of limit analysis predicts a collapse mechanism at the junction between nozzle and sphere. The corresponding stress distribution which is attained after 17 loadings increments is also indicated in the same figure. The results demonstrate that the limit analysis procedures yield but a conservative lower bound for the limit pressure.

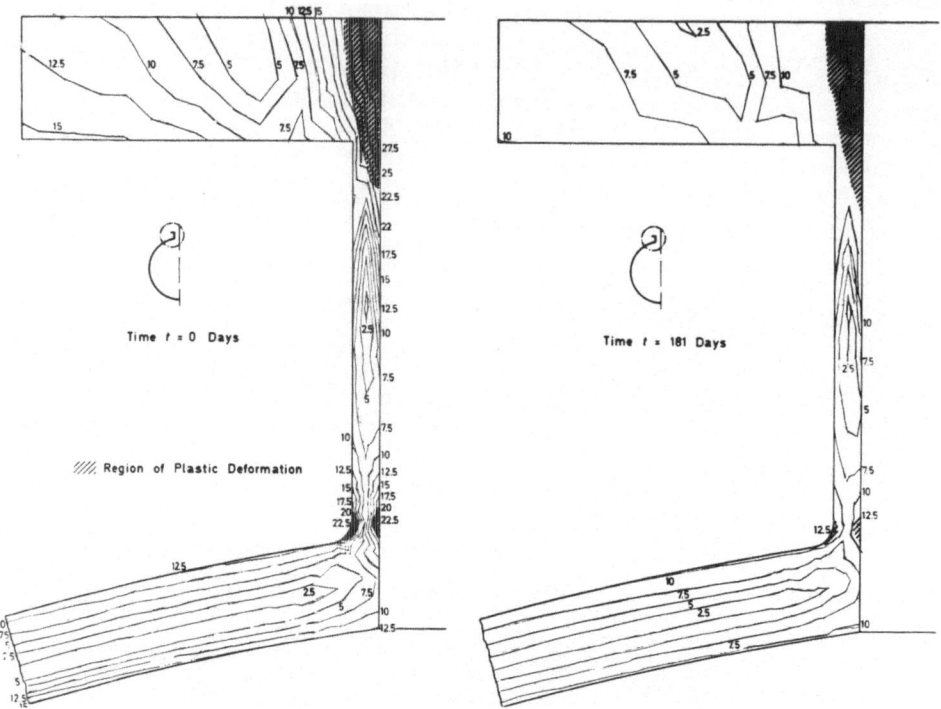

Fig. 28.   Spherical shell with nozzle under steady state temperature loading.
Equivalent stress distribution $\bar{\sigma}$ [kp/mm²].

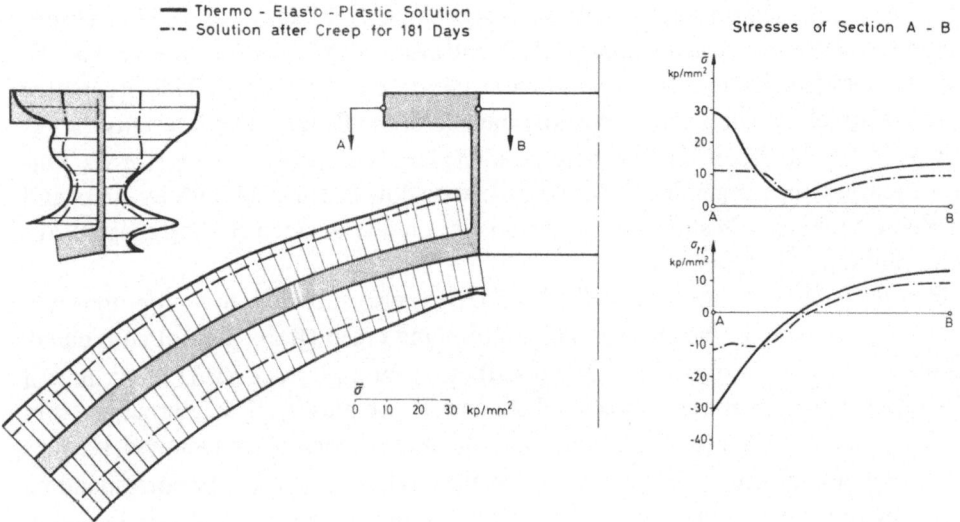

Fig. 29.   Spherical shell with nozzle. Influence of creep on stress distribution.

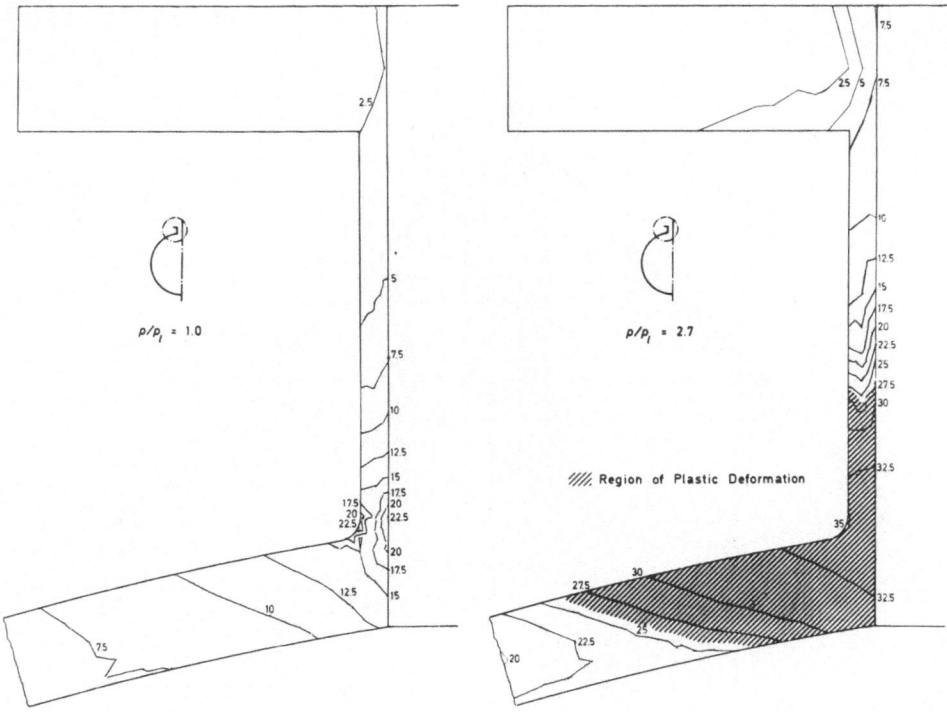

Fig. 30. Spherical shell with nozzle. Internal pressure p. Equivalent stress distribution $\bar{\sigma}$ [kp/mm^2].

3.2.4.2. *Re-Entry vehicle nose cone*. The second example deals with the analysis of a re-entry vehicle nose cone subjected to severe temperature gradients. The geometry and idealisation with axisymmetric solid elements are shown in Figure 31. The ring inserted in the structure consists of a highly anisotropic material, whereas the rest of the structure is made of steel. The final temperature distribution is given in Figure 32. The initial load method was used for the thermo-elasto-plastic analysis in conjunction with temperature dependent material properties which are shown in Figure 33. Again the temperature distribution was incremented in a step-wise procedure. Although the analysis was not completed, some of the results are given here for temperatures corresponding to 10% of the final values. The corresponding deformations are shown in Figure 34 while Figure 35 illustrates the distribution of equivalent stresses [35].

These two examples demonstrate that an important class of problems can be solved by the initial load technique. The nonlinearities due to plasticity and creep are accounted for by 'equivalent' load perturbations without updating the stiffness properties. This method is readily implemented in a finite element computer program if kinematically equivalent loads due to initial stresses or strains can be computed. It is only necessary to change the sequence of operations in order to account for the incrementation of the loading and the possible iteration within each load step. This technique is presently available in the integrated finite element software package

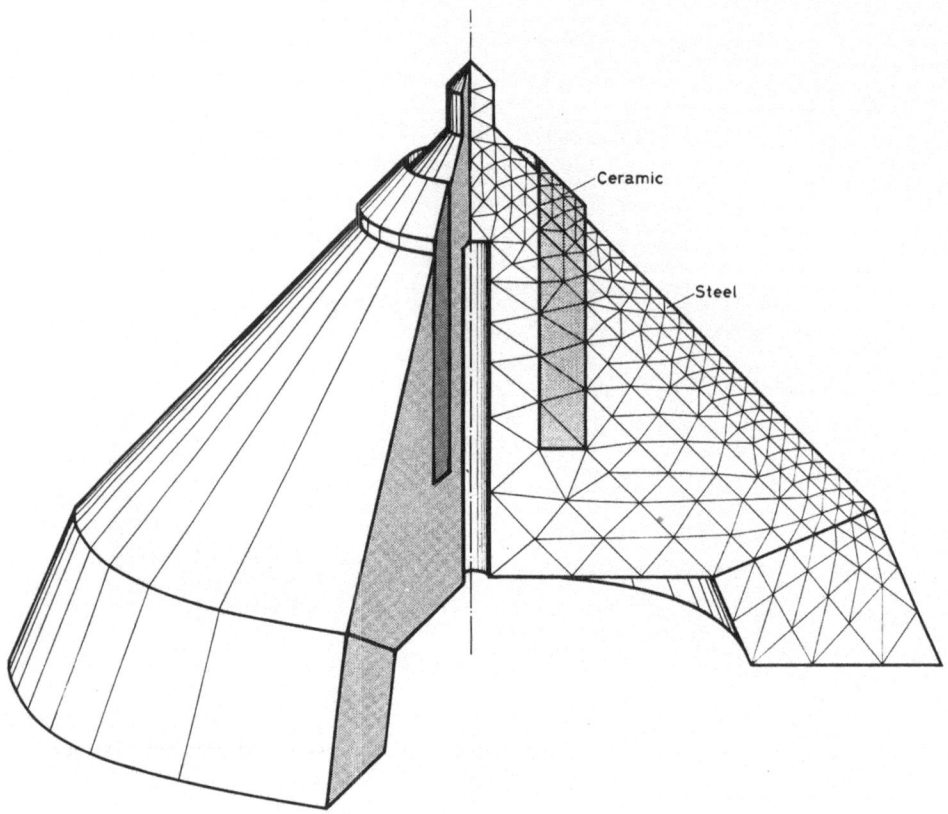

Fig. 31.   Re-entry vehicle nose cone idealisation by Axisymmetric ring elements TRIAX 6.

ASKA III-1 [32], [77] for the solution of thermo-elasto-plastic problems including creep effects.

3.3. LINEAR ELASTO-DYNAMICS

The rapid evolution in air- and spacecraft technology has made transient response analysis increasingly important. For this reason we turn our attention now to the formulation and solution of the equations of motion with particular emphasis on eigenvalue techniques for large scale linear problems.

3.3.1. *The Equations of Motion*

Let us commence our survey by reviewing first the finite element formulation of the equations of motion. Thereafter, direct and modal solution techniques are discussed for the transient response analysis of complex structures.

3.3.1.2. *Formulation.* Using the principle of d'Alembert inertia forces are interpreted

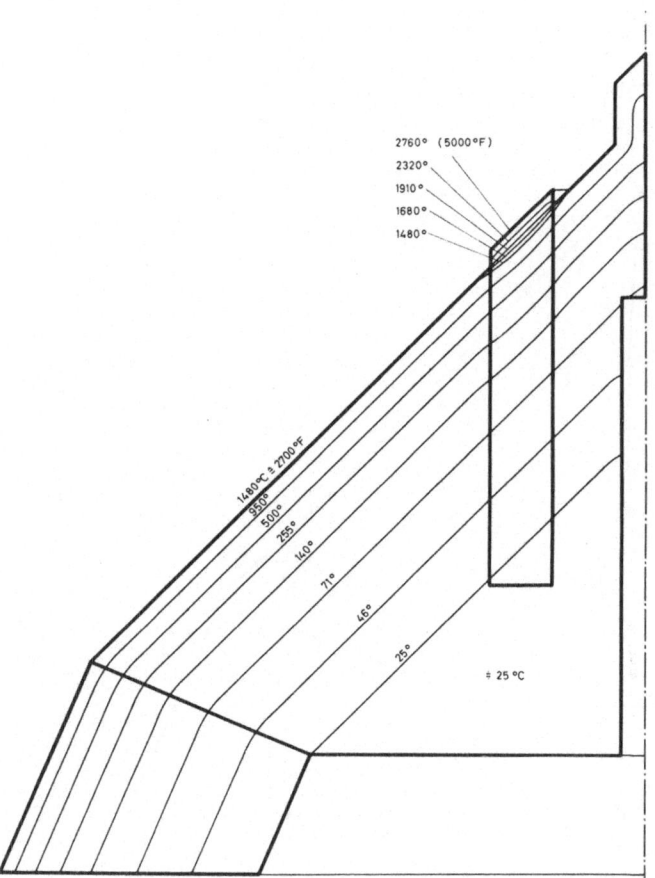

Fig. 32.   Re-entry nose cone. Temperature distribution [°C].

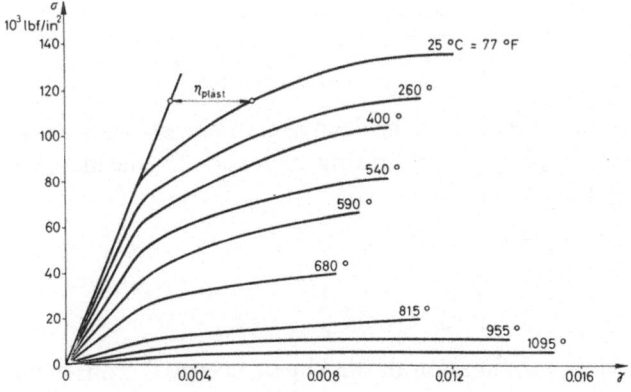

Fig. 33.   Temperature dependent stress-strain properties.

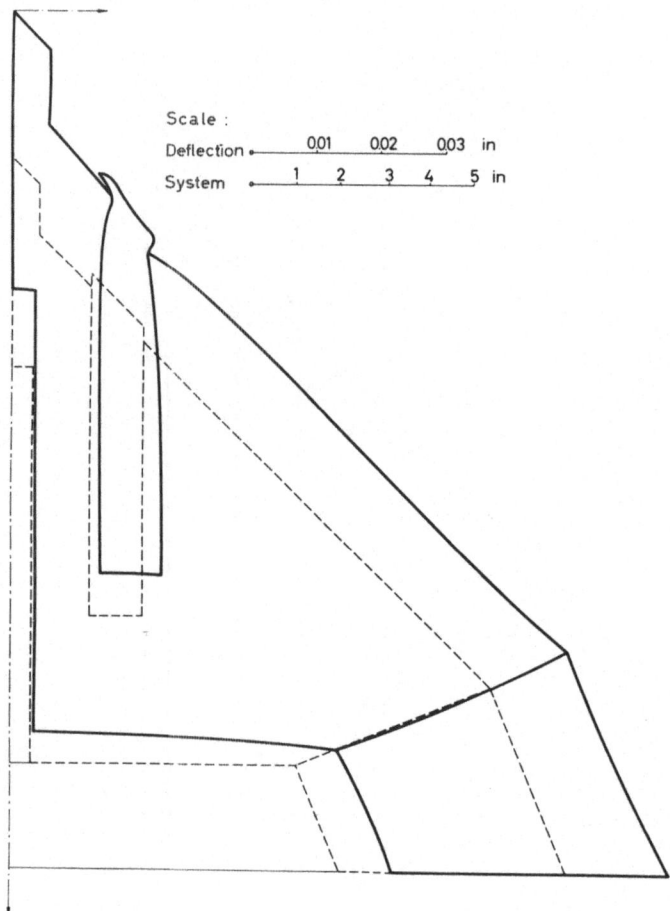

Fig. 34.   Re-entry nose cone. Deflected shape of structure.

as distributed volume loads

$$\mathbf{p}_v = -\mu\ddot{\mathbf{u}},\tag{46}$$

where μ denotes the mass density. Recalling the definition of kinematically equivalent nodal loads in Equation (11) we obtain the following expression for the mass distribution within a finite element [36], [37]

$$\mathbf{m} = \int_v \boldsymbol{\omega}^t \mu \boldsymbol{\omega} \, dv.\tag{47}$$

We remark that this spatial approximation of the kinetic energy is 'consistent' with the kinematic assumption for the strain energy or the equivalent stiffness. Rayleigh's minimum principle states in this case that the frequencies of the idealised structure yield upper bounds of the exact values.

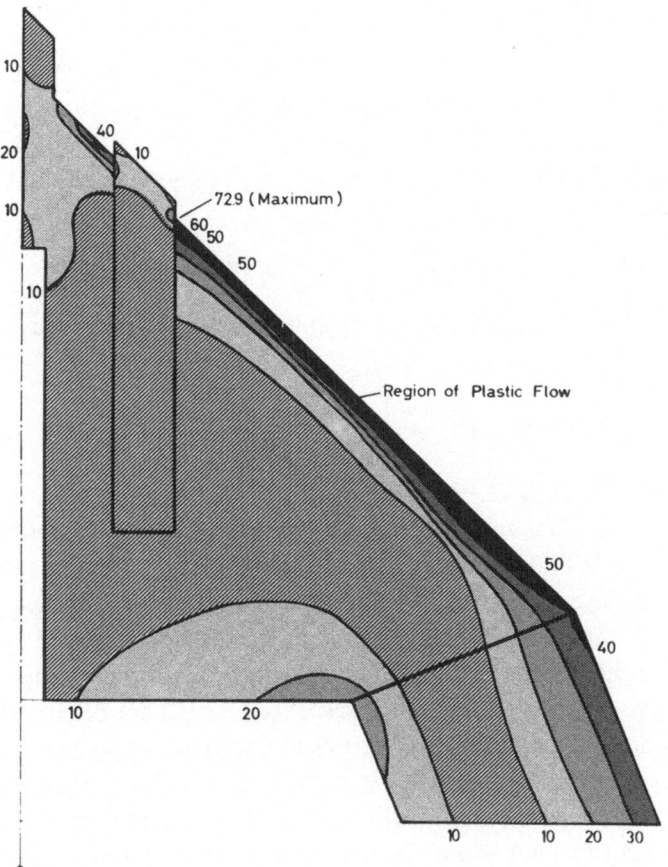

Fig. 35.   Re-entry nose cone. Equivalent stress distribution $\bar{\sigma}\,[10^3\ \mathrm{lbf/in^2}]$.

In practice, the consistent mass is often diagonalised by some lumping procedure to decouple inertia effects within an element. This technique reduces considerably the computational effort during the transformation of the generalised eigenvalue problem to standard form. Naturally, the frequency bound of the solution is lost due to the lumping procedure. Moreover, considerable distortions are introduced with higher order elements as lumping after tributary area concepts introduces erroneous weighting factors.

Viscous energy dissipation is accommodated on the material level by a generalisation of the constitutive law in Equation (9). Considering materials of the Kelvin type the stress is in this case a function of strain as well as strain rates

$$\boldsymbol{\sigma} = \mathbf{E}\boldsymbol{\varepsilon} + \mathbf{F}\dot{\boldsymbol{\varepsilon}}. \tag{48}$$

The dissipation matrix $\mathbf{F}$ describes the damping characteristics of the constitutive law. The corresponding finite element approximation yields an expression for the element damping forces in terms of nodal velocities

$$\mathbf{c} = \int_v \dot{\varepsilon} \mathbf{F} \dot{\varepsilon} \, dv. \tag{49}$$

The associated element damping matrix $\mathbf{c}$ may be in general unsymmetric depending on the properties of the 'dashpot'-matrix $\mathbf{F}$.

Including the inertia and damping effects in the expression of virtual work we obtain the following equations of motion for an element

$$\mathbf{m}\ddot{\varrho} + \mathbf{c}\dot{\varrho} + \mathbf{k}\varrho = \mathbf{P}. \tag{50}$$

Assembling the individual element contributions the dynamic behaviour of the idealised structure derives from

$$\mathbf{M}\ddot{\mathbf{r}} + \mathbf{C}\dot{\mathbf{r}} + \mathbf{K}\mathbf{r} = \mathbf{R}. \tag{51}$$

There are two sources of nonlinearities involving time as well as displacements, velocities, accelerations and their respective gradients. Time variable processes occur, e.g. during extreme flight conditions such as launch or re-entry. The stiffness properties change in the case of heating due to the temperature dependence of the material properties; the mass distribution varies with the fuel consumption and ablation; the airforces depend mainly on the flight velocities. Time-independent nonlinearities occur e.g. due to geometric and material effects which are discussed in Section 3.2. Aside from non-linear stiffness properties the airforces may also be coupled to displacements as well as time derivatives leading to aero-elastic phenomena [38].

In the following we restrict the discussion to linear structural dynamics in which $\mathbf{M}, \mathbf{C}, \mathbf{K}$ remain constant and the excitation force is a function of time only, $\mathbf{R} = \mathbf{R}(t)$. We emphasise that a number of factors influence the damping characteristics of the structure besides material damping. Therefore, we commonly resort to proportional (Rayleigh) damping if an explicit form of $\mathbf{C}$ is needed, especially since sophisticated experimental evidence is missing

$$\mathbf{C} = \alpha\mathbf{M} + \beta\mathbf{K}. \tag{52}$$

In this case no additional storage is required for $\mathbf{C}$ and orthogonality is preserved in respect to the modeshapes of the undamped free vibration problem. Note that only the two parameters α and β are at our disposal which can be adjusted to attain a specified level of damping for a limited frequency range. In this case all other frequency components are over-critically damped. Alternatively, the concept of proportional damping can be extended using a Caughey series expansion for constructing an orthogonal $\mathbf{C}$ with specified damping properties for a wide range of the frequency spectrum [39]. Considering that modal damping factors are better supported by experimental evidence modal response analysis becomes mandatory since in this case no explicit form of $\mathbf{C}$ is necessary.

3.3.1.3. *Response solution.* Two techniques are at our disposal when making transient response calculations: Modal Analysis and Direct Integration. Both methods differ only in the coordinate basis to which the motion is referred.

In the mode superposition method the finite element basis is replaced by ortho-normal coordinates as obtained from the free vibration analysis having two objectives in mind:

(a) To reduce the dynamic system simply by omitting frequency components which are of little interest.

(b) To simplify the transient response analysis by decoupling the equations of motion.

It is clear that this orthogonalisation of the equations of motion requires a large computational effort. In general four steps can be distinguished:

(1) Static condensation of the finite element basis (dynamic idealisation).

(2) Spectral analysis of the free vibration problem.

(3) Transformation and integration of the equations of motion.

(4) Transformation back into the finite element basis for determining the transient stress response.

The orthogonalisation is discussed in Sections 3.3.2 and 3.3.3 in context with the solution of large scale eigenvalue problems.

The direct integration of the equations of motion avoids this costly orthogonalisation. After recasting the equations of motion into incremental form, the essential operation involves the propagation of the solution at time t to time $t + \Delta t$ using a step-forward integration scheme. To this end numerical analysts have proposed explicit finite difference techniques as well as implicit schemes. With specific reference to finite element dynamics three methods are commonly used:

(1) Newmark's Beta Methods [40],

(2) Wilson's Theta Methods [41], [42],

(3) Finite Elements in Time [31], [43], [44].

The principal difference of these techniques is related to the assumption on the time variation of displacements, velocities and accelerations during each time-interval. Accuracy and stability of the algorithm are of prime importance for the successful solution of the coupled set of equations of motion. Space limitations prevent us from going into details, we only remark that the finite time elements with cubic approximation of the inertia forces and iteration within each time step yield extremely accurate and economic results when applied to the non-linear dynamic analysis of cable structures [44]. Moreover, for the special case of linear systems ref. [44] proposes both conditionally and unconditionally stable algorithms, which prove very efficient.

Comparing mode superposition and direct time integration we recognize that in the first case the effort is spent mainly on the orthogonalisation of the equations of motion reducing the transient response analysis to a trivial task. On the other side the main effort is spent on obtaining reliable results from direct integration of the coupled equations of motion. Considering that practically all existing step by step schemes exhibit some form of numerical damping and distortions of the spectrum (period elongations), it becomes quite clear that modal analysis provided, at least in the past, more accurate results than the direct time integration methods. Moreover, the frequency spectrum is in most cases of great engineering interest and forms a prerequisite for random response analyses. Since no explicit formulation of the damping matrix is needed, modal damping can be used directly which is also more likely to be supported by experimental evidence. These advantages of modal analysis disappear for nonlinear problems and general (non-orthogonal) damping. In these cases we might consider repeated modal decompositions for different time steps, but in general the computational effort will be prohibitive when compared with that of direct integration.

We conclude that modal analysis is well suited for the long duration response of linear structures, while direct integration should be used for the short duration response of linear as well as non-linear problems. Referring once more to [44] we note that for large systems it may turn out that the direct integration method, based on finite elements in time, is in many cases the more efficient and economical.

3.3.2. *Modal Analysis*

In the following the spectral decomposition of the equations of motion is discussed in regard to free vibration analysis, static condensation and transient response analysis. The solution of the resulting large scale eigenvalue problem will be dealt with in Section 3.3.3.

3.3.2.1. *Free vibration analysis.* The equations of an undamped dynamic system

$$\mathbf{M\ddot{r}} + \mathbf{Kr} = \mathbf{0}$$

have the solution

$$\mathbf{r} = \mathbf{r}_0 e^{i\omega t} \quad \text{where} \quad \omega = \text{real}. \tag{54}$$

We can recast the resulting eigenvalue problem in either of the following forms

$$\mathbf{MX} = \mathbf{KX\Lambda} \quad \text{where} \quad \Lambda = \begin{bmatrix} 1 \\ \overline{\omega_i^2} \end{bmatrix} \tag{55}$$

$$\mathbf{KY} = \mathbf{MY\Omega} \quad \text{where} \quad \Omega = \begin{bmatrix} \omega_i^2 \end{bmatrix}. \tag{56}$$

There are two reasons in favour of the first formulation.

(1) Numerical Accuracy

The Bauer-Fike theorem on conditioning states that for symmetric matrices the change in eigenvalues due to perturbations is bounded by any normal norm of the perturbation matrix [45]. In other words, if the original dynamic matrix is slightly perturbed, the relative change in the large eigenvalues is small in comparison to changes in the small eigenvalues. In structural dynamics we are mainly interested in the lowest frequencies; hence they are determined more accurately if we use the first formulation, eq. (55).

(2) Physical Justification

In practice, static analysis normally precedes the dynamic solution in which case positive definiteness of the stiffness can be guaranteed. This property forms the basis for a stable computation of the dynamic matrix, e.g. by Choleski decomposition.

For this reason we refer in the following to the first formulation. Only in the case that the mass matrix is diagonal does the second formulation offer computational advantages since the transformation of the generalised eigenvalue problem to standard form is then a trivial operation.

3.3.2.2. *Static condensation.* In general, the initial static idealisation exhibits too many degrees of freedom to permit an efficient spectral analysis. As a result a number of reduction techniques have been proposed to deduce a more economic 'dynamic' idealisation from the 'static' one [46], [47].

In the following we turn to the implementation of 'static condensation' in the finite element software package ASKA [19]. The nodal degrees of freedom are thereby divided into four families, local, suppressed, prescribed and external (for substructuring). This clear organisation of freedom families provides a vehicle to deal conveniently with kinematically prescribed degrees of freedom, e.g. ground accelerations for earthquake analysis, and statically condensed degrees of freedom for the reduction of the dynamic system. The substructure capability can be readily used to separate master from dependent degrees of freedom. Interpreting 'externals' as 'masters' we can recast the equations of motion into the following form

$$\begin{bmatrix} \mathbf{M}_{DD} & \mathbf{M}_{DM} \\ \mathbf{M}_{MD} & \mathbf{M}_{MM} \end{bmatrix} \begin{bmatrix} \ddot{\mathbf{r}}_D \\ \ddot{\mathbf{r}}_M \end{bmatrix} + \begin{bmatrix} \mathbf{K}_{DD} & \mathbf{K}_{DM} \\ \mathbf{K}_{MD} & \mathbf{K}_{MM} \end{bmatrix} \begin{bmatrix} \mathbf{r}_D \\ \mathbf{r}_M \end{bmatrix} = \begin{bmatrix} \mathbf{R}_D \\ \mathbf{R}_M \end{bmatrix}. \tag{57}$$

The dependent degrees of freedom $\mathbf{r}_D$ can be determined in terms of the 'master' freedoms if inertia effects are neglected. Time-invariance of the transformation is assured if no excitation forces act on the dependent degrees of freedom, $\mathbf{R}_D = \mathbf{0}$. If $\mathbf{K}_{DD}$ is non-singular the following transformation relates dependent and master degrees of freedom

$$\mathbf{r}_D = -\mathbf{K}_{DD}^{-1}\mathbf{K}_{DM}\mathbf{r}_M = \mathbf{S}_D\mathbf{r}_M. \tag{58}$$

The original finite element basis is next condensed to the master degrees of freedom level and leads to the new dynamic idealisation

$$\mathbf{M}^*\ddot{\mathbf{r}}_M + \mathbf{K}^*\mathbf{r}_M = \mathbf{R}_M(t), \tag{59}$$

where $\mathbf{M}^* = \mathbf{S}^t\mathbf{M}\mathbf{S}$; $\mathbf{K}^* = \mathbf{S}^t\mathbf{K}\mathbf{S}$ are fully populated matrices and

$$\mathbf{r} = \mathbf{S}^t\mathbf{r}_M \quad \text{with} \quad \mathbf{S} = \begin{bmatrix} \mathbf{S}_D \\ \hdotsfor{1} \\ \mathbf{I} \end{bmatrix}. \tag{60}$$

Expanding these transformations we obtain for the condensed stiffness and mass matrices

$$\begin{aligned} \mathbf{K}^* &= \mathbf{K}_{MM} + \mathbf{K}_{MD}\mathbf{S}_D \\ \mathbf{M}^* &= \mathbf{M}_{MM} + \mathbf{M}_{MD}\mathbf{S}_D + (\mathbf{M}_{MD}\mathbf{S}_D)^t + \mathbf{S}_D^t\mathbf{M}_{DD}\mathbf{S}_D. \end{aligned} \tag{61}$$

If the dependent degrees of freedom are massless $\mathbf{M^*} = \mathbf{M}_{MM}$.

We remark that the eigen-spectrum does not remain invariant during static condensation since no similarity transformations are involved. Caution should be exercised when selecting the master degrees of freedom, as they determine the amount of distortion in the eigen-spectrum. Note the analogy between static condensation and the Rayleigh-Ritz technique for the reduction of degrees of freedom. From Rayleigh's minimum principle we can claim that the frequencies of the condensed system form upper bounds to the original problem if kinematically equivalent mass matrices are used. However, nothing can be said about the actual approximation error of the eigenvalue spectrum which depends solely on the choice of the 'Ritz basis' in the present case the master degrees of freedom.

3.3.2.3. *Transient response calculation.* Mode shapes of the free vibration analysis form the basis for the transient response calculation by mode superposition. The transformation of the finite element nodal coordinate basis to the orthonormal eigenvector basis serves two objectives, a reduction of the number of degrees of freedom and a simplification of the time response analysis of the equations of motion due to uncoupling. Transformation of basis

$$\mathbf{r} = \mathbf{X} \; \boldsymbol{\eta} \; .$$
$$\text{\scriptsize{(}n \times 1\text{)} \quad \text{\scriptsize{(}}n_X \times 1\text{)}}$$
$$(62)$$

The dynamic system is simply reduced if only a small number of mode shapes $n_X < n$ are considered for the response analysis. Assuming linearity of the original finite element basis we invoke the principle of superposition for transforming the equations of motion

$$\boldsymbol{\Lambda}\ddot{\boldsymbol{\eta}} + \bar{\mathbf{C}}\dot{\boldsymbol{\eta}} + \boldsymbol{\eta} = \mathbf{X}^t \mathbf{R}(t),$$
$$(63)$$

where the transformation of the damping is given by

$$\bar{\mathbf{C}} = \mathbf{X}^t \mathbf{C} \mathbf{X}.$$
$$(64)$$

For orthogonal damping this transformation yields a diagonal damping matrix, resulting in an uncoupled set of second order differential equations. In this case the dynamic response of the ith mode involves the solution of

$$\ddot{\eta}_i + 2\zeta_i \omega_i \dot{\eta}_i + \omega_i^2 \eta_i = P_i(t).$$
$$(65)$$

Here the energy loss mechanism of the structure is defined in terms of the effective damping ratio ζ_i of the mode under consideration. Each equation of motion can be solved either directly by numerical integration of the resulting convolution integrals or by step forward integration procedures. For special forcing functions the time integration can also be carried out analytically.

For non-zero initial conditions the following relationships have also to be considered

$$\mathbf{X}\boldsymbol{\eta}(0) = \mathbf{r}(0)$$
$$\mathbf{X}\dot{\boldsymbol{\eta}}(0) = \dot{\mathbf{r}}(0).$$
$$(66)$$

Note that the inverse of $\mathbf{X}$ has rank n_X, hence there is no unique solution for $\boldsymbol{\eta}(0)$ and $\dot{\boldsymbol{\eta}}(0)$. However, the relevant strain and kinetic energy expressions can be minimized by the following transformation [50]

$$\boldsymbol{\eta}(0) = \mathbf{X}^t \mathbf{K} \mathbf{r}(0)$$
$$\dot{\boldsymbol{\eta}}(0) = \mathbf{X}^t \mathbf{K} \dot{\mathbf{r}}(0). \tag{67}$$

For $n = n_X$ this transformation degenerates into the inverse of $\mathbf{X}$ which can also be confirmed from the orthogonality properties of the eigenvectors

$$\mathbf{X}^{-1} = \mathbf{X}^t \mathbf{K} \quad \text{and} \quad \mathbf{X}^{-1} = \boldsymbol{\Lambda}^- \mathbf{X}^t \mathbf{M}. \tag{68}$$

3.3.3. *Eigenvalue Techniques for Large Scale Problems*

In the following a direct and an iterative method are compared for the eigensolution of large scale Hermitian matrices.

3.3.3.1. *Direct method* (Tridiagonalisation via Householder). The eigensolution of the direct method involves the following five steps
  (1) Static condensation $\mathbf{M}^* \mathbf{X}^* = \mathbf{K}^* \mathbf{X}^* \boldsymbol{\Lambda}^*$.
  (2) Transformation of the generalised eigenvalue problem to standard form $\mathbf{D}^* - \boldsymbol{\Lambda}^* \mathbf{I}^*$.
  (3) Householder reduction to tridiagonal form $\mathbf{T}_3^*$.
  (4) **QR**-Eigenvalue solution of tridiagonal matrix $\boldsymbol{\Lambda}^*$.
  (5) Inverse-Iteration for computation of mode shapes $\mathbf{X}^*$.
Table II summarizes the different steps of transformation together with a count of major operations. It clearly indicates that by far the largest amount of computational effort arises from the static condensation while only a small fraction is spent on the formation of the dynamic matrix $\mathbf{D}$ together with its reduction to tridiagonal form. Hence in the case of large scale matrices the actual eigenvalue solution does not consume a significant part of the overall computing time. In addition, we should note that the eigenspectrum of the original basis may be distorted to a considerable extent by static condensation and this may even affect the lowest range which is of particular interest for the subsequent response analysis.

3.3.3.2. *Iterative method* (Simultaneous Vector Iteration). In essence, this technique is an extension of the power method but with iteration on a subspace instead of a single vector. In this case, no intermediate transformations are necessary as opposed to the direct method in which static condensation and transformation to standard form precede the actual eigenvalue solution. Simultaneous subspace iteration is a very effective technique for the computation of the eigenspectrum. This technique was originally proposed in [51], [52] and extended to large scale matrices in [53], [54]. A detailed error analysis is given in reference [55] together with a discussion of the convergence properties. An elegant formulation and a geometric interpretation of the subspace inverse iteration has been presented in [60] and [61].

TABLE II

Direct Householder method

EIGENVALUE PROBLEM·

$$M X - K X \Lambda \xrightarrow[\text{condensation}]{\text{static}} M^* X^* - K^* X^* \Lambda^*$$

OPERATION	TRANSFORMATION	COMPUTATIONAL STEP	OPERATION COUNT
Formation of Mass and Stiffness (Master and Dependent DOF)		K , M	
Static Condensation (Reduction to Master DOF)	$M^* - $ $K^* - $	$K^* = S^t K S$ $M^* = S^t M S$	$\sim \frac{1}{2} n_p (b^2 + 8 n_m b + 4 n_m^2)$
Dynamic Matrix (Transformation to Standard Form)	$D^* = $	$K^* = U^t U$ $D^* = U^t M U^{-1}$	$\sim \frac{5}{6} n_m^3$
Householder Transformation (Reduction to Tridiagonal Form)	$T_j^* = $	$P_i^t D_i P_i \rightarrow T_j^*$	$\sim \frac{2}{3} n_m^3$
Q–R Transformation (Reduction to Diagonal Form → Eigenvalues)	$\Lambda^* = $	$T_k^* = Q_k R_k$ $T_{k+1}^* = Q_k^t T_k Q_k \rightarrow \Lambda^*$	$\sim 6 n_m$ per iteration
Inverse Iteration (Computation of Eigenvectors)		$T_j^* X_{k+1}^* = X_k^* \Lambda^*$	$\sim 4 n_m^2 n_s$ per iteration

The purpose of the simultaneous vector iteration is the evaluation of an eigensubspace **V** which satisfies

$$MV_k = KV_{k+1}\Lambda_{k+1} \quad \text{with} \quad k \to \infty. \tag{69}$$

The efficiency of this method derives from the fact that it is much easier to establish a starting subspace which is 'close' to **X** than to determine good approximations of single eigenvectors which is done in the power method. Moreover, costly deflation or purification processes are avoided if more than one eigenvector are desired.

Table III summarizes the different steps of simultaneous vector iteration together with a count of major operations. The computational effort is primarily determined by the initial triangular factorisation and by the repeated evaluation of new subspaces. Operations on the reduced projection matrix and orthogonalisation of subspaces are hardly of any influence. The rate of convergence of the iteration process depends on three factors, the separation of eigenvalues, the choice of the starting subspace and the order of the subspace.

3.3.3.3. *Comparison of the direct and iterative eigenvalue methods.* The performance of the two eigenvalue techniques is compared in Table IV. Central processing times on a CDC 6600 are given for different examples which were obtained using the DYNAN (ASKA IIa) package of ASKA [54], [58].

TABLE III
Simultaneous vector iteration

EIGENVALUE PROBLEM:

$$\mathbf{M X} = \mathbf{K X \Lambda} \xrightarrow[\text{iteration}]{\text{subspace}} \mathbf{M V}_k = \mathbf{K V}_{k+1} \bar{\mathbf{\Lambda}}_{k+1}$$

OPERATION	TRANSFORMATION	COMPUTATIONAL STEP	OPERATION COUNT
Formation of Mass, Stiffness and Starting Subspace		$\mathbf{M}, \mathbf{K}, \mathbf{V}_o$	
Triangular Factorisation		$\mathbf{K} = \mathbf{U}^t \mathbf{U}$ $\mathbf{X} = \mathbf{U} \mathbf{V}_o$	$\sim \frac{1}{2} n b^2$ $\sim n b n_s$
Simultaneous Vector Iteration (Computation of new Subspace)		$\mathbf{V}_{k+1} = \mathbf{U}^{-t} \mathbf{M} \mathbf{U}^{-1} \mathbf{V}_k$	$\sim 4 n b n_s$
Dynamic Reduction (Projection of Subspace)		$\mathbf{B}_{k+1} = \mathbf{V}_{k+1}^T \mathbf{V}_k$	$\sim \frac{1}{2} n_s (n_s + 1)$
Jacobi Transformation (Computation of Eigenvalues and Eigenvectors)		$\mathbf{B}_{k+1} = \mathbf{Q}_{k+1} \mathbf{\Lambda}_{k+1} \mathbf{Q}_{k+1}^T$	$\sim n_s^3 + n_s^4$
Computation of Starting Subspace (Projection and Orthogonalisation)		$\tilde{\mathbf{V}}_{k+1} = \mathbf{V}_{k+1} \mathbf{Q}_{k+1}$ $\tilde{\tilde{\mathbf{V}}}_{k+1} = \tilde{\mathbf{V}}_{k+1} \bar{\mathbf{\Lambda}}_{k+1}^{-\frac{1}{2}} \mathbf{R}_{k+1}^{-1}$	$\sim n n_s^2$ $\sim n n_s^2 + \frac{1}{2} n_s^3$

The total execution time is composed of three contributions
(1) Formation of **K** and **M**.
(2) Formation of **K*** and **M*** by static condensation (for the direct method)
(3) Solution of mode shapes and frequencies **X** and **Λ**.

The first example, the spectral analysis of the housing of a ship crew, involves 7281 DOF which are condensed to 122 'master' freedoms. It is remarkable that the actual eigenvalue analysis requires only one per cent of the total solution time.

The second example deals with the free-free vibration analysis of the FS-28 aircraft which will be discussed further in Section 3.4. The original finite element idealisation involves 5515 DOF which are reduced by static condensation to 102 'master' freedoms. Again the 'preliminary' operations require 98% of the total solution time.

The third example involves the spectral analysis of a delta-wing using the fifth order plate bending elements TUBA 6. Three different mesh lay-outs have been used to set off the idealisation error from other effects. A comparison of the direct Householder method and the simultaneous vector iteration clearly indicates that the latter technique is extremely efficient for large size problems if only a small number of mode shapes and frequencies are desired. For small and medium size problems the direct method proves superior even when few frequencies are required.

TABLE IV
Examples of computation times for direct and iterative eigenvalue methods (CDC 6600)

Example	Order of System n	Maximum Half-Bandwidth b	Method of Solution	Order of Condensed System n_M	Number of Eigenvectors n_s ( ) Order of Subspace	CP-Time in Seconds			Error in Natural Frequencies
						K,M	K^*,M^*	Λ,X	
Shaft of Ship Screw (Shell, Beam and Solid Elements)	7281	850	direct	122	30	1940	5315	83	
Aircraft FS-28 (Shell and Beam Elements)	5515	300	direct	102	10	1558	1388	64	
Delta Wing (Plate-bending Elements)	Fine Mesh 1408	150	iterative	-	20 (40)	166	-	1163	"reference values"
			iterative	-	3 (6)	166	-	155	-
			direct	66	3	176	163	17	$f_3 = 0.03$ %
			direct	66	20	176	163	23	$f_{20} = 3.8$
			direct	121	20	178	270	80	$f_{20} = 0.8$
	Medium Mesh 341	100	iterative	-	20 (45)	41	-	197	$f_{20} = 0.07$
			iterative	-	3 (6)	41	-	24	$f_3 = 0.04$
			direct	30	5	43	15	5	$f_5 = 0.7$
			direct	30	20	43	15	6	$f_{20} = 13.$
	Coarse Mesh 176	60	iterative	-	20 (39)	22	-	98	$f_{20} = 0.3$
			iterative	-	3 (6)	22	-	16	$f_3 = 0.08$
			direct	28	5	22	7	4	$f_5 = 0.7$
			direct	28	20	22	7	6	$f_{20} = 8.5$
			direct	-	20	22	-	215	$f_{20} = 0.3$

However, the accuracy of the higher mode shapes is found to deteriorate sharpely due to static condensation.

In the case of the coarse mesh, the direct technique has also been applied to the original system omitting the step of static condensation. A comparison with the iterative technique indicates the efficiency of the direct method if we consider that now all 176 eigenvalues are obtained. Moreover, 0.3% error in the 20th frequency demonstrate that the idealisation error remains negligible in the case of the highly refined plate bending elements TUBA 6 whose interpolation is based on a quintic expansion.

3.3.3.4. *Future developments.* The previous comparison clearly indicates the shortcomings of the two eigenvalue techniques which are presently available in the DYNAN package. For this reason the ISD is extending and improving the two methods along two lines:

(1) Extension of the direct Householder method to 'out of core' matrices. At the moment the dynamic system has to be reduced in ASKA to 'in core' size by static condensation. However, within a short space of time it will be possible to solve eigenvalue problems involving up to 1000 deg of freedom without reduction. In this case the dynamic matrix is first transformed to 'hyper-tridiagonal' form using Householder similarity transformations and subsequently tridiagonalized. It is clear that this technique retains all information of the original eigenspectrum as opposed

to the static condensation procedure. This development will be available by summer of 1973.

(2) Improvement of the simultaneous vector iteration to accelerate convergence. We recall that the rate of convergence is strongly influenced by the choice of the starting subspace. Using an idea developed in the thesis of ref. [55] we can use static condensation as a special form of the Rayleigh-Ritz technique to construct a trial basis which is then corrected by simultaneous vector iteration. Preliminary investigations indicate that this choice of starting subspace doubles the rate of convergence. Moreover, it is now possible to look at the direct and iterative approaches from a similar point of view. We recall, that the direct method solves the wrong eigenproblem (due to static condensation) 'exactly', while the iterative scheme utilizes this approximate basis from static condensation as starting value from which a portion of the exact eigenspectrum is obtained. This improvement of simultaneous vector iteration will be also available by summer of 1973 in an extended dynamic packet ASKA IIb.

3.3.4. *Examples*

In the following three examples are presented to illustrate different aspects in structural dynamics. The free-free vibration analysis of the FS-28 power glider involves an eigenvalue problem with 5000 degrees of freedom and the separation of rigid body from deformation modes. The time response analysis of a rocket shows specific dynamic features when subjected to propellant cut-off, stage separation and thrust build up. In conclusion, the dynamic response of a pressure vessel under short duration pulse serves to demonstrate the effects of static condensation and dynamic reduction.

3.3.4.1. *Free-free vibration of FS-28 power glider.* The static and dynamic idealisation are shown in Figure 36 for the left half of the aircraft. The finite element discretisation of the structure involves shell, beam and stringer elements with 5515 deg of freedom and a consistent mass matrix. The dynamic system is reduced first to 102 master degrees of freedom by static condensation and subsequently the rigid body modes are separated from deformation modes. Recall that a certain level of intuition is required to capture the relevant portions of the original eigenspectrum with the new dynamic idealisation. In our case, the distribution of master nodes with translational degrees of freedom follows directly the configuration of the different aircraft components. Ultimately, the direct Householder technique is used for the solution of the eigenvalue problem.

Figures 37, 38 and 39 present the resulting mode shapes and frequencies which are compared with experimental results. Measured and computed frequencies agree rather well for mode shapes which involve empennage and wing bending [57].

3.3.4.2. *Transient response of rocket.* In this example the longitudinal dynamic response of a rocket is presented for the load cases: propellant cut-off, stage separation and thrust development. A lumped parameter model (mass-spring system) serves

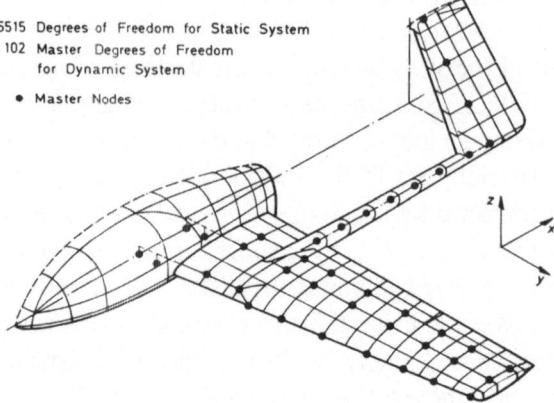

5515 Degrees of Freedom for Static System
102 Master Degrees of Freedom
     for Dynamic System

 • Master Nodes

Fig. 36.   FS-28 power glider. Finite element idealisation.

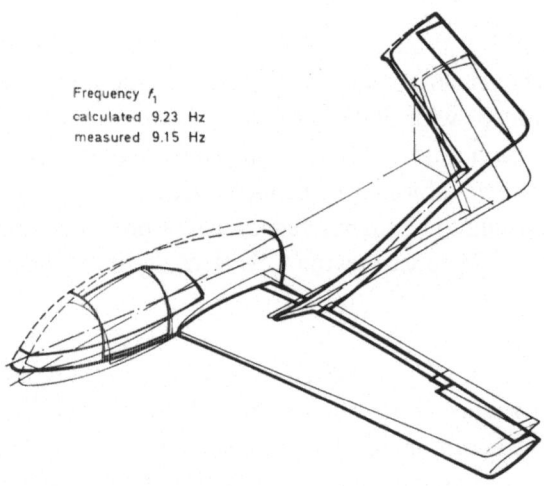

Frequency f_1
calculated 9.23 Hz
measured 9.15 Hz

Fig. 37.   FS-28 power glider. First symmetrical mode.

to idealise stiffness and mass properties of its individual structural components. The free-free response analysis is performed using mode superposition with 32 deg of freedom and no damping. For a comprehensive description of this example see ref. [56]. Figure 40 shows the time variation of longitudinal accelerations due to propellant cut-off corresponding to a sudden drop in thrust at the second stage. The initial conditions are $\ddot{r}(0) = 4$ g and $\dot{r}(0) = 0$ while $r(0)$ is determined from the equation of motion at time zero. We remark that the resulting total accelerations include free body and deformation response. Since the cylindrical fuel tanks are stiff in comparison to the other structural components the values of the second stage correspond to the free body accelerations of the centre of gravity. In contrast, the

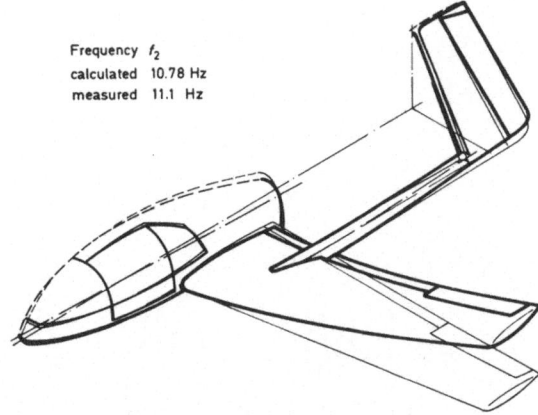

Fig. 38.   FS-28 power glider. Second symmetrical mode.

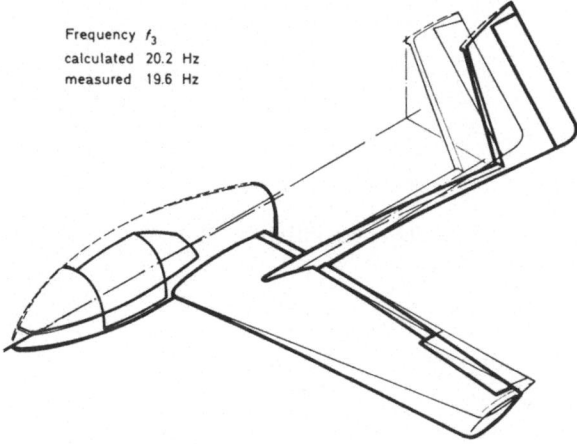

Fig. 39.   FS-28 power glider. Third symmetrical mode.

upper part of the third stage is excited up to ± 20 g due to the low frequency suspension.

Figure 41 illustrates the time variation of longitudinal accelerations as a result of stage separation. Initial displacements $r(0)$ are specified as static deformations due to internal pressure while initial velocities and accelerations remain zero. This loading case corresponds to a relatively slow decrease in internal pressure required for stage separation. Due to weak coupling hardly any high frequency components are transferred to the heavy portions of the third stage.

Experimental results are used to define the build-up of thrust as forcing function for the third stage. Figure 42 presents the resulting time variation of longitudinal accelerations for quiescent initial conditions. Note the large amplitude excitation of the rocket motor which is caused by the small stiffness of its suspension. Any

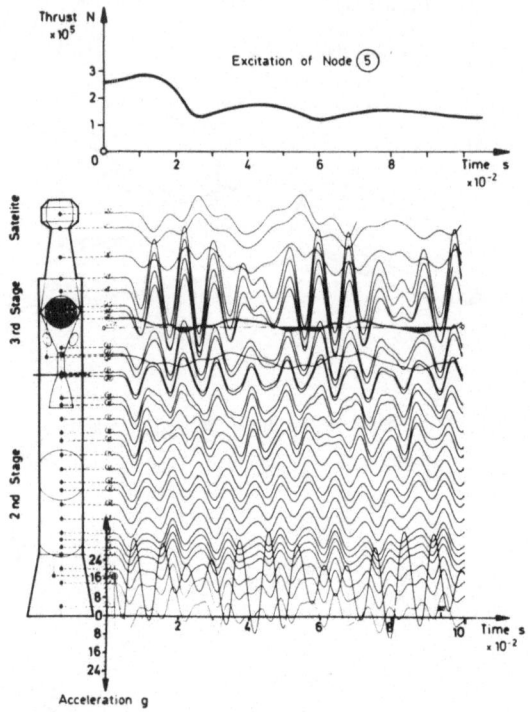

Fig. 40.   Time response of rocket due to propellant cut-off. Total accelerations (rigid body + elastic).

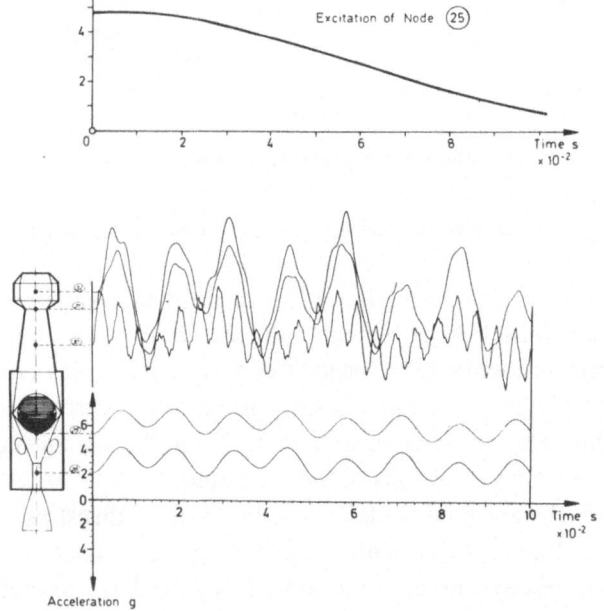

Fig. 41.   Time response of rocket at stage separation. Total accelerations (rigid body + elastic).

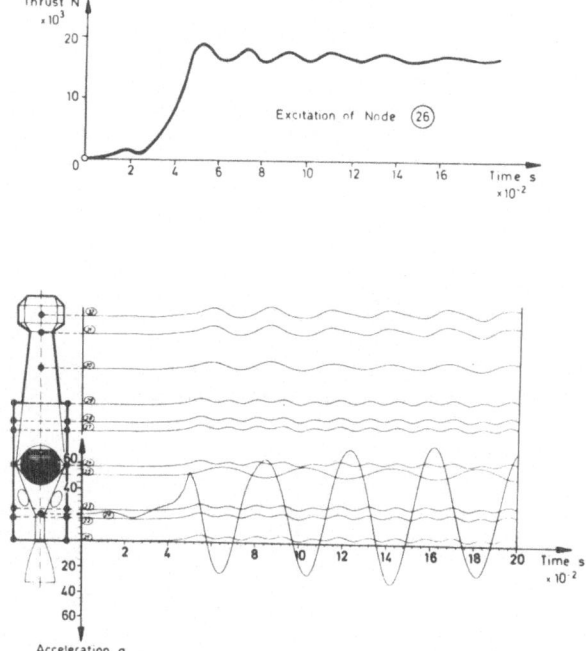

Fig. 42. Time response of rocket due to thrust-surge. Total accelerations (rigid body + elastic).

coupling with the fuel pressure could lead to pogo problems. However, these are unlikely to occur in our case due to the large pressure difference between tank and motor.

3.3.4.3. *Transient response of spherical shell with nozzle.* The last example concerns a pressure vessel with radial nozzle subjected to a short-duration pressure pulse. The spherical shell is idealised with the axisymmetric curved shell element SABA 2 [71], [72] shown in Figure 43. Consistent mass matrices are utilized to discretize the mass distribution of the shell structure; an additional point mass at node 8 accounts for the inertia of auxiliary devices such as adjacent piping [59].

The effect of static condensation is summarized in Table V, in which the eigenvalue spectrum of the original 81 DOF system is compared with that of the condensed 24 DOF system with master nodes having translational degrees of freedom only. Note the excellent agreement in the first three eigenvalues for which the upper bound properties of the condensed system can be easily verified. The other portion of the eigenvalue spectrum demonstrates the strong distortion due to static condensation. Certain modes are missing altogether while others are newly introduced. Obviously the choice of master DOF is crucial for a successful response analysis which depends solely on the eigenvalue spectrum of the dynamic system excited by the frequency content of the forcing function.

The undamped time response is now investigated using mode superposition for

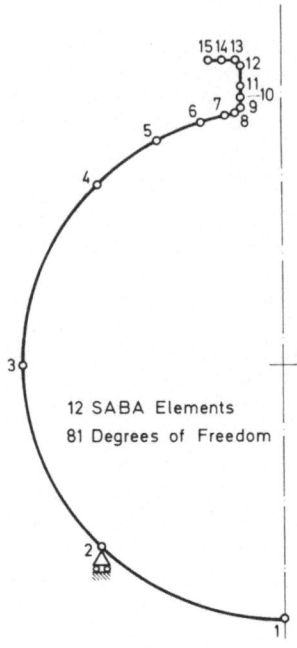

Fig. 43.   Spherical vessel with radial nozzle. Finite element idealisation.

TABLE V
Natural frequencies of spherical vessel with nozzle

Mode Number	Original System (81 degrees of freedom)	Statically Condensed System (24 translational degrees of freedom)	Distortion of Frequencies
1	68.9321	68.9322	0. %
2	334.1111	336.9653	+ 0.8
3	519.0951	524.7533	+ 1.1
4	675 5620	-	-
5	709.4587	709.5437	+ 0.01
	-	730.7976	-
6	742.2543	829.5707	+ 11.8
7	820 1799	-	-
	-	912.4607	-
	-	996.5227	-
8	887 9811	1014.7535	+ 14.3
	:		:
80	840082 32		
81	2002891.96		

a trapezoidal impulse of internal pressure. The computations are carried out in two steps for the frequency spectrum of the original system assuming quiescent initial conditions. The first analysis considers all mode shapes which correspond to the extreme frequency ratio of 1 to 30000. In the second case, the dynamic system is reduced by using only the first twenty mode shapes which correspond to a frequency ratio of 1 to 60. Figure 44 shows the short-time displacement response of two characteristic nodal points. Note the difference in high frequency components between the original and reduced system. For the equivalent stress, the short-time response is illustrated in Figure 45. The spatial distribution of stresses is shown in Figure 46,

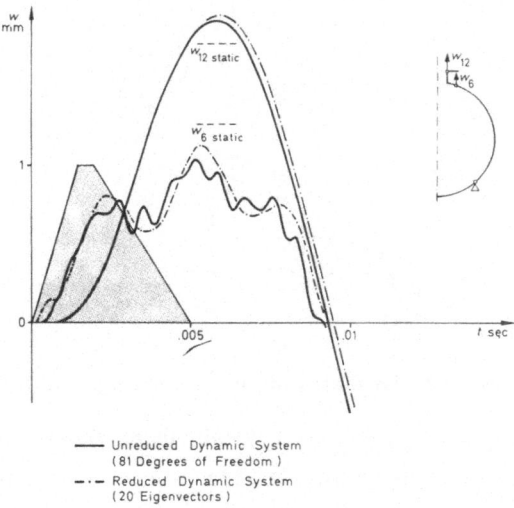

Fig. 44.   Spherical shell with nozzle. Short-time displacement response due to pressure pulse.

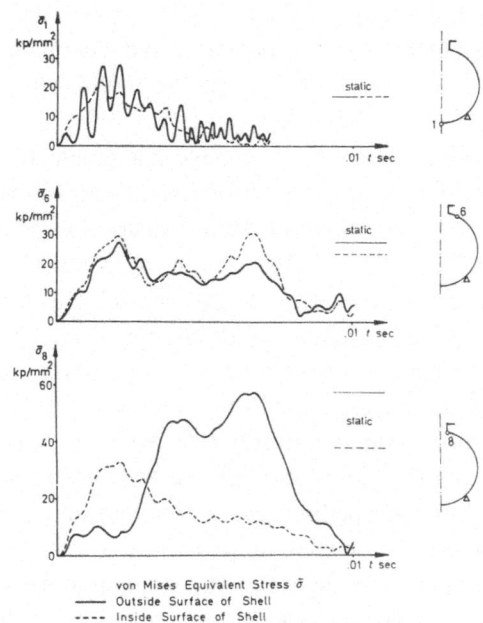

Fig. 45.   Spherical shell with nozzle.

where the envelope of equivalent stresses is plotted for the time interval 0 to 0.015 s. The stress results are based on solutions of the reduced dynamic system.

Alternatively, the equations of motion have been integrated directly using the procedures of Runge-Kutta four point method, Newmark $\beta = \frac{1}{4}$ method [40], Wilson $\theta = 2.0$ method [41] and finite elements in time [43]. For a time step of 10^{-5} s these

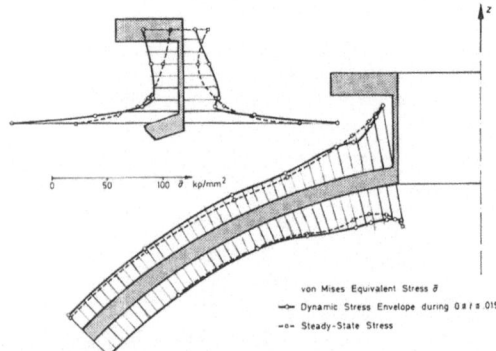

Fig. 46.   Spherical shell with nozzle dynamic stress envelope due to pressure pulse.

results can't be distinguished from the solution using mode superposition on the original frequency spectrum; see, however [44].

4. Concluding Remarks

The preceding exposition attempts to describe the general configuration and mission of the Space Shuttle and some of the analytic capabilities developed at the ISD. The finite element method proves a powerful and flexible computer-oriented tool with the help of which structural integrity can be assured under severe environments and conditions.

Due to space limitations we could not engage in assessing the remaining ingredients of analysis, the proper description of environmental conditions. Here we only remark that disciplines such as aerodynamics, heat transfer and acoustics play an integral part in the proper formulation of the analytic problem. Coupled interaction phenomena require additional research and further sophistication in the analysis techniques. For instance, let us consider the development and use of high thrust liquid propulsion systems which have introduced an instable dynamic interaction involving launch vehicle structure thrust oscillations and pressure pulsation of the liquid propellant. This phenomenon, commonly referred to as pogo, requires an accurate representation of both the liquid propellant and elastic structure free-vibration characteristics, the liquid propellant feedline system and the characteristics of the propulsion system. The solution techniques are similar to those employed for control systems in that they operate in the frequency domain and employ either open-loop (e.g. Nyquist) or closed loop (e.g. eigenvalue) approaches. The primary objective is to ensure the design integrity of the vehicle either in that unstable vibrations are not experienced or in that the candidate suppression device will eliminate or minimize such instabilities [64], [65].

We have demonstrated that the methods of determining the free-vibration characteristics of a shuttle type vehicle are relatively well in hand. Indeed the refined finite elements available today and the powerful eigenvalue methods allow for a highly accurate determination of the natural mode shapes and their associated frequencies of large and complex structures in a cost effective manner. Likewise, the requirements for transfer functions for the feedline and propulsion systems are relatively well defined. However, an improved description of the sloshing model is required, to account for the coupling due to the asymmetry of the Space Shuttle Vehicle. The finite element method has already demonstrated its flexibility and scope in evaluating both the ducted flow and the film lubrication problems [6]. Studies are currently conducted to obtain similar results of a highly accurate nature for a sloshing model including a reliable determination of the tank bottom pressure. This latter requirement is necessary for those cases in which the pressure drop due to the feedline outlet is of sufficient magnitude as to require modification of closed tank free-vibration characteristics. Demands for improved accuracy also stem from potential use of active suppression devices in which closed loop systems are employed based on signals measured upstream of the pump inlet. Closed loop analysis requires the determination of complex eigenvalues and eigenvectors. Although complex eigen-solution techniques are currently in use [63], their level of efficiency and reliability requires further refinement prior to their adaption in a production environment.

In conclusion we remark that two areas are critical for the successful analysis of complex structures such as the Space Shuttle Vehicle, the development of a data-bank and the display of data with the help of Interactive Graphics [69]. We have seen that the shuttle involves a substantial volume of information for input, output and intermediate computations. The data sources include engineering drawings, technical reports, tables and handbooks, and computer output resulting from prior analysis. The task of ensuring that the model under evaluation represents the appropriate 'as-designed' configuration is indeed most formidable in view of the large volumes of data and wide variety of sources. The solution of this problem is critical throughout the design process especially as authorized changes to requirements and/or designs are introduced and appropriate management visibility and control of the data base is essential.

The role of Interactive Graphics is strongly identified with the basic concept of the finite element method as a means of minimizing manpower requirements, decreasing elapsed time for project completion and minimizing actual computer usage. The ability to visualize graphical displays of input and output in a real time environment affords the engineer ready access to the vast volume of information stored in the computer devices and eliminates wasted time while waiting for printed results or micro film plots. Of more importance is the assurance that the input configuration is valid. This feature minimizes invalid computer runs and, coupled with the other inherent system advantages, represents substantial cost savings which should at least offset the costs associated with setting up and maintaining an Interactive Graphics environment.

References

[1] Argyris, J. H., 'Energy Theorems and Structural Analysis', *Aircraft Engineering* **26** (1954), 347, 383; **27** (1955), 42, 80, 125, 145; and as book (Butterworths, London, 1960), 5th ed., Plenum Press, New York, 1971.

[2] Argyris, J. H., 'Recent Advances in Matrix Methods of Structural Analysis', *Progress in Aeronautical Sciences*, Pergamon, London, 1964.

[3] Argyris, J. H., 'Continua and Discontinua', Opening address to 1st Conference on Matrix Methods of Structural Mechanics, Dayton, Ohio, 1967.

[4] Argyris, J. H. and Scharpf, D. W., 'Some General Considerations of the Natural Mode Technique', Part I and II, *Aeron. J. Roy. Aeron. Soc.* **73** (1969).

[5] *ISD–ISSC Proc. of the Symp. on the Finite Techniques*, University of Stuttgart, 1969.

[6] Argyris, J. H., 'The Impact of the Digital Computer on Engineering Sciences', Twelfth Lanchester Memorial Lecture, *The Aeron. J. Roy. Aeron. Soc.* **74** (1970).

[7] Bushell, D., Almroth, B. O., and Brogan, F., 'Finite Difference Energy Method for Non-Linear Shell Analysis', *Int. Journal Comp. Struct.* **1** (1971).

[8] McLay, R. W., 'Completeness and Convergence Properties of Finite Element Displacement Functions – A General Treatment', *AIAA Paper* 67-143 (1966).

[9] Walz, J. E., Fulton, R. E., and Cyrus, N. C., 'Accuracy and Convergence of Finite Element Approximations', *Proc.2nd Int. Conf. on Matrix Methods of Structural Mechanics*, Dayton, Ohio, 1968.

[10] Pian, T. H. H. and Tong, P., 'Basis of Finite Element Methods for Solid Continua', *Int. J. Num. Meth. Eng.* **1** (1969).

[11] Argyris, J. H. and Fried, I., 'The LUMINA Element for the Matrix Displacement Method', *Aeron. J. Roy. Aeron. Soc.* **72** (1968), 514–517.

[12] Argyris, J. H., Fried, I., and Scharpf, D. W., 'The HERMES 8 Element for the Matrix Displacement Method', *Aeron. J. Roy. Aeron. Soc.* **72** (1968).

[13] Ergatoudis, I., Irons, B. M., and Zienkiewicz, O. C., 'Curved Isoparametric Quadrilateral Elements for Finite Element Analysis', *Int. J. Solids Struct.* **4** (1968).

[14] Dunne, P. C., 'Complete Polynomial Displacement Fields for Finite Element Methods', *Aeron. J. Roy. Aeron. Soc.* **72** (1968).

[15] Buck, K. E., 'Rotationskörper unter beliebiger Belastung', in K. E. Buck, D. W. Scharpf, E. Stein und W. Wunderlich (eds.), *Finite Elemente in der Statik*, W. Ernst und Sohn, München, 1972.

[16] Argyris, J. H. and Scharpf, D. W., 'The SHEBA Family of Shell Elements for the Matrix Displacement Method', Part I and II, Natural Definition of Geometry, Strains and Stiffness, *The Aeron. J. Roy. Aeron. Soc.* **72**, 1968.

[17] Argyris, J. H., Balmer, H., Doltsinis, J. St., and Willam, K. J., 'Finite Element Analysis of Thermomechanical Problems,' *Paper presented at the 3rd Int. Conf. on Matrix Methods of Structural Mechanics* Wright-Patterson Air Force Base, Ohio, 1971. To be published in proceedings.

[18] Argyris, J. H., Faust, G., Roy, J. R., Szimmat, J., Warnke, E. P., and Willam, K. J., 'Finite Elemente zur Berechnung von Spannbeton-Reaktordruckbehaltern,' ISD Report No. 137 (1973). Published also in DAfStb (1973).

[19] Schrem, E. and Roy, J. R., 'An Automatic System for Kinematic Analysis', ASKA Part I, *IUTAM Symp. on High Speed Computing of Elastic Structures*, Liege, 1970.

[20] Schrem, E., 'Computer Implementation of the Finite Element Procedure', *Proc. of ONR Symp. on Num. and Computer Methods in Struct. Mech.* University of Illinois, Urbana, 1971.

[21] Baumann, E., and Rowe, J., 'Some ASKA Applications on the B-1 Bomber and Space Shuttle', Private Communication, North American Rockwell, Downey, Cal., 1972.

[22] Washizu, K., *Variational Methods in Elasticity and Plasticity*, Pergamon Press, 1968.

[23] Oden, J. T., *Finite Elements of Nonlinear Continua*, McGraw-Hill Book Co., 1972.

[24] Argyris, J. H. and Roy, J. R., 'General Treatment of Structural Modifications', *ASCE* **89** (1972)

[25] Prager, W., 'Non-isothermal plastic deformation', *Proc. Kininkl. Nederl. Akad. van Wetenschappen, Amsterdam* **61** (1958).

[26] Naghdi, P. M., 'Stress-Strain Relation in Plasticity and Thermo-Plasticity', in '*Plasticity*', *Proc. 2nd Symp. on Naval Struct. Mech.*, Pergamon, 1960.

[27] Argyris, J. H., 'Elasto-Plastic Matrix Displacement Analysis of Three-Dimensional Continua', *J. Roy. Aeron. Soc.* **69** (1965).

[28] Argyris, J. H., Scharpf, D. W., and Spooner, J. B., 'Die elastoplastische Berechnung von allgemeinen Tragwerken und Kontinua', *Ingenieur-Archiv* **37** (1969). Extended English Version was published also in *Proc. 3rd Conf. on Dimensioning*, Budapest, 1969.

[29] Zienkiewicz, O. Z., Valliapan, S., and King, I. P., Elasto-Plastic Solutions of Engineering Problems, Initial Stress, Finite Element Approach', *Ing. Num. Meth. Eng.* (1969).

[30] Argyris, J. H. and Scharpf, D. W., 'Methods of Elastoplastic Analysis', *Proc. of the Symp. on Finite*

Element Techniques, Fourth Int. Ship Structures Congress (ISSC) Stuttgart (1969), also published in *ZAMP* (1972), 517–552.

[31] Argyris, J. H. and Chan, A. S. L., 'Applications of Finite Elements in Space and Time', *Ingenieur-Archiv* **41** (1972), 235–257.

[32] Balmer, H. and Doltsinis, J. St., 'ASKA Part III-1 Material Nonlinearities, Lecture Notes with Example Problems', ISD Report No. 132, University of Stuttgart, 1972.

[33] Rabotnov, Y. N., *Creep Problems in Structural Members*, North Holland Publ. Co., 1969.

[34] Greenbaum, G. A. and Rubinstein, M. F., 'Creep Analysis of Axisymmetric Bodies Using Finite Elements', *Nuclear Eng. and Design* **7** (1968).

[35] Mareczek, G., 'Elastoplastische Berechnung eines Wiedereintrittskörpers unter extrem hoher Temperaturbelastung', Diplomarbeit am ISD, Stuttgart, 1967.

[36] Archer, J. S., 'Consistent Mass Matrix for Distributed Systems', *Proc. ASCE* **89** (1963).

[37] Argyris, J. H., 'Some Results on the Free-Free Oscillations of Aircraft Type Structures', *Revue Française de Mécanique* **15**, 3e trimestre (1965).

[38] Bisplinghoff, R. L., Ashley, H., and Halfman, R. L., *Aeroelasticity*, Addison-Welsey Publ. Co., Reading, 1957.

[39] Wilson, E. L. and Penzien, J., 'Evaluation of Orthogonal Damping Matrices', *Int. J. Num. Methods* **1** (1972).

[40] Newmark, N. M., 'A Method of Computation for Structural Dynamics', *Proc. ASCE* **85** (1959)

[41] Clough, R. and Wilson, E. I., 'Dynamic Finite Element Analysis of Arbitrary Thin Shells', *Proc. Conf. at Palo Alto* (1971).

[42] Wilson, E. L., Farhoomand, I., and Bathe, K. J., 'Nonlinear Dynamic Analysis of Complex Structures', *Int. J. Earthquake Engineering and Structural Dynamics* **1** (1972).

[43] Argyris, J. H. and Sharpf, D. W., 'Finite Elements in Time and Space', *Aer. J. Roy. Aeron. Soc.* **73** (1969) 1041–1044, and *Nuclear Engineering and Design* **10** (1969).

[44] Argyris, J. H., Dunne, P. C., and Angelopoulos, T., 'Non-Linear Oscillations Using the Finite Element Technique', ISD Report 136, University of Stuttgart, 1972; also published in *Comp. Methods in Appl. Mech. of Engineering* (1973).

[45] Wilkinson, J. H., *The Algebraic Eigenvalue Problem*, Clarendon Press, Oxford, 1965.

[46] Guyan, R., 'Reduction of Stiffness and Mass Matrices', *AIAA J.* **3** (1965).

[47] Uhrig, R., 'Reduction of the Number of Unknowns in the Displacement Method Applied to Kinetic Problems', *J. Sound and Vibration* **4** (1966).

[48] Hurty, W. C. and Rubinstein, M. F., *Dynamics of Structures*, Prentice Hall Inc., Englewood Cliffs, New Jersey, 1964.

[49] Hurty, W. C., 'Introduction to Modal Synthesis Techniques, *paper presented at the Winter Annual Meeting of the ASME*, Washington, D.C., 1971.

[50] Malejannakis, G. A., 'Anwendung der Matrizenverschiebungsmethode auf erzwungene Sschwingungen proportional gedämpfter elastischer Systeme', ISD Report No. 99 (1971).

[51] Bauer, F. L., 'Das Verfahren der Treppen-Iteration und verwandte Verfahren zur Lösung algebraischer Eigenwertprobleme', *ZAMP* **8** (1957).

[52] Jennings, A., 'A Direct Iteration Method of Obtaining Latent Roots and Vectors of a Symmetric Matrix', *Proc. Camb. Phil. Soc.* **63** (1957).

[53] Brönlund, O. E., 'Eigenvalues of Large Matrices', *Proc. of the Symp. on Finite Element Techniques*, Fourth Int. Ship Struct. Congress (ISSC), Stuttgart, 1969.

[54] Braun, K. A., Brönlund, O. E., Bühlmeier, J., Dietrich, G., Frick, G., Johnsen, T. L., Kiesbauer. H. T., Malejannakis, G. A., Straub, K., and Vallianos, G., DYNAN Lecture Notes with Computational Examples', ISD Report No. 109, University of Stuttgart, 1971.

[55] Brönlund, O. E., 'Die simultane Verbesserung einer beliebigen Anzahl genäherter Eigenvektoren von hermiteschen Matrizen', Dr. -Ing. Thesis, University of Stuttgart, 1972.

[56] Argyris, J. H., Brönlund, O. E., Kayser, L. T., Malejannakis, G. A., and Straub, K., Längsdynamik der 2./3. Stufe der ELDO-Rakete Europa IFG', ISD Report No. 86, Stuttgart, 1971.

[57] Kiessling, F., 'Praktische Anwendung der Matrizenverschiebungsmethode auf die Schwingungsanalyse eines Versuchsflugzeugs', Diplomarbeit am ISD (1971).

[58] *DYNAN User's Reference Manual*, ISD Report No. 97 (1971).

[59] Argyris, J. H., Buck, K. E., Scharpf, D. W., and Willam, K. J., 'Linear and Nonlinear Methods of Structural Analysis', *First Int. Conf. on Struct. Mechanics in Reaktor Technology*, Berlin 1971, also published in *Nuclear Engineering and Design* **19** (1972).

[60] Bathe, K. J., 'Solution Method for Large Generalized Eigenvalue Problems in Structural Engineering', SESM Report 71-20, Dept. of Civil Eng. University of California, Berkely, 1971.

[61] Clough, R. W. and Bathe, K. J., 'Finite Element Analysis of Dynamic Response', *paper presented at the 2nd US Japanese Seminar* (1972).

[62] Fried, I., 'Discretization and Computational Errors in Higher-Order Finite Elements, *AIAA* **9** (1971).

[63] Brönlund, O. E. und Bühlmeir, J., 'Einige Verfahren zur Berechnung von Eigenwerten und Eigenvektoren von nicht-hermiteschen Matrizen unter besonderer Berücksichtigung von struktur-dynamischen Problemen', ISD Report No. 128, Universität Stuttgart, 1972.

[64] Rubin, S., 'A General Study of the POGO', *paper preseuted at the 23rd Congress of the IAF*, Vienna, 1972.

[65] Valid, R., Ohayon, R., and Berger, H., 'The Computation of Elastic Tanks Partially Filled with Liquids for the Prevision of POGO Effects', *paper presented at the 23rd Congress of the IAF*, Vienna, 1972.

[66] Buck, K. E., 'Zur Berechnung der Verschiebungen und Spannungen in rotationssymmetrischen Körpern unter beliebiger Belastung', Dr.-Ing. Thesis, Universität Stuttgart, 1970.

[67] Gloudeman, J. F., 'Zur numerischen Berechnung der linearen und nicht-linearen Differentialgleichungen mit hermiteschen Interpolationspolynomen', Dr.-Ing. Theses, Universität Stuttgart, 1970.

[68] Willam, K. J., 'Finite Element Analysis of Cellular Structures', Ph. D. Thesis, University of California, Berkeley, 1969.

[69] Grieger, I., 'INGA, Interaktive graphische Analyse, Benutzerhandbuch', ISD Report No. 135, University of Stuttgart, 1973.

[70] Scharpf, D. W., 'Die Frage der Konvergenz bei Berechnung elastoplastisch-deformierbarer Tragwerke und Kontinua', Dr.-Ing. Thesis, Universität Stuttgart, 1969.

[71] Chan, A. S. L. and Firmin, A., 'The Analysis of Cooling Towers by the Matrix Finite Element Technique', *Aeron. J. Roy. Aeron. Soc.* **74** (1970).

[72] Argyris, J. H., Buck, K. E., Lochner, N., and Scharpf, D. W., 'Matrix Displacement Analysis of Plates and Shells, A general Formulation of the Linear Theory', ISD Report No. 103, University of Stuttgart, 1971.

[73] Argyris, J. H. and Lochner, N., 'On the Application of the SHEBA Element'. *Comp. Meth. in Appl. Mech. and Eng.* **1** (1972).

[74] Argyris, J. H., Haase, M., and Malejannakis, G. A., 'Natural Geometry of Surfaces with Specific Reference to the Matrix Displacement Analysis of Shells', ISD Report No. 134, University of Stuttgart, 1973. Also published in *Comp. Methods in Appl. Mech. and Eng.* (1973).

[75] Bergan, P. G. and Clough, R. W., 'Convergence Criteria for Iterative Process', *AIAA TN* (1972).

[76] Fuchs, G. v., Roy, J. R., and Schrem, E., 'Hypermatrix Solution of Large Sets of Positive Definite Linear Matrices,' *Comp. Methods in Applied Mech. and Eng.* **1** (1972).

[77] Balmer, H., Doltsinis, J. St., and König, M., 'Elastoplastic and Creep Analysis with the ASKA Program System', to be published in *Comp. Methods in Appl. Mech. and Eng.* (1973). ·

EARTH AND OCEAN PHYSICS APPLICATIONS PROGRAM (EOPAP)

F. O. VONBUN

NASA, Goddard Space Flight Center, Greenbelt, Md., U.S.A.

This paper describes, in abbreviated form, NASA's proposed Earth and Ocean Physics Applications Program (EOPAP).

The solid Earth and the oceans have provided the base for man's activities on this planet through its history. Both have supplied him with the necessary substance for his life and have further stimulated his aspirations. At the same time however, both have acted as severe task masters, plaguing him with natural disasters and influencing nearly every move in his daily life. Earthquakes, tidal waves and storms have wreaked havoc to mankind, taking a large toll of life and property. These phenomena are of an unmanageable scale and until recently have all been unpredictable. Space capabilities have provided, for instance, the ability to predict the paths of potentially disastrous hurricanes and accompanying tidal waves.

One of the aims of this program is to apply yet another kind of space capability to help us to predict earthquakes, storm surges, tidal waves, and the condition of the surface of the world's oceans. Man in the process of creating his society, is at the same time creating his own disasters, which may, if not halted, ultimately dwarf those of nature. Pollution of the oceans and the depletion of its food resources are already real threats at present. Space techniques will, and can, play a major role in helping to map ocean currents, circulations and ocean management capability for handling problems associated with pollution, food resources, shipping and climate.

This new program is being designed to properly blend geophysics and oceanography with techniques developed under the space program to achieve practical applications of benefit to man. As you know, the keynote of this congress is: 'SPACE FOR WORLD DEVELOPMENT'. To translate this into a somewhat different form, this may be stated as follows: 'USE WHAT WE HAVE LEARNED IN THE LAST DECADE AND APPLY IT DURING THE COMING YEARS FOR THE BENEFIT OF MANKIND'.

In the following, the EOPAP is outlined in some detail. In particular, are shown the Program Goals or Objectives, the Experiments, the Spacecraft and the schedules needed to achieve these objectives.

In Figure 1 and 2, four major goals are outlined: two in Earth Dynamics and two in Ocean Dynamics. The accomplishment of these goals, which evolved over the last few years, is the essence of this program. If achieved, they will help to obtain answers to pressing problems of our environment on the only planet man can live on – the Earth. The purpose here is to identify, develop, demonstrate, and utilize relevant space measurements and the necessary analytical techniques to provide data and their usage for contributing significantly to the development and validation of predictive

L. G. Napolitano et al (eds.), Astronautical Research 1972, 239–245. All Rights Reserved
Copyright © 1973 by D Reidel Publishing Company, Dordrecht-Holland

- ● EARTHQUAKE HAZARD ASSESSMENT AND ALLEVIATION
  - ● <u>UTILIZE</u> SPACE CAPABILITY TO PROVIDE IMPORTANT INFORMATION FOR EARTHQUAKE HAZARD ASSESSMENT AND ALLEVIATION (FAULT MOTION, POLAR MOTION, EARTH ROTATION VARIATIONS, TECTONIC MOTION)
  - ● <u>DEVELOP</u> MATHEMATICAL MODELS TO PREDICT PROBABLE TIME, LOCATION AND INTENSITY OF EARTHQUAKES

- ● GLOBAL SURVEYING AND MAPPING
  - ● <u>UTILIZE</u> SPACE CAPABILITY TO EXTEND GEODETIC CONTROL TO REMOTE AREAS AND THE OCEAN FLOOR (0.5m VERT., 10m HORIZ.)
  - ● <u>PROVIDE</u> REFERENCE FIELDS (GRAVIMETRIC, MAGNETIC) OF THE EARTH FOR USE IN SURVEYING, GEOPHYSICAL STUDIES, AND POSSIBLE ASSISTANCE IN SEARCH FOR MINERAL RESOURCES

Fig. 1.   Major goals – Earth dynamics.

- ● OCEAN CURRENTS AND CIRCULATION
  - ● <u>UTILIZE</u> SPACE CAPABILITY TO DETERMINE SEA SURFACE TOPOGRAPHY AND THE SEPARATION BETWEEN THE SEA SURFACE AND THE GEOID TO IMPROVE MODELS OF OCEAN CIRCULATION AND CURRENTS
  - ● <u>APPLY</u> MODELS FOR CIRCULATION AND CURRENT TO PROBLEMS ASSOCIATED WITH POLLUTION, FOOD RESOURCES, SHIPPING AND CLIMATE

- ● OCEAN SURFACE CONDITION MONITORING
  - ● <u>UTILIZE</u> SPACE CAPABILITY FOR SYNOPTIC MEASUREMENTS OF SEA STATE, SURFACE WIND AND ITS DIRECTION, STORM SURGES....
  - ● <u>APPLY</u> THE ABOVE TO HELP SHIPPING, FISHING, WEATHER FORECASTING

Fig. 2.   Major goals – Ocean dynamics.

models for earthquake hazard alleviation, ocean surface conditions, and ocean circulation.

In order to make this program successful, both experiments together with their respective models as well as special spacecraft and or flight missions are needed. Figures 3 and 4 outline major experiments in Earth and Ocean Dynamics together with the metric tracking systems which are required.

In parallel to these developments the necessary theories, mathematical models, and the data handling problems are being worked on to assure that these tools are available as soon as the missions are flown, or the experiments are performed in order to really 'use' the data obtained. To be more specific, models are being developed which use space tracking data for Earth and Ocean Dynamics. That is, models to predict (in the final version) the probability of position and occurrence of large earthquakes, using such additional input data as plate motion, polar motion, and Earth rotation. Further, under development are mathematical models to determine the Earth's gravity and magnetic fields, as well as to use space tracking data for geodetic control (station locations, continental shelf survey, etc.).

For the area of Ocean Dynamics, development of models is underway to study and determine seastate, sea surface winds, sea surface topography, ocean circulation and currents using the on-board radar altimeter, microwave seatterometer, microwave radiometer as well as the spacecraft orbital parameters.

- EARTHQUAKE HAZARD ASSESSMENT AND ALLEVIATION
  - SAN ANDREAS FAULT EXPERIMENT (SAFE) (LASER, POSS VLBI)
  - PLATE MOTION EXPERIMENT (VLBI, LASER)
    EAST–WEST COAST, US–JAPAN, US–GERMANY & OTHERS
  - SOLID EARTH TIDES (LASER, ATS–G)
  - EARTH MOTION DETERMINATION: POLAR MOTION;
    UT–1, MOTION IN SPACE (LASER, VLBI, RADIO)

- GLOBAL SURVEYING AND MAPPING
  - GRAVITY FIELD DETERMINATION, (LASER, SATELLITE TO SATELLITE
    TRACK (SST) ATS–NIMBUS GEOS...., GRADIOMETER) N.G.S.P.
  - GEOID AND GRAVITY FINESTRUCTURE (ALTIMETER,LASER, RADAR)
  - STATION LOCATION DETERMINATION (LASER, USB, RADAR, OPTICAL,
    DOPPLER,)
  - MAGNETIC FIELD DETERMINATION (MAGNETOMETER,SCALAR, VECTOR)
  - INTERNATIONAL GEOD. EXP., ISAGEX, EPSOC, ...

Fig. 3.   Experiments – Earth dynamics.

- OCEAN CURRENT AND CIRCULATION
  - OCEAN SURFACE TOPOGRAPHY; GEOID, TRENCHES; SLOPE,
    (ALTIMETER, SST, LASER)
  - GENERAL CIRCULATION (LAGRANG TRACERS, ALTIMETER, SST)
  - OPEN OCEAN TIDES, TSUNAMIES (ALTIMETER, SST)

- OCEAN SURFACE CONDITION MONITORING
  - SEA STATE, WAVE DIRECTION (ALTIMETER, SPEC. MODE)
  - SURFACE WINDS (ALTIMETER, SCATTEROMETER)
  - STORM SURGES (ALTIMETER, SST)

Fig. 4.   Experiments – Ocean dynamics.

Further, theoretical work is planned and ongoing to improve (two orders of magnitude) our capability to determine spacecraft orbits, reduce our tracking systems errors as well as the errors associated with the propagation media, the troposphere and the ionosphere.

Figure 5 depicts a planned flight mission schedule over the next decade showing the major spacecraft which are essential to this program.

LAGEOS (Laser Geodetic Satellite) is a completely passive spacecraft (Figure 6) consisting of a heavy core (580 kg) covered with laser corner reflectors, to be launched into a polar, circular orbit of 3000 to 4000 km. This spacecraft's main purpose is to help solve problems in the area of Earth dynamics. The design of the spacecraft is such that measurement accuracies of 2 to 5 cm can be obtained using, say, subnanosecond pulsed laser systems.

SEASATS-1 (Sea Satellite 1) is a spacecraft tailored toward oceanographic measurements. It will carry a radar altimeter (50 cm accuracy), a microwave scatterometer, a microwave radiometer, an imaging radar, an infrared sensor, an SST* (30 cm, 0.003 cm s^{-1}), laser corner cubes, and will be three-axis stabilized. Its weight will be in the order of 400 kg, its orbit will be circular, polar with a height of ~ 500 to ~ 700 km.

GEOPAUSE (a spacecraft far out of the Earth's gravity 'noise', thus the word geopause) is a spacecraft to act as a high precision orbiting tracking station needed for all EOPAP missions (except LAGEOS). Its major tracking system will be an SST (2 to 3 cm in range and 0.003cm s^{-1} to 0.001 cm s^{-1} in range rate), and it will further carry laser retroreflectors and an accelerometer (to measure 10^{-12} g's). Its weight will be in the order of 350 kg, its orbit polar, circular and ~ 30000 km high.

GRAVSAT (Gravity Satellite) is a spacecraft to probe the Earth's gravitational field. It will carry an SST (2 to 3 cm range, 0.003 to 0.001 cm s^{-1} range rate) to work together with the Geopause. Its weight will be in the order of 1000 kg, its orbit very low ~ 250 to 300 kg, polar and circular. It will carry a drag compensating system (satellite in a satellite).

SEASATS-2, a spacecraft of advanced capability (compared to SEASATS-1) with its aim to act as a prototype of a final operational sea surveillance spacecraft.

In order to solve problems in the area of Earth and Ocean Dynamics, intersite distances have to be determined to, say, 2 to 3 cm, the ocean's topography (variations of the surface) has to be determined to say 10 cm, polar motion and UT-1 have to be determined to 2 to 10 cm, wave heights have to be measured between 1 to 15 m and surface wind speeds and directions are needed to an accuracy of approximately 20% in magnitude and less than 20° in direction.

All these are quantities of at least two orders magnitude smaller than past satellite techniques could provide. Thus, to make a proper judgment possible, several experiments have been carried out in the past and/or are underway at present.

About a year ago, we successfully finished a polar motion experiment. Using, in

---

* Satellite-to-Satellite Tracking.

LAGEOS		S	Δ							
SEASATS-1		S		Δ						
GEOPAUSE			S			Δ				
GRAVSATS			S			Δ				
SEASATS-2							S			Δ
CAL. YEAR	72	73	74	75	76	77	78	79	80	81	82

S START
Δ LAUNCH

Fig. 5.   EOPAP spacecraft.

SOLUTION		SURVEY DIFFERENCE
1	408699.20 m	+ 43 cm
2	408698.87 m	+ 10 cm
3	408699.33 m	+ 56 cm
4	408699.44 m	+ 67 cm
5	408698.91 m	+ 14 cm
GROUND SURVEY	408698.77 m	

Fig. 6.   Godlas-senlas chord distance.

essence, only one laser and the Beacon Explorer-C spacecraft (equipped with laser corner reflectors) we did determine the distance between two points on the ground to say 30 cm (precision), as shown in Figure 6 (GODLAS – Goddard Laser, and SENLAS – a laser positioned at Sneca, N.Y., ~400 km from Goddard). Figure 7 shows the results obtained from this experiment.

We further finished chord distance determination experiments using the Very Long Baseline Interferometry (VLBI) technique. Precisions in the order of 50 cm and less were obtained using baselines of 5000 km in length (Massachusetts – California). This is a conservative figure by not taking any statistical improvements into account yet.

Figure 8 depicts some of our results in summary form. As can be seen, we have demonstrated with past and present experiments that the needed data, as mentioned before, can indeed be obtained during the next few years. The program as outlined is

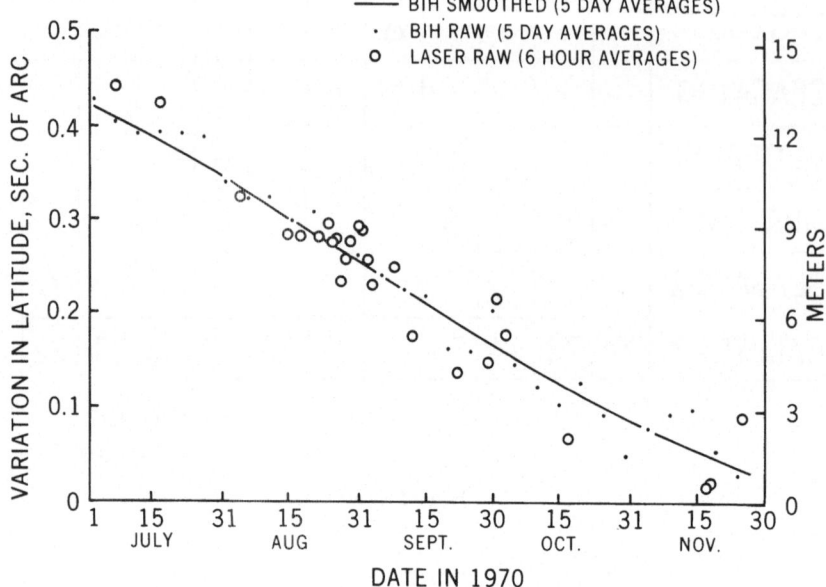

Fig. 7. Variation in latitude of Goddard laser

POLAR MOTION

CONVENTIONAL ASTRONOMY – 40 STATIONS	1 m	5 DAY AVERAGE
SATELLITE DOPPLER TRACKING – 12 STATIONS	1 m	2 DAY AVERAGE
SATELLITE LASER TRACKING – 1 STATION 1 COMPONENT	1 m	6 HOUR AVERAGE

INTERSITE DISTANCE

SATELLITE CAMERA TRACKING	10 m	GLOBAL
SATELLITE LASER TRACKING	0.3 m	400 km
VLBI	0.5 m	4000 km

GRAVITY FIELD – LUNAR MASCONS

HI–LO RANGE RATE	0.03 cm/s	1 mgal

SEA STATE

AIRBORNE ALTIMETER	1.5 m

SURFACE WINDS

AIRBORNE SCATTEROMETER	20%	1–15 m/s

Fig. 8. Recent Earth and ocean physics advances.

really feasible and its outcome is not only proven by theoretical error analyses which were done of course, but also, and this is very important, founded on experimental evidence. This should not indicate that the road of the future will be an easy one. On the contrary, because large improvements (one to two orders of magnitude) in the reduction of metric errors have to be made, the task will be difficult. Nevertheless, we are confident that the goals as outlined can be achieved.

LATEST RESULTS FROM THE EARTH RESOURCES PROGRAM

(Summary of Discussion Session)

WILLIAM NORDBERG

NASA, Goddard Space Flight Center, Greenbelt, Md., U.S.A.

The results summarized here deal exclusively with observations from the first Earth Resources Technology Satellite (ERTS-1) and with the interpretation of these observations in terms of surveying and mapping Earth resources over large areas. This summary is based on nine reports presented to the IAF Congress on 14 October 1972. The titles and authors of these reports are listed below as References.

The ERTS-1 was launched on 23 July 1972 and with minor exceptions, has transmitted multi-spectral images of approximately 6.5×10^6 km² of the Earth's surface every day since 25 July 1972. Sunlight reflected by the Earth is imaged and measured by two sensor systems, namely, a set of three Return Beam Vidicon (RBV) television cameras and a Multi-Spectral Scanner (MSS), in the 0.5–0.6, 0.6–0.7, and 0.7–0.8 μm wavelength intervals. The MSS alone provides photometric images in one additional interval of 0.8–1.1 μm. Measurements from each sensor are recorded on two wideband video tape recorders when the satellite is not within transmission range of a receiving station. The satellite also carries a Data Collection System (DCS) which acquires messages from randomly transmitting platforms, distributed over North America, and retransmits these messages to a central analysis facility. These messages consist of measurements of such parameters as water quality, rainfall, snow depth, and seismic activity which are made by the platforms directly and automatically at remote and inaccessible sites.

The optical measurements made by the RBV and MSS systems are transmitted to three receiving stations in the U.S.A., located in Maryland, California, and Alaska, and to a Canadian station in Saskatchewan during each daytime pass of the satellite over these stations. A fifth station is planned to be in operation in Brazil. During nighttime passes, measurements recorded on the tape recorders, over other parts of the world, are transmitted to the three U.S.A. stations. RBV and MSS observations made over the 6.5×10^6 km² every day amount to about 10^{11} digital bits which are converted to about 1350 photographic images daily at the NASA Data Processing Facility (NDPF) in Greenbelt, Maryland. Approximately ten copies of each of these images are reproduced photographically and distributed to several agencies of the U.S. Government who conduct experiments to determine the applicability of ERTS observations in various operations related to resource management such as, cartographic mapping; survey of land use, crops, timber and soils; assessment of water quality and supply; exploration of mineral resources; and detection of environmental hazards. Both the U.S. Departments of Commerce and Interior archive all photographs at their respective data centers. ERTS-1 photographs may be purchased from at least one of these archives at nominal cost by *anyone*, simply by writing to the EROS

L. G. Napolitano et al (eds.), Astronautical Research 1972, 247–250 All Rights Reserved
Copyright © 1973 by D. Reidel Publishing Company, Dordrecht-Holland

Data Center at Sioux Falls, South Dakota. Canadian pictures may also be purchased from the archives in that country.

In addition, NASA has reached agreements with about 330 investigators who analyse the ERTS-1 observations for specific applications in the fields of agriculture and forestry, geography, geology, hydrology, oceanography, meteorology, and other sciences dealing with the quality of people's environment. These investigators are affiliated with federal and local governments, universities, or private industrial organizations. Over one hundred of them are from countries other than the United States.

In order to observe most areas of the world, the satellite is in a nearly polar orbit and could make observations anywhere, during about 14 passes in a southwesterly direction, from 80° north to 80° south, every day. Passes occur at approximately the same local time (about 0930 at the Equator) at each location to assure uniform illumination of the imaged scenes. The fields of view of the MSS and RBV cameras are the same and encompass a strip, 185 km wide along the satellite's path. Images are taken daily along selected portions of such strips, about 1800 km apart at mid latitudes. The orbit is adjusted such that a strip observed on one day is contiguous with the strip observed on the previous day. As each strip advances by about 180 km from a pass on one day to the same pass on the next day, the entire world could be covered by observations once every 18 days. However, power and data transmission capacities of the spacecraft, as well as processing throughout of the NDPF are limited such that contiguous images every 18 days are possible only over North America. Over the rest of the world strips amounting to a total length of about 26000 km (and of course 185 km wide) can be taken every day. These strips are distributed over areas which are forecast to be cloud free and in which at least one of the 330 investigators has agreed to conduct his analysis. Depending on contrast and geometric shape, objects larger than 50 to 100 m in diameter may be easily identified in the ERTS-1 images. The images are sufficiently free of distortion so that they can serve readily as maps on scales up to 1:250000. This combined with the ability of ERTS to view very large areas repeatedly, makes this stallite a most effective and economic tool to map features not previously detected as well as variations in environmental characteristics (Risley).

Each of the spectral intervals of the sensors permits the discrimination of different sets of features which are related to Earth resources. For example, the 0.5–0.6 μm interval provides for the greatest transparency of water such that shoals, sand bars, coastal shelves, and similar underwater features which might relate to ocean or lake dynamics or pose hazards to navigation can be mapped. The 0.6–0.7 μm interval provides for somewhat lesser transmission in water, but is especially useful to map suspended material and sediments. It also shows the greatest contrast between areas covered by vegetation and those denuded naturally or by people's activities, and is therefore most useful for mapping cultural patterns and land use. The two infrared intervals take advantage of the very low reflectivity of any kind of surface water and the relatively high reflectivity of vegetation, depending on type and state. This has led to the production of excellent maps of watersheds, drainage patterns, wetlands and swamps. In particular, in the Amazon region, maps were revised drastically on the basis of even these very early ERTS observations (Mendonca). In the same area, those rivers and tributaries, primarily the Amazon, which carry heavy silt and sediments could be clearly distinguished from those which were clear and free of turbidity, namely, the Rio Negro and all but one of its tributaries, by analyzing the 0.6–0.7 μm interval. Geological formations could also be delineated on the basis of vegetation changes in this and other regions by analyzing the 0.7–0.8 and 0.8–1.1 μm intervals.

In the 0.6–0.7 μm as well as in the infrared intervals one can clearly map linear features which relate to geological structure and one can recognize boundaries, between geological formations (Porter). Geological evidence of glacial surges was found in the Canadian arctic, and various geological stages of river flow were mapped in that region (Morley). Even in thoroughly explored areas of the world, between 10–20% of these linear features had not been mapped before. In unexplored areas the number of such features relating to geological structure first observed by ERTS is even greater. The mapping of these structural features is of significance to mineral exploration, the location of ground water and possibly the recognition of hazards to construction.

The most powerful analyses made, however, from ERTS-1 observations are those which combine all four spectral bands so that features may be identified and mapped on the basis of their varying spectral characteristics throughout these intervals. For example, snow cover of the Arctic Ocean ice and of the North American mountain chains was not only mapped for its areal extent and variations with topographic height, but incipient melting conditions could also be recognized (Davis). This was based on the fact that snow appears increasingly darker at longer wavelengths as melting and water amount increase. This type of analysis when made over a period of time can lead to the computation of the amount of water runoff and therefore of water supply in large watersheds.

Of great significance to environmental surveys is the finding that burnt forest and agricultural areas stand out clearly in the multi-spectral images and they can be mapped very precisely. This makes it possible not only to assess fire damage very economically, but to plan more efficiently for reforesting and reclamation of the affected areas (Thorley).

Similar findings have been made for areas affected by strip mining or floods (Porter). Water pollution resulting from the dumping of acid wastes by barges in New York harbor was detected and mapped by analyzing MSS images in all four spectral intervals (Davis).

Multi-spectral analysis of images in California and in the Southwestern U.S. permitted easily the classification of agricultural crops, grazing lands, soils and various land use practices by the population. Almost instantly, a land use map of the entire San Joaquin County in California was produced from the first ERTS-1 picture of that area (Thorley). Automatic classifications of land use and agricultural features were made employing computerized 'clustering' techniques (Landgrebe). These showed that about 13–18 different classes of vegetation, soil and land use features can be recognized in one typical ERTS-MSS image covering an area of 185 × 185 km.

Finally it was pointed out that ERTS pictures could provide observations extremely useful to the analysis of hydrodynamic processes in harbors, inlets, lakes or even in the open oceans. The spatial structure of vortices and other flow patterns displayed in ERTS pictures could be analyzed similarly to laboratory experiments, except that in this case, nature itself is staging the experiment (Napolitano). Particularly prominent displays of such hydrodynamic features were shown for the sedimentation plumes associated with the Mackenzie River in the Beaufort Sea (Morley) and with tidal patterns in Long Island Sound, Delaware Bay and other regions of the Atlantic coast of the Northeastern U.S. (Davis).

References

Davis, William Dr, NOAA/ESSA, Rockville, Maryland – 'Hydrographic and Atmospheric Surveys with ERTS-1'.

Landgrebe, David Dr, LARS/Purdue Univ., Indiana – 'Automatic Classification of Soils and Vegetation with ERTS-1'.

Mendonca, Fernando de Dr, Brazilian Institute for Space Research – 'Preliminary Results of ERTS-1 Observations of Amozonia'.

Morley, L. W. Dr, Dir. Canada Ctr. for Remoting Sensing – 'Application of ERTS-1 Observations to Resource Surveys in Canada'.

Napolitano, Prof., Inst. of Aerodynamics, Naples, Italy – 'Hydrological Applications of ERTS-1 Observations'.

Nordberg, William Dr, NASA/Goddard Space Flight Center, Greenbelt, Md. – 'The Earth Resources Technology Satellite-1'.

Porter, J. Robert Jr, Earth Satellite Corp., Washington, D.C. – 'Preliminary Analysis of ERTS-1 Data for Geologic, Forestry, and Land Use Applications'.

Risley, Edward, U.S. Geological Survey, Washington, D.C. – 'Geological Findings, Land Use Surveys, and Geographic Mapping with ERTS-1'.

Thorley, Gene, Forestry Remote Sensing Lab., Berkeley, California – 'ERTS-1 Applications to California Resource Inventory'.

DEVELOPMENTS IN APPLICATIONS OF REMOTE SENSING TO HYDROLOGY

ANDREW ADELMAN, REUBEN AMBARUCH, and JOHN W. SIMMONS

IBM, Federal Systems Division, Electronics Systems Center, Huntsville, Ala., U.S.A.

Abstract. The progress and initial results in the first major part of a streamflow forecasting project, utilizing aircraft (and later space) derived imagery for watersheds of the Tennessee River Valley, are presented. The initial study phase, aimed at assessing the feasibility of applying remote sensed data to prediction of watershed performance, is approximately thirty percent complete. Demonstration of feasibility will lead to a means of prediction of the hydrological behavior of ungaged watersheds using remote sensed data to minimize time, effort, and cost of achieving these predictions, as well as provide good prediction capability with fewer ground instruments in existing installations.

Catchments have been chosen for calibration and simulation from the Tennessee Valley region, for which a wealth of climatological and streamflow historical data as well as extensive aerial photographic and ground survey coverage are available. An initial set of 14 catchments chosen for analysis represents five distinct physiographic provinces in the region and vary in size from 6.9 to 365 km^2.

A computer program based on the well-known Stanford Watershed Model IV and a companion parameter-optimization program are used to calibrate model parameters for the selected catchments, based on historical data. The accuracy of the models is shown by comparison of synthesized with observed streamflows. Methods of determining model parameters from physical characteristics, observable or inferable from remotely acquired photographs, without the necessity of historical streamflow data, are discussed.

Extension of the effort includes validating applicability of the models to catchments in other regions of the world. Sensitivity of parameter estimation and simulation accuracies to lower resolution in sensor data, such as Earth observation data acquired from space, will be assessed. Detailed procedures for applying the models will be prepared and made available for potential use in all parts of the world.

1. Introduction

The application of remote sensing to hydrology was suggested by Dr Peter A. Castruccio at the International Astronautical Congress, Brussels, 1971. From this suggestion has evolved an application development project of three major parts:

(1) A proven hydrological simulation model has been modified for use in the Tennessee Valley, a highly instrumented and extensively surveyed region of the U.S.A. Model parameters are derived from geomorphological characteristics of selected basins as determined from remote sensed images (aerial photographs), with a minimum of survey data and a priori knowledge of subsurface properties (i.e., soils). The model inputs are hourly recorded precipitation data and evaporation statistics. Historical streamflow records are used to evaluate the accuracy of simulated streamflows produced by the model. This part of the project is presently in process, testing accuracy of the model of 35 catchments of the Tennessee Valley.

(2) A second major instrumented hydrological system, geographically and climatically different from the Tennessee Valley, in another part of the world, is chosen for this second part. The same methodology and criteria are used to verify transferability of the model.

(3) In the final part, using the model and techniques previously developed, the hydrological responses to hypothetical precipitation of uninstrumented watersheds

in various parts of the world are predicted. The knowledge, experience, and techniques developed will enable preparation of user directions for applying the model in any part of the world.

This report summarizes progress and achievements to date in Part 1 of the project: the simulation models selected and modified for the study; the methods used to derive model parameter values from aerial photographs; and the results of several years of streamflow prediction in those river basins which have been simulated to date, compared to actual streamflow measurements. Results to date conservatively indicate that short and long term predictions will be routinely achievable with tolerances of ±25% or less without historical data.

2. Basic Project Concept

That aspect of hydrology known as Streamflow Forecasting undertakes to predict the outflow from a given catchment, as a function of time, in response to a given precipation event under given initial conditions. This capability is vital to effective planning for urban/industrial development, flood control, hydroelectric power, navigation, and water resources management.

Figure 1 depicts the cross section of a somewhat idealized rural catchment and iden-

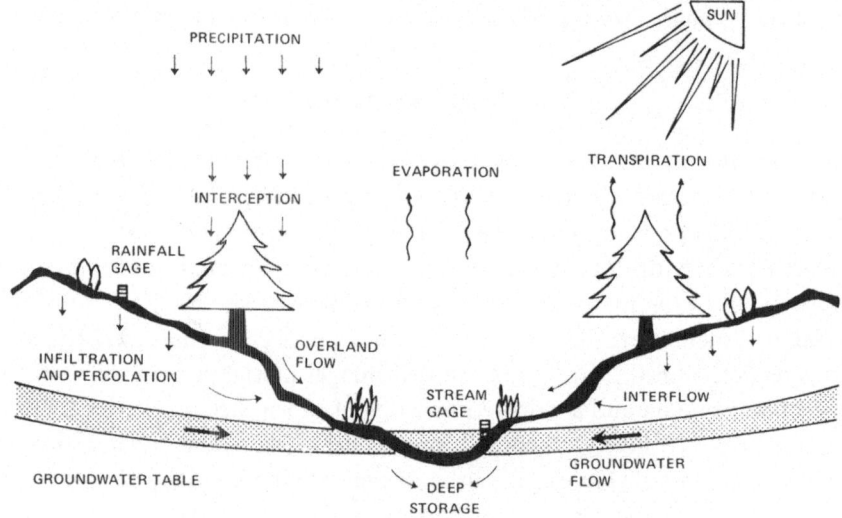

Fig. 1. Cross section of idealized rural catchment.

tifies the principal phenomena at work the rainfall-runoff relationship. The input (precipitation) is partially intercepted by vegetation and water retention areas. Moisture reaching pervious surfaces divides between overland flow, infiltration, and evaporation. Through subsurface processes, interflow and ground water flow contribute ultimately to streamflow, with some losses due to transpiration through plant life.

All the phenomena involved in the hydrological cycle are widely and well understood qualitatively, and several empirical relationships have been developed. Integrating them into comprehensive parametric models has been achieved by several investigators. Of these, the Stanford Watershed Model [5] is probably the best known. In order to apply any model to a given catchment, it is necessary that it be calibrated. That is, several years of streamflow are simulated using actual precipitation data, synthesized flows are then compared to actual recorded flows, model parameters (as well as seasonal factors) are adjusted by hydrologists and the process repeated until acceptable simulation accuracy has been achieved. More recently, self-calibrating models [12] have been developed which use trial and error routines to optimize model and seasonal parameters, but the calibration still requires historical streamflow data.

Current methods of streamflow forecasting require installation and maintenance of a system of instruments so that data may be collected for several years. The basin of a major river must be divided into hundreds of watersheds of manageable size and each one provided its own stream gage. Although the installation and maintenance expenses may not objectionable, waiting for several years of data collection to initiate a a new development may be intolerable. There is real need for a method of predicting the hydrological behavior of a previously ungaged catchment, quickly (within two months, say, rather than three to five years) and with resonable accuracy.

As previously suggested by others [e.g. 16], it would be advantageous to be able to establish simulation model parameters, either directly or inferentially, from observations of basin morphology and climate, taking advantage of all available prior knowledge of the area. Earth observation spacecraft make observations using a variety of sensors, covering large areas in a short time. The objective of this project is to assess the extent to which the data provided by Earth observation missions can be used to reduce the time, effort, and cost of predicting the hydrological behavior of a given catchment.

The study aims to devise a set of universally applicable prediction model parameters in terms of photographically observable characteristics (such as areas, elevations, and land use) and inferable characteristics (such as soil depth and porosity). These parameters and their related observables are quantified and optimized (by methods which will be explained later in this report) for 25 watersheds selected from all six physiographic provinces of the Tennessee Valley. This set of optimized paramater values is the one which yields simulated streamflows which are highly correlated with actual recorded flows for periods of six to ten years. Sensitivity of simulation accuracy to variations in each parameter is determined in selected basins. Optimized model parameters are next correlated with observable physiographic characteristics and the appropriate 'translation' tools (graphs, tables, nomograms) developed. The relationships thus determined are then tested on a set of ten watersheds. Required spatial and spectral resolution of images, appropriate sensors, and data acquisition methods are also determined.

3. Major Project Elements

The major elements of the project are (1) a watershed simulation model, (2) a calibration model, (3) selected watershed for use as test sites, (4) a historical data base, and (5) a physiographic data base.

The Stanford Watershed Model [4, 5] is probably the best known of the parametric hydrological models and, in all its modifications, is probably the most widely used. Since it was originally published in 1962, several reports have appeared in literature describing modified versions and applications [1, 6, 7, 8, 10, 13, and others]. The model uses a moisture accounting system to synthesize a continuous hydrograph from (1) recorded climatological data, precipitation, evaporation and (for snowmelt situations) temperature; (2) measurable watershed characteristics such as drainage area and fraction of the watershed in impervious surfaces; and (3) parameters used in the computational process which are known to vary in magnitude among watersheds but have not been quantitatively tied to specific measurable watershed properties. For example, one parameter indexes the capacity of the soil of the watershed as a whole to retain water.

The model accounts for all moisture entering the watershed until it leaves by stream-flow, evapotranspiration, or subsurface outflow. A series of relations, each based on empirical observation or theoretical description of a specific hydrologic process, is used to estimate rates and volumes of moisture movement from one storage category to another, in accordance with current storage states and the calibrated watershed parameters. The model routes channel inflow from the point where it enters a tributary channel to the downstream point for which a hydrograph is required. A subroutine exists for including snow in the accounting but is not needed in the present study.

In a recent research program a version of the Stanford Watershed Model IV, in FORTRAN IV, was developed for use with a self-calibrating streamlined version of the model. These models are referred to as the Kentucky Watershed Model (KWM) and OPSET (because it generates the OPtimum SET of parameters). The availability and utility of these models and reports describing them [8, 12, 15] led to their use in this project.

Watersheds chosen for calibration and simulation in the study are those of the Tennessee Valley Region in the southeastern United States. Its proximity to the state of Kentucky makes it specially attractive as a region in which to apply OPSET and KWM. Additionally, the Tennessee Valley is extensively instrumented, and copious historical climatological and streamflow data are available. It contains approximately 560 rainfall stations, both hourly and daily, and almost 1000 streamflow records have been accumulated in the past several decades. Other climatological data such as temperature, evaporation, wind, and humidity have been carefully collected and catalogued. Complete topographic map coverage is available for the area, consisting of 7.5 min quadrangle maps derived from aerial photographs at a scale of 1:24000. The Tennessee Valley has also been completely surveyed by aerial photography, and black and white photographs are available for every watershed, mainly at a scale of

1:24000. Some photographic coverage is also recently available in pseudocolor infrared taken from high altitude aircraft, at a scale of 1:63000. Within the Tennessee Valley 55 watersheds have been identified as suitable to the purposes of the project, representing all six of its physiographic provinces. It is estimated that 35 watersheds will be needed for confidence in the results of the correlation process. Of these, 25 can be used for calibration and 10 for tests of simulation accuracy.

The historical data base for the operation of OPSET and KWM consist of digitized precipitation and evaporation data. Additional inputs required by OPSET are daily discharge and selected flood hydrograph data for the years for which the model parameters are to be calibrated. Hourly and daily precipitation data are available in the form of precipitation tables and precipitation data tapes, the latter being more common since 1960. Precipitation data from printed tables is digitized and stored in puched card format from which it may be read, along with data from precipitation tapes, for use in a program for merging and synchronizing data from both hourly and daily gages, in order to provide the most accurate record of hourly precipitation data available, stored in a precipitation tape, which is subsequently transferred to the master watershed data base. Evaporation records from several pan evaporation stations in the region are digitized, with pan evaporation factors and seasonal adjustment constants, and incorporated into the data base.

The sources of physiographic data are topographical maps, photographs, and soil charts and catalogs pertaining to the area under study. Other data, such as geological maps and survey reports are also used where available, but derivation of watershed and model parameters through interpretation and analysis of photographs is preferable as a developmental step toward the application of data sensed more remotely. In a research study at the University of Kentucky, Ross [15] investigated the correlation of model parameter values with measurable physical characteristics of the watersheds used in that particular study. The procedures that he suggests for determining OPSET input parameters, and the correlations that he observed between output parameters and watershed characteristics, form the basis of this study methodology.

The model parameters which are fixed inputs for both OPSET and KWM are readily derived from observable basin characteristics. Six self-explanatory examples appear in Table I. The 13 model parameters normally estimated by OPSET for operation of KWM are the most difficult to measure, directly or indirectly. This is particularly true of those parameters whose quantification depends upon knowledge of subsurface conditions, particularly soil type distribution and depth. In his research in Kentucky, Ross found several useful correlations between parameters and watershed characteristics, some indeterminate correlations, and some which are apparently effective in too few of the observed watersheds to be correlated. The present study will extend the analysis to the selected basins of the Tennessee Valley.

Determination of the feasibility of applying remotely sensed data to streamflow forecasting depends upon completion of a series of error and sensitivity analyses concerning (1) the effects on streamflow simulation accuracy of deviations in each

TABLE I

Quantification of model parameters from remotely sensed observables (examples)

PARAMETER NAME	CODE NAME	UNIT	DEFINITION	QUANTIFICATION
DRAINAGE AREA	AREA	SQ MI	TOTAL AREA WITHIN THE WATERSHED BOUNDARY	PLANIMETER MEASUREMENT FROM TOPOGRAPHIC MAPS AND/OR PHOTOGRAPHS IMAGERY MUST BE SUCH AS TO PERMIT IDENTIFI-CATION OF RIDGE LINES
IMPERVIOUS SURFACE FRACTION	FIMP	--	FRACTION OF TOTAL WATERSHED AREA COVERED BY PAVING, ROOFTOPS, ETC., WHOSE RUNOFF CONTRIBUTES DIRECTLY TO A STREAM	ESTIMATED FROM REMOTELY DERIVED IMAGES (AERIAL PHOTOS IF AVAILABLE) USUALLY ZERO FOR RURAL AREAS EXCEPT WHERE THERE ARE LARGE AREAS OF EXPOSED ROCK
WATER SURFACE FRACTION	FWTR	--	FRACTION OF TOTAL WATERSHED AREA COVERED BY WATER SURFACES AT NORMAL LOW FLOW	ESTIMATED FROM REMOTELY DERIVED IMAGES (AERIAL PHOTOS IF AVAILABLE) VIRTUALLY ZERO FOR WATERSHED CONTAINING NEITHER LAKES NOR SWAMPS
VEGETATIVE INTERCEPTION MAXIMUM RATE	VINTMR	IN /HR	MAXIMUM RATE OF RAINFALL INTERCEPTION BY WATERSHED VEGETATION	ESTIMATED FROM LAND COVER, DERIVED FROM IMAGE INTER-CEPTION VALUE ASSIGNED IS 0 10 FOR GRASSLAND, 0 15 FOR MODERATE FOREST, 0 20 FOR HEAVY FOREST, OR SOME WEIGHTED AVERAGE IN A WATERSHED HAVING MORE THAN ONE TYPE OF VEGETATIVE COVER SIMULATED FLOWS ARE NOT VERY SENSITIVE TO THIS PARAMETER
OVERLAND FLOW SURFACE SLOPE	OFSS	FT /FT	AVERAGE SLOPE OF OVERLAND RUNOFF PATHS	SELECT AT RANDOM SEVERAL POINTS WITHIN THE WATERSHED MEASURE THE SLOPE OF EACH IN A DIRECTION PERPENDICULAR TO THE RECEIVING CHANNEL AVERAGE THE SLOPES
OVERLAND FLOW MANNING'S "N"	OFMN OFMNIS	--	ROUGHNESS COEFFICIENT FOR OVERLAND RUNOFF, PERVIOUS AND IMPERVIOUS SURFACES	DETERMINE LAND USE/COVER, ESTIMATE ROUGHNESS COEFFICIENTS FROM PUBLISHED TABLES [CHOW, REF 3]

model parameter, (2) the relationships between the accuracy with which physiographic characteristics are determined and the accuracy with which model parameters are quantified and (3) accuracy (resolution) of the sensor used to observe the basin. Those analyses have recently been initiated.

4. Interim Results and Conclusions

In the presently active part of the project, the preponderant share of the effort has been given to construction of the historical and physiographic data bases. Two watersheds of the Tennessee Valley, designated 'White Hollow' and 'Little Chestuee', have been calibrated and simulated. Table II shows the results of the continuous simula-

TABLE II

KWM continuous simulation accuracy

WHITE HOLLOW			LITTLE CHESTUEE		
WATER YEAR	CORRELATION (%)		WATER YEAR	CORRELATION (%)	
	MONTHLY	DAILY		MONTHLY	DAILY
1957	97 0	89 9	1950	98 5	91 6
1958	96 7	91 7	1951	98 5	95 5
1959	92 3	84 7	1952	97 9	76 6
1960	92 0	78 5	1953	96 5	82 1
1961	96 5	90 9	1954	98 5	95 4
1962	97 7	92 1	1955	99 2	88 8
1963	99 2	95 8	1956	98 8	95 8
1964	96 1	79 3			

tions; fidelity of simulated to observed flow is indicated by monthly and daily statistical correlation coefficients. The results are acceptable for every year, unless one is interested in accurate simulation of peak flows, with respect to both peak flow rate and time of occurrence of peak flow rate. Examples of simulated and observed flows for one month for each basin are shown in Figures 2 and 3. Work done to date indicates that (1) the models as they were used in Kentucky and modified by IBM are applicable to the Tennessee Valley after calibration and (2) recorded precipitation data adequately represent the actual rainfall over the basins for the purposes of this study.

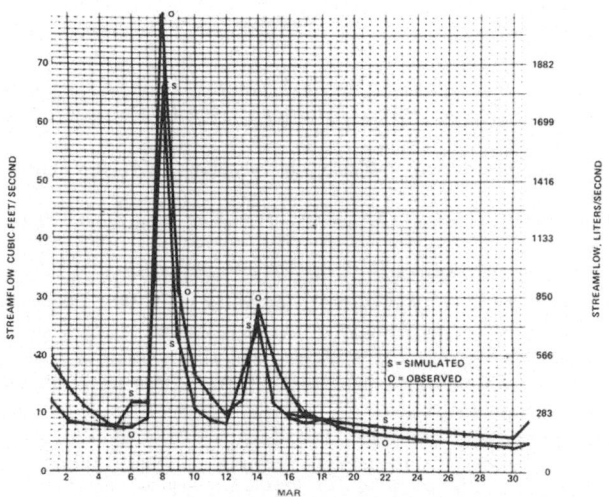

Fig. 2.   One month simulated and observed flow, White Hollow, March 1961.

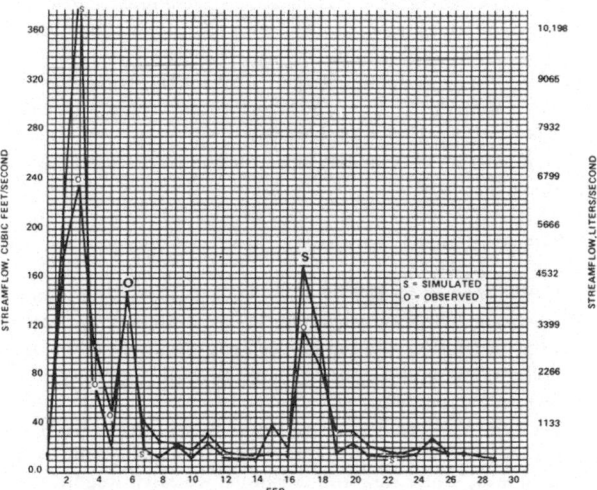

Fig. 3.   One month simulated and observed flow, Little Chestuee, February 1956.

Using the correlations between model parameters and watershed characteristics reported by Ross for several Kentucky watersheds, a complete set of Model Parameters was estimated for the White Hollow watershed and used in the KWM model to simulate eight years of streamflow. Except for two years, the daily and monthly correlation coefficients ranges from 70.7 to 97.6%. This trial treatment of a known basin as if it were ungaged is not a conclusive test of application to remote, ungaged areas, but the result is definitely encouraging. It indicates that for many purposes it is feasible to predict the hydrological performance of an ungaged basin with nothing more than space or aerial imagery and sparse ground samples.

References

[1] Anderson, E. A. and Crawford, N. H., *The Synthesis of Continuous Snowmelt Hydrographs on a Digital Computer*. Stanford, California: Stanford University, Department of Civil Engineering, Technical Report No. 36, 1964.
[2] Castruccio, P. A., 'Use of Remote Sensing in Hydrology', XXII International Astronautics Congress, Brussels, 1971.
[3] Chow, Ven Te (ed.), *Handbook of Applied Hydrology*. New York: McGraw-Hill Book Company, 1964.
[4] Crawford, Norman H. and Linsley, R. K., *The Synthesis of Continuous Streamflow Hydrographs on a Digital Computer*. Stanford, California: Stanford University, Department of Civil Engineering, Technical Report No. 12, July 1962.
[5] Crawford, Norman H. and Linsley, R. K., *Digital Simulation in Hydrology: Stanford Watershed Model IV*. Stanford California: Stanford University, Department of Civil Engineering, Technical Report No. 39, July 1966.
[6] *Hydrocomp International*, 'HSP Operations Manual', Palo Alto California, 1969.
[7] *Hydrocomp International*, 'Simulation of Continuous Discharge and Stage Hydrographs in the North Branch of the Chicago River', Palo Alto, California, 1970.
[8] James, L. Douglas, *Economic Analysis of Alternative Flood Control Measures*. Lexington: University of Kentucky, Water Resources Institute, Research Report No. 16, 1968.
[9] James, L. Douglas, *An Evaluation of Relationships Between Streamflow Patterns and Watershed Characteristics through the Use of OPSET: A Self-Calibrating Version of the Stanford Watershed Model*. Lexington: University of Kentucky, Water Resources Institute, Research Report No. 36, 1970.
[10] Ligon, James T., Law, Albert G. and Higgins, Donald H., *Evaluation and Application of a Digital Hydrologic Simulation Model*. Clemson, South Carolina: Clemson University, Water Resources Research Institute, Report No. 12, November 1969.
[11] Linsley, Ray K., Kohler, Max A. and Paulhus, Joseph L. H., *Hydrology for Engineers*. New York: McGraw-Hill Book Co., 1958.
[12] Liou, Earnest Y., *OPSET: Program for Computerized Selection of Watershed Parameter Values for the Stanford Watershed Model*. Lexington: University of Kentucky, Water Resources Institute, Research Report No. 34, 1970.
[13] Lumb, Alan Mark., *Hydrologic Effects of Rainfall Augmentation*. Stanford, California: Stanford University, Department of Civil Engineering, Technical Report No. 116, November 1969.
[14] Root, R. R. and Miller, L. D., *Identification of Urban Watershed Units Using Remote Multispectral Sensing*. Fort Collins, Colorado: Colorado State University Environmental Resources Center, Completion Report Series No. 29, June 1971.
[15] Ross, Glendon A., *The Stanford Watershed Model: The Correlation of Parameter Values Selected by a Computerized Procedure with Measurable Physical Characteristics of the Watershed*. Lexington: University of Kentucky, Water Resources Institute, Research Report No. 35, 1970.
[16] Thomas, D. M. and Benson, M. A., 'Centralization of Streamflow Characteristics from Drainage Basin Characteristics', Open File Report, U.S. Department of Interior, Geological Survey, 1969.
[17] *U.S. Dept. of Agriculture*, 'Aerial-Photo Interpretation in Classifying and Mapping Soils', *Soil Conservation Service Agriculture Handbook 294*, October 1966.

RESULTS AND EVALUATION OF THE EOLE
OPERATIONAL LOCATION SYSTEM

GÉRARD BRACHET

Département Calcul d'Orbites, Division Mathématiques et Traitement, CNES, France

Abstract. The EOLE location and data collection system has been operational since early September 1971.
It is essentially a satellite which tracks and collects data from a fleet of meteorological balloons and ground transponders.

During each satellite interrogation of a transponder, range and range rate are measured and stored in the satellite memory. These are later transmitted to a ground telemetry station and processed in the CNES Brétigny computing center.

The data processing system for transponder location computation is described. The main difficulties which have been encountered are emphasized particularly for the location of fast moving transponders such as meteorological balloons.

Preliminary results of a detailed evaluation of the system capability are given, with particular emphasis on the location accuracy.

The contribution of the main sources of errors are analyzed: orbit computation, range and range rate accuracy, epoch determination.

Future developments of EOLE type location system are briefly described.

1. Principle, Objectives and Achievements of the EOLE Location System

The EOLE system was designed originally as a data collection and location system oriented towards the tracking of a large number of meteorological balloons flying at constant pressure level (200 mb) in the southern hemisphere. The objective was a better understanding of the atmospheric flow pattern and therefore did not require very accurate location capability (Morel, 1972).

The system is outlined in Figure 1 (Muller, 1972):

(1) The transponders interrogation program is established on the ground and fed into a special memory in the satellite.

(2) According to this program, the satellite interrogates transponders sequentially at given times on 460 MHz.

(3) If a transponder recognises its code number, it answers back by transmitting data from its own sensors on 400 MHz.

In this process, range and two way range rate are measured by the satellite and stored in its 131 kbits core memory together with the data transmitted by the transponder (Benac, 1972).

The same process is repeated on other transponders with a time cycle of 625 ms. During one pass of the satellite above a transponder, 3 to 30 interrogations are made, providing enough information for the transponder's position to be computed either in a pure geometrical mode or after a least squares statistical data processing.

(4) On command from a ground telemetry station, the on-board memory is transmitted to Earth and recorded.

(5) The information is then transmitted to the CNES Brétigny computing center

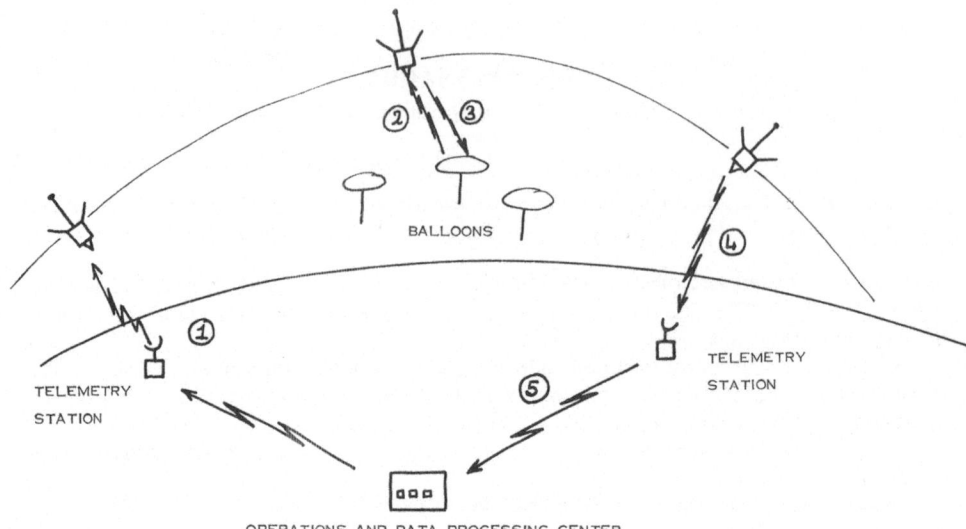

Fig. 1.   EOLE system description. 1 – interrogation program is transmitted to satellite. 2 – balloon interrogation. 3 – balloon answer. 4 – in board data are transmitted down to telemetry station. 5 – data are transmitted to Brétigny and processed.

where it is processed. (Dargent, 1972). Results are generally available within 24 to
36 h, but 7 h is achieved for some special experiments.

The EOLE satellite is a 80 kg gravity gradient stabilized spacecraft with its 400
MHz cone-shaped antenna pointing towards the ground (Figure 2), launched on
August 16 1971. Its orbit is 690 km perigee, 900 km apogee, with a 50° inclination and
a 1 h 40 min revolution period.

A total of 479 meteorological balloons (Figure 3) were launched from 3 special
launching stations in Argentina at 33°, 39° and 53° latitude south, during the period
Aug.–Nov. 1971.

The average life time was 150 days, giving a satisfactory coverage of the southern
hemisphere.

Other applications of the EOLE data collection and location system were soon
started; they included:

– Tracking of transponder-carrying floating buoys in northern Pacific and in
southern Indian Ocean.

– Data collection from transponders aboard merchant navy ships, including the
SS FRANCE.

– River flow control and collection of ground meteorological data from remotely
located ground transponders in Latin America and India.

– Tracking of sailboats. Figure 4 shows the route followed by Pen Duick 4, during
its trip to Europe in August 1972, as plotted from the EOLE location results.

– One of the most spectacular result of the EOLE ground transponder tracking
experiment is shown on Figure 5, where the route of an Iceberg is plotted from

Fig. 2.   The EOLE satellite.

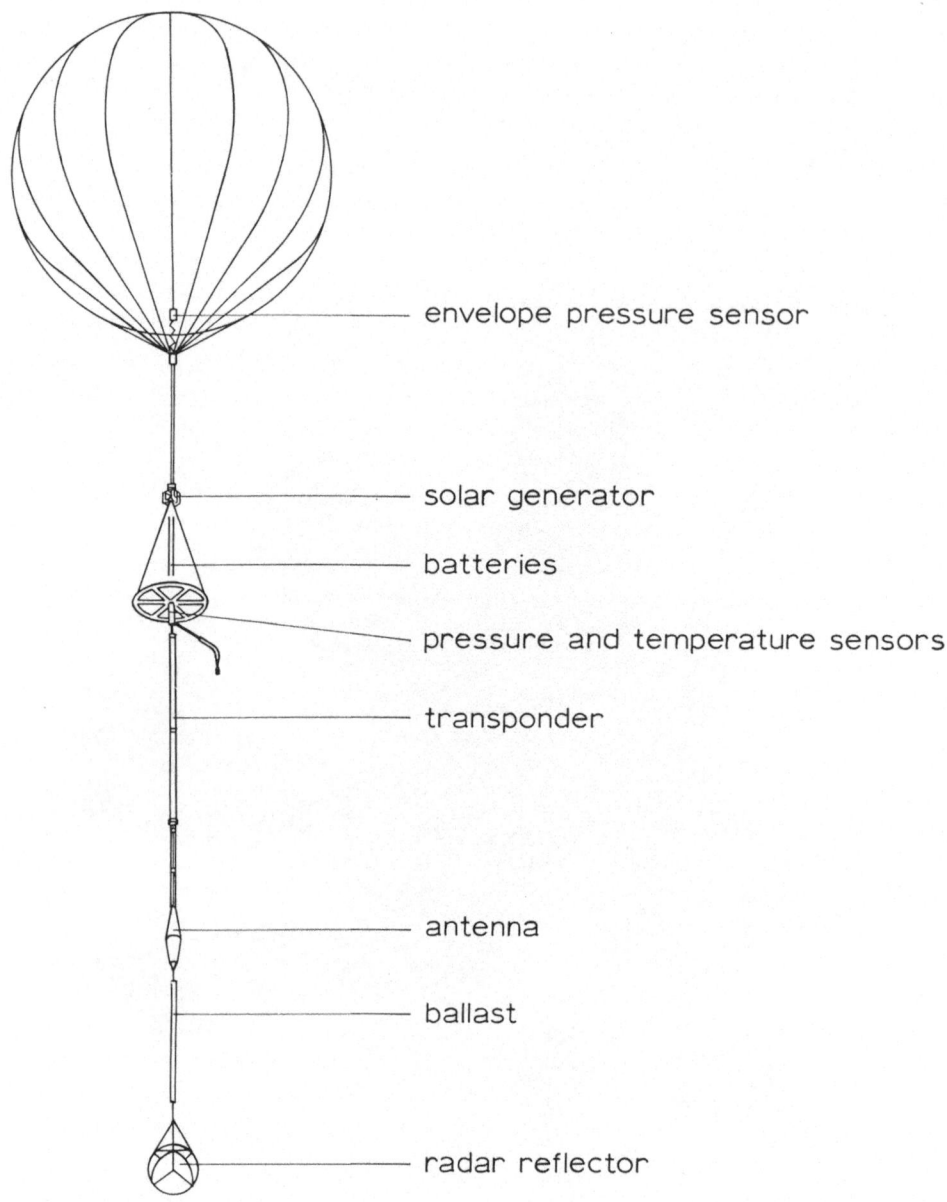

envelope pressure sensor

solar generator

batteries

pressure and temperature sensors

transponder

antenna

ballast

radar reflector

Fig. 3.   EOLE balloon.

February 1972 up to September 1972 following the coast of the Antartic continent by 65° of latitude south.

In September 1972, a few balloons are still being located, and other experiments are in process, such as tracking of floating buoys in the Gulf Stream (NOAA) and in northern atlantic (Fisheries Laboratory of Lowestoft, U.K.).

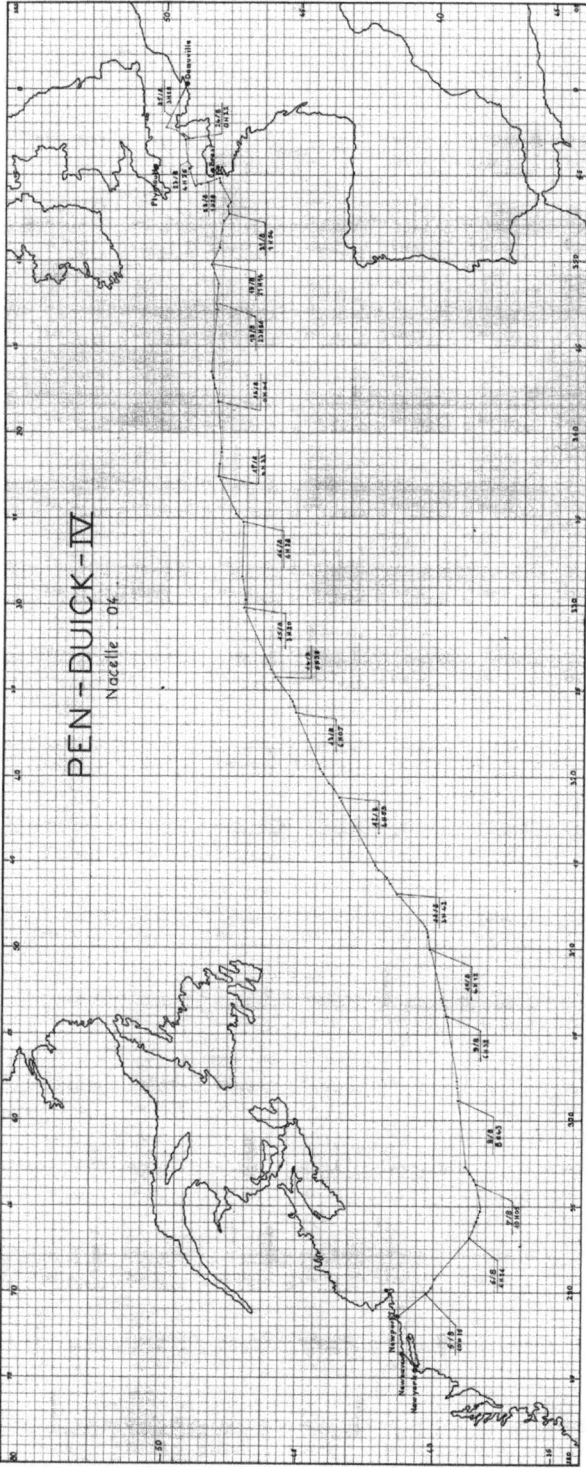

Fig. 4. Transatlantic route followed by Pen Duick 4, August 1972.

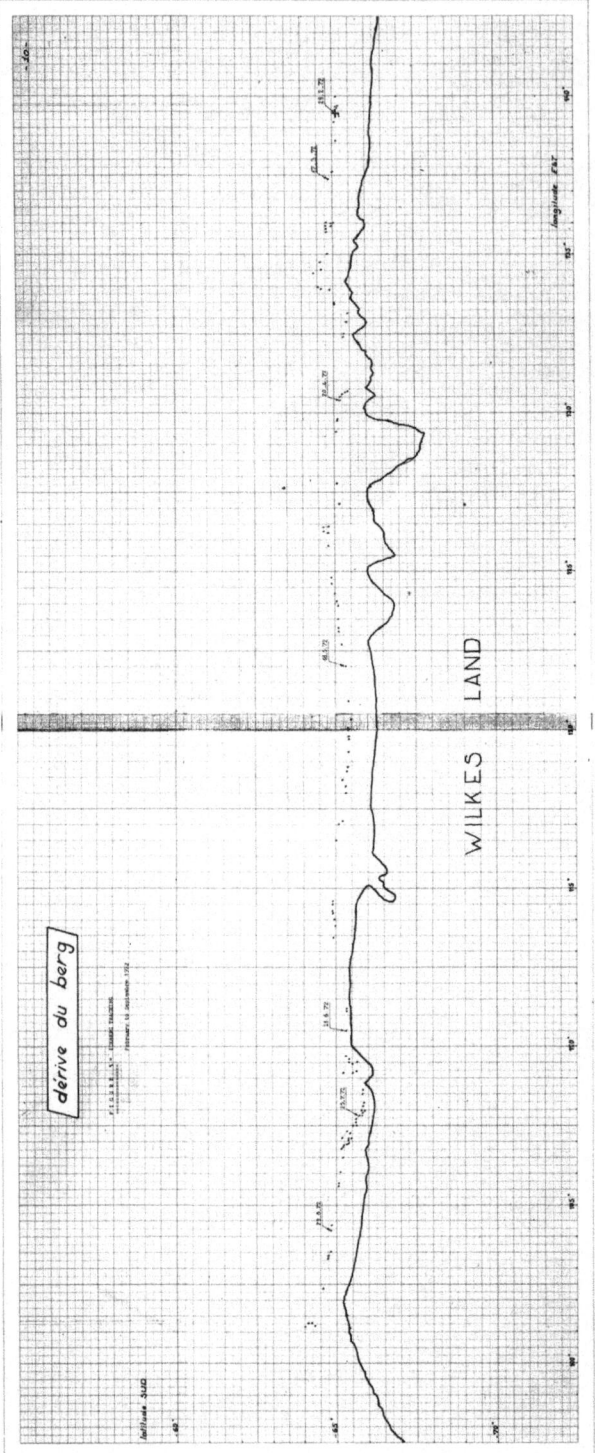

Fig. 5.   Iceberg tracking, Febrary to September, 1972.

The EOLE system has thus proved to be a complete success and has demonstrated its ability to provide support for a wide range of applications, such as oceanography, glaciology, meteorology, navigation, and other environment sciences.

2. Location Accuracy Analysis

Each range and range-rate measurement from the satellite to the transponder leads to a position if the altitude of the transponder is assumed to be known with an accuracy comparable to the range accuracy.

This position is found geometrically as the intersection of the surface of constant altitude above the ellipsoïd with the range sphere centered at the satellite position and the Doppler cone, whose axis is along the satellite velocity and the half angle θ given by $\cos \theta =$ range rate/satellite velocity (Figure 6).

However, this geometric solution cannot give the transponder velocity and its accuracy is of course very dependent on the random errors in the individual measurements.

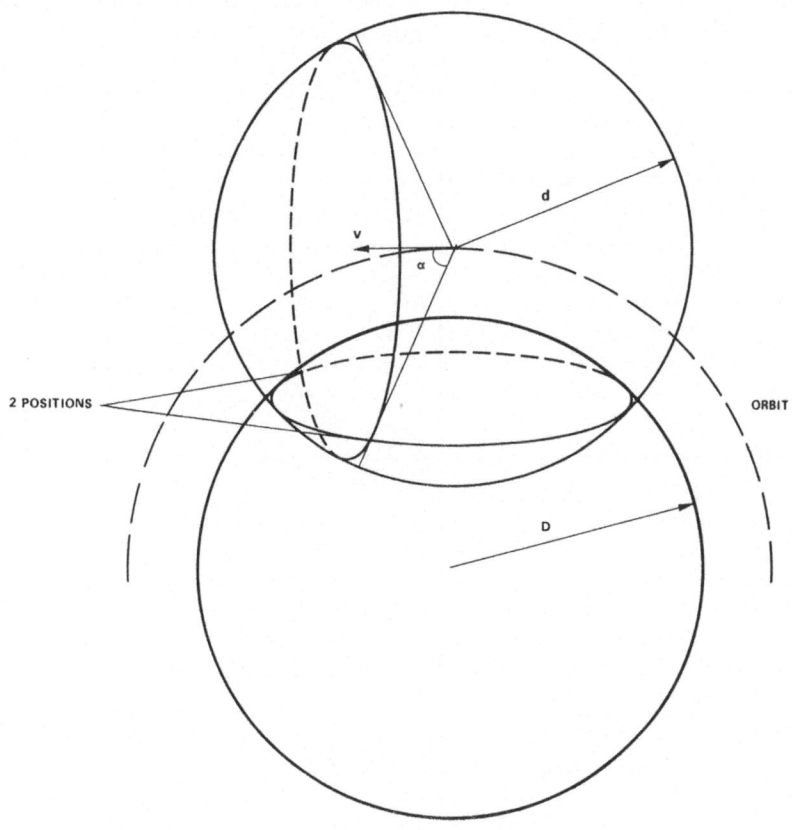

Fig. 6.   Location system geometry.

This is why a least square adjustment of the transponder position is preferred when the number and distribution of measurements during one pass allows it.

Figures 7 and 8 show the distribution of the computed positions of fixed ground transponders located in the CNES Pretoria telemetry station (South Africa) and in the Marine Science Laboratory in Victoria (British Columbia, Canada). These two examples have been chosen because of the very different geometry of the satellite passes above the station, the first being at $-25.5°$ latitude, and the other at $+48.5°$. The rms position accuracies are respectively 1.3 and 2.5 km

2.1. ORBIT ACCURACY

The orbit was computed every day using tracking data collected during the previous three days by the two CNES interferometric tracking stations in Kourou and Pretoria, using Brouwer's analytical first order model (Brouwer, 1969), where short periodic perturbations in J_2, long periodic perturbations in J_2, J_3, J_4 and J_5, and secular terms due to J_2, J_2^2 and J_4 are taken into account. Figure 9 shows the along-track and the in-plane and out-of-plane cross track differences with a more accurate orbit computed from the same tracking data but using a much more sophisticated force model. The along-track errors of the simple model sometimes reach 1.7 km, while the cross-track components are always better than 1.2 km.

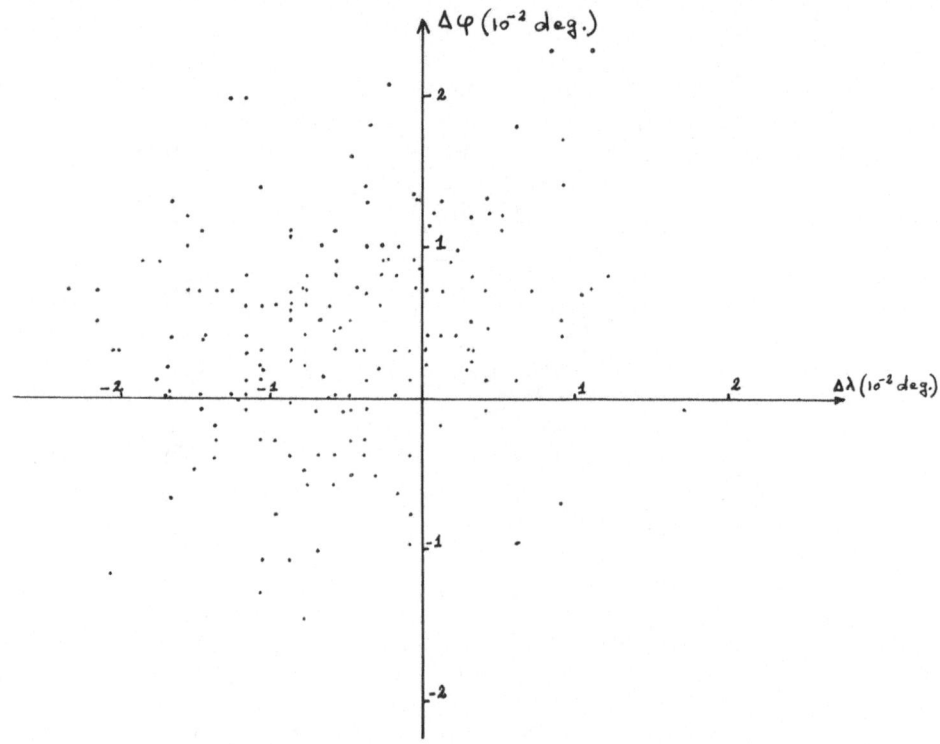

Fig. 7.   Pretoria ground transponders EOLE locations.

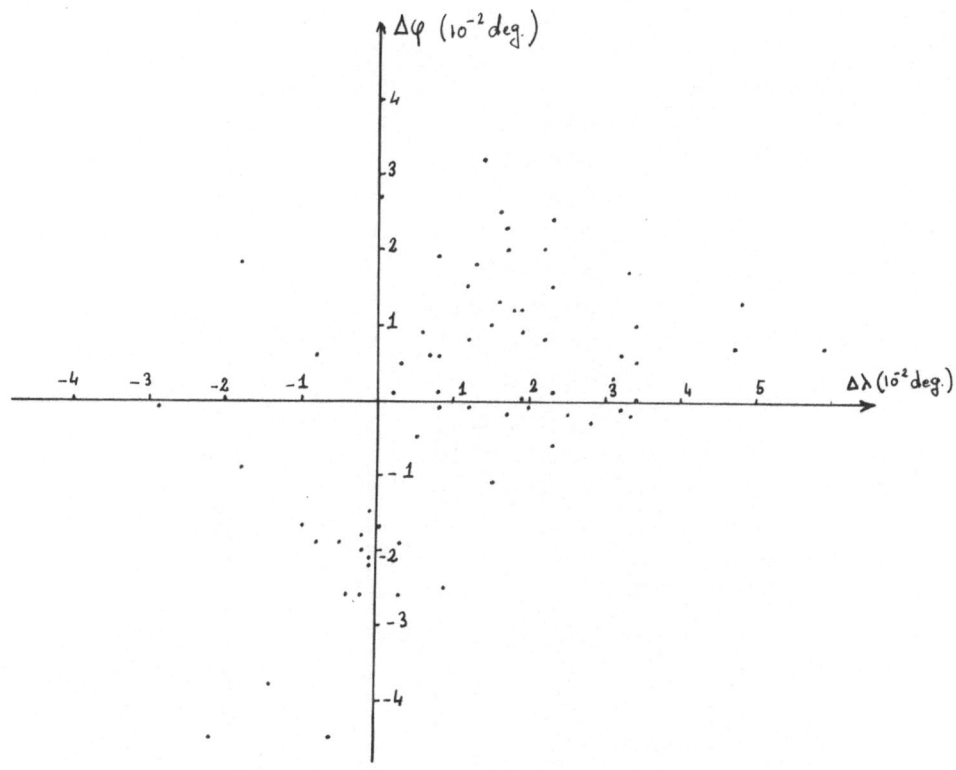

Fig. 8.   Victoria buoy No. 20 EOLE locations.

2.2. Transponder tracking accuracy

Using the satellite position as a reference, the location accuracy of fixed or slowly moving transponder is a function of
- the range measurement accuracy,
- the range-rate measurement accuracy,
- the timing accuracy of the measurements,
- the transponder altitude accuracy.

The range rate is measured via a two way Doppler system on 460 MHz (down link) and 400 MHz (up link), integrated over 125 ms. The sources of error are therefore only the satellite oscillator frequency, and the propagation errors (ionospheric and tropospheric refraction).

Range is obtained by phase modulation of the carrier frequency on both links. Three modulating frequencies are used: 48, 2304 and 2688 Hz, corresponding to 3125, 65.1 and 55.8 km respectively. Here the source of errors are satellite phase measurement (a function of signal level) and transit time in its electronics, transponder transit time and propagation.

The effect of tracking data and altitude errors vary with the geometry of the pass above the transponder, which is generally represented by the 'ground track geocentric

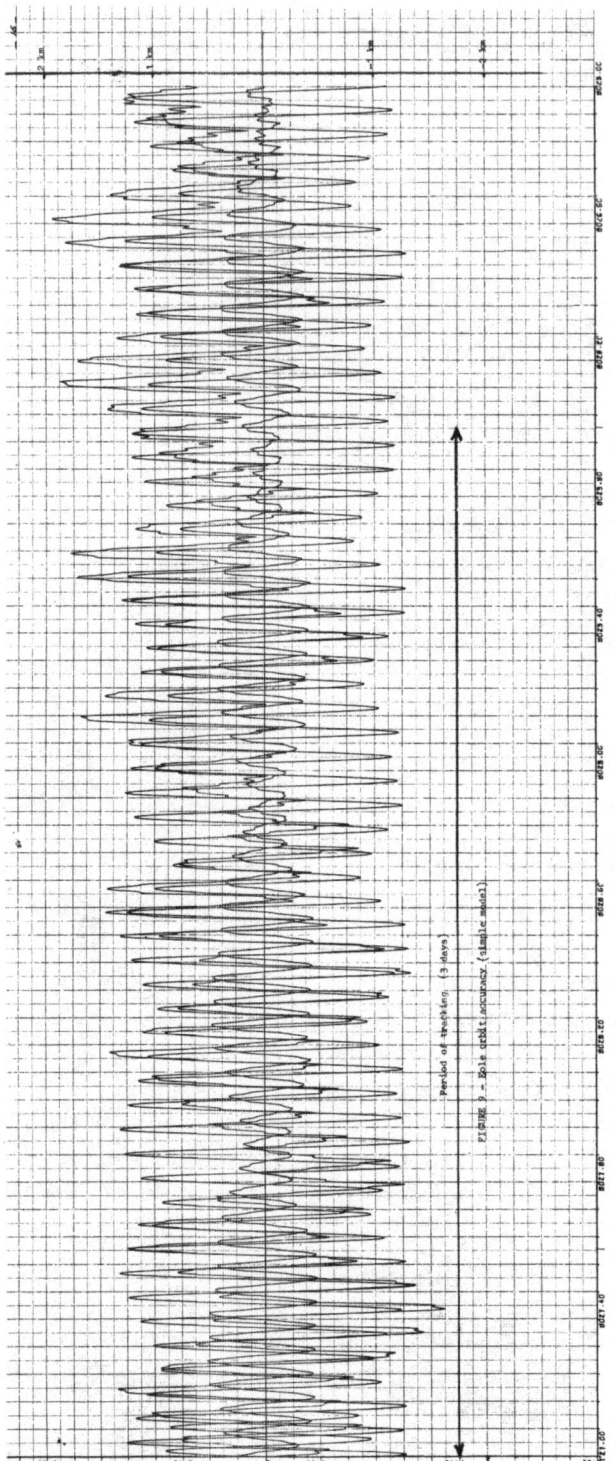

Fig. 9. EOLE orbit accuracy (simple model).

distance' (ψ), although another important parameter is the distribution pattern of measurements along the pass.

Figure 10 shows that the dispersion of locations from one satellite pass, for the 10 buoys in the Marine Science Laboratory of Victoria, B.C., plotted as a function of geocentric distance to the ground track. This dispersion originates *only* from differences in range accuracy from one buoy to the other, as this is the only parameter which is transponder dependent, all others being satellite dependent only.

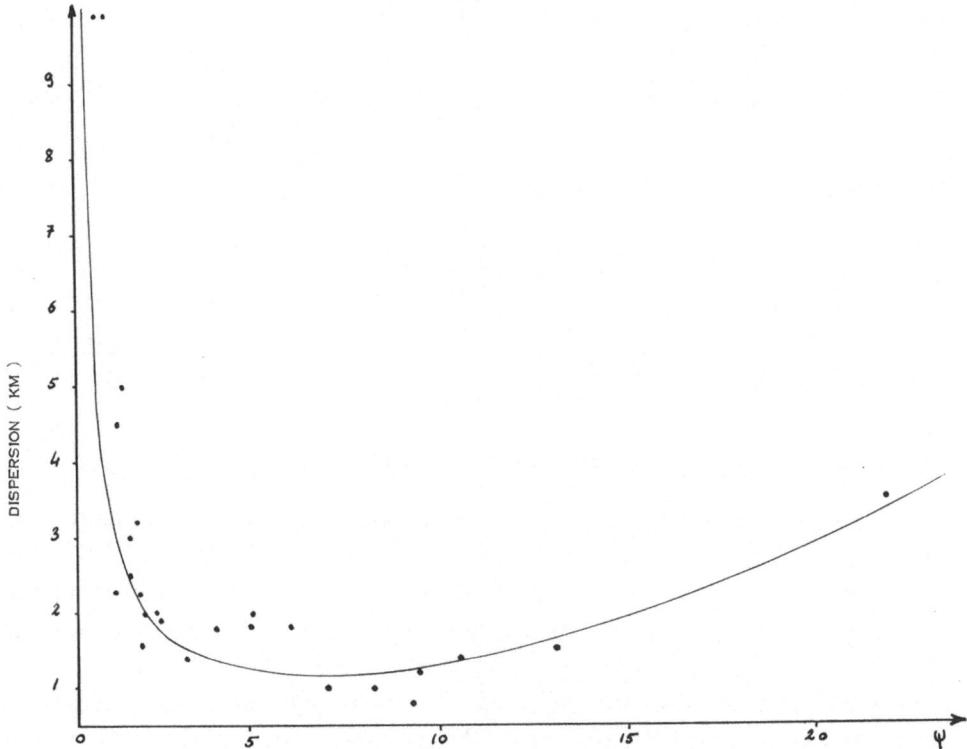

Fig. 10.   Victoria buoys location dispersion as a fonction of ground track distance ψ.

Further analysis showed that calibration differences in the transponder transit time can reach 800 m, so that errors of this order of magnitude can have effects 2 or 3 times larger on the location errors.

Figure 11 shows the dispersion of purely geometric solutions for the same buoy for two different passes. Again the effect of ground track distance is clearly visible.

2.3. LOCATION OF FAST MOVING OBJECTS

In the case of rapidly moving objects, a number of new problems arise, concerning both the data processing itself and the final location accuracy:

– No approximate position is available to allow an iterative adjustment of the position.

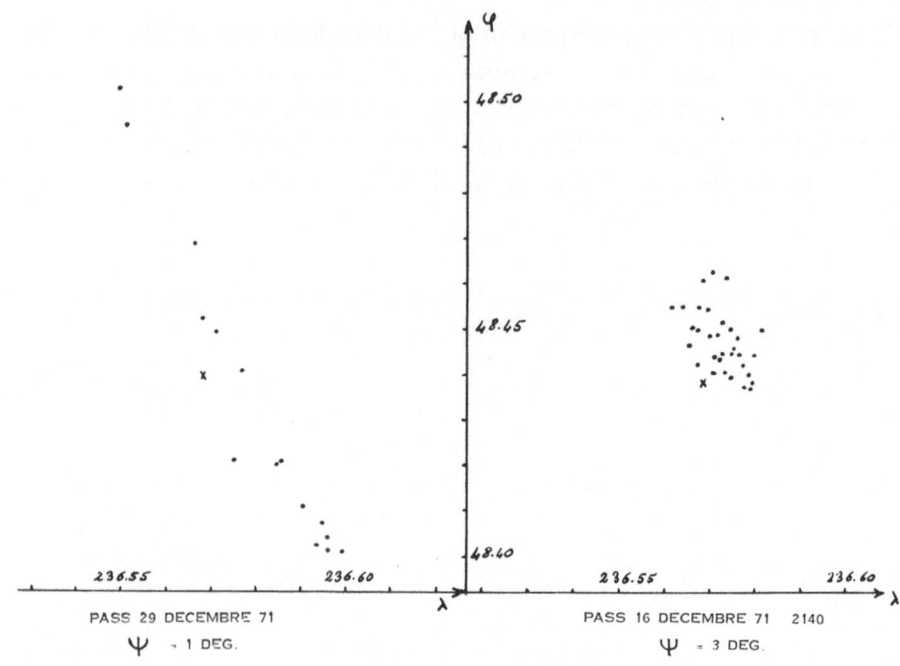

Fig. 11.   Geometric locations for buoy No. 27.

– The Doppler measurement is very significantly affected by the balloon velocity itself.

The first of these two problems could be solved by a purely geometric method, but the second introduces inaccuracies, forcing a choice among the three solutions given below:

(a) Geometric locations are computed from each range and range rate measurements with no wind. An average wind is then computed from the first and last positions. This wind is introduced in the geometric method and the process is iterated (Internal/wind).

(b) A theoretical wind model is introduced in the data processing system and the geometric locations are computed.

(c) An average wind computed from locations on two successive passes is introduced as in (b) above (Meteorological wind).

The choice is then made according to two criteria:

– continuity of the polygonal contour which should be almost a straight line,
– length of each segment which should be approximately proportional to the time elapsed from one measurement to the next.

Figure 12 shows the results of these three methods for three typical balloons. Clearly method (a) is never acceptable while the theoretical wind model (b) gives better results for two of the balloons, and method (c) is always the best, which is after all a good proof that our knowledge of wind flow pattern was not that good before the EOLE experiment!

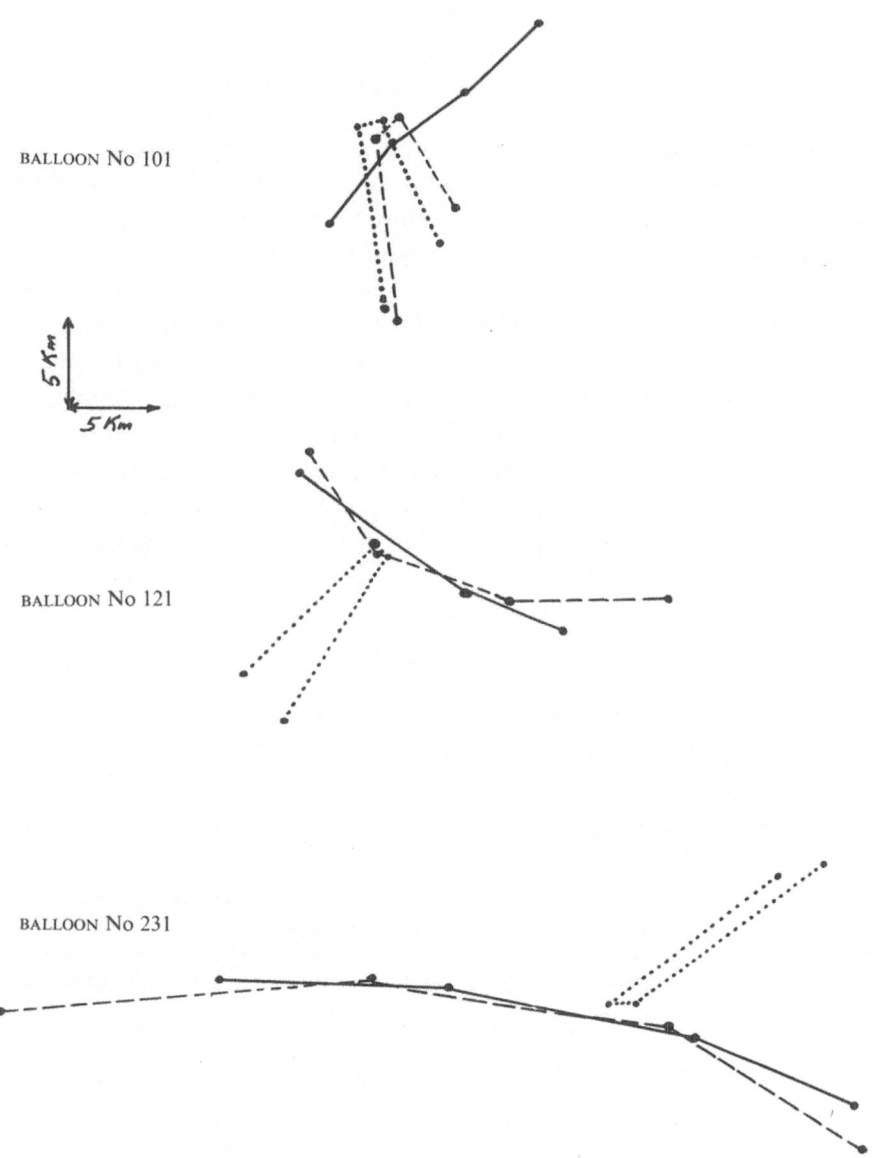

Fig. 12.   Geometric solutions for balloon positions. .... Internal wind; ---- wind model; —— 'actual' wind.

Approximate solutions having then been obtained, it is possible to use them as initial values for an iterative least square adjustment of the position and velocity.

However, since even a good geometric solution gives always two solutions symmetric with respect to the satellite's ground track, both are used in the iterative process and the final solution is the one giving the smallest residuals.

Finally, the balloon position being a fairly sensitive function of the wind model, a corrective matrix is provided, thus allowing improvements in the positions as wind models improve during the EOLE scientific analysis.

3. Conclusions – Future Developments

Many projects which are now in an advanced stage are based on the EOLE principle of design, including the platform tracking system in the TIROS N satellite, with a capability of locating 2000 balloons or ground platforms, using one way Doppler data collected during two successive passes of the satellite.

But it is, not surprisingly, geodetic and navigation sciences which may benefit most of EOLE type systems: The CNES GEOLE project (for Geodetic EOLE) is based on such an idea, with the goal of 10 m location accuracy for one pass, and 1 m in 24 h (Husson and Thieriet, 1970). Improvement would be obtained from higher frequency measurements (2 GHz), high altitude orbit and accurate trajecto-graphy through a network of fixed ground transponders as a reference. A test of this system will be performed in late 1975 on a more local scale with the DIALOGUE satellite, to be launched on a lower, 700 km orbit but carrying the GEOLE tracking subsystems (Thieriet, 1972).

References

Benac, P.: 1972, 'Experience EOLE, chaîne de localisation et de recueil des données', COSPAR XVth General Assembly, Madrid.

Brouwer, D.: 1969, 'The solution of the problem of artifical satellite theory without drag', *Astron. J.* **64**, 1274.

Dargent: 1972, 'Data processing for the EOLE project', International Telemetry Conference, Los Angeles.

Husson, J. C. and Thieriet, D.: 1970, 'A Geodetic Support System, the GEOLE Project', *Int. Hydrographic Rev.* **XLVII**, No. 2.

Morel, P.: 1972, 'Latest results in space meteorology: Experiments in Satellite Platform location an Data relay techniques', COSPAR XVth General Assembly, Madrid.

Muller, J.: 1972, 'Le déroulement de l'expérience EOLE', COSPAR XVth General Assembly, Madrid.

Thieriet, D.: 1972, 'The Dialogue Project', CNES Internal Report No. CNES/PR/AM-DA-72-T-109.

KEY TECHNOLOGICAL CHALLENGES OF THE
EARTH RESOURCES TECHNOLOGY SATELLITE PROGRAM

I. S. HAAS, B. T. BACHOFER, and G. P. FISHMAN

Earth Observatory Programs, Space Division, General Electric Company, Philadelphia, Pa. 19101, U.S.A.

1. Introduction

The first Earth Resources Technology Satellite was launched on July 23, 1972, and by this day in October has completed more than 1100 orbits of the Earth. The spacecraft is observing 200 scenes per day and transmitting data to three prime ground stations in the United States and one in Canada. This data is being processed at NASA's ground data processing facility for subsequent distribution on a worldwide basis. The job of making ERTS successful has required meeting and solving a number of key technological challenges. Because time does not permit discussing all of them, we have selected four major topics for today's meeting: orbit and coverage parameters, spectral characteristics, system performance and photographic interpretation and information extraction.

2. Orbit and Coverage

In order to maximize the usefulness of the multispectral data collected from ERTS, the conditions under which the data is collected must be systematically controlled to produce repeated coverage under nearly constant observation conditions. Figure 1 itemizes the various requirements and constraints which affect the selection of the

REQUIREMENT/CONSTRAINT	ORBITAL PARAMETER	NOMINAL/VALUE
• SPACECRAFT GROUND COMMUNICATION TIME • SENSOR RESOLUTION	ALTITUDE (SEMI-MAJOR AXIS)	917.0 KM (7294.7 KM)
• CONSTANT IMAGE SCALE	ECCENTRICITY	0
• REPETITIVE COVERAGE, SIDELAP AND COVERAGE CYCLE DURATION	NODAL PERIOD	103.3 MIN (18 DAY REPEAT CYCLE)
• INTRACK OVERLAP, AND PICTURE CENTER CONTROL	PREDICTED EPHEMERIDES	CONTROL WITHIN 2 SEC (~13 KM)
• ILLUMINATION OF SCENE • MINIMIZE CLOUDS/HAZE	NODE TIME	9:30–10:00 AM
• CONSTANT ILLUMINATION CONDITIONS	INCLINATION	99.1 DEG (SUN SYNC)

Fig. 1. Orbit and coverage considerations.

L. G. Napolitano et al. (eds.), Astronautical Research 1972, 273–280. All Rights Reserved
Copyright © 1973 by D. Reidel Publishing Company, Dordrecht-Holland

ERTS orbital parameters. The altitude of 917 km results primarily from a tradeoff between the amount of communication time between the spacecraft and the fixed ground stations, which increases as the altitude increases, vs the sensor resolution, which decreases as the altitude increases. In order to produce imagery at a constant scale, a fixed altitude, and therefore, a circular orbit is required. Thus, typically the orbital eccentricity should be near zero.

The nodal period of 103.3 min will produce complete Earth coverage in 18 days, or 251 orbits. So that the scenes collected in the next 18-day period will approximately repeat the scenes collected during the previous 18-day period, precise control of the nodal period is required.

In the along-track direction, picture center is controlled by precise execution of turn-on and turn-off commands to the payload devices. These turn-on and turn-off times are computed from predicted ephemerides, which are calculated from tracking data from the spacecraft. Picture center control in the along-track direction is maintained within a tolerance of about 13 km because of command execution granularity of up to 2 s.

A subsatellite swath of about 185 km wide is viewed by the sensors as the spacecraft moves over the surface of the Earth, as shown in Figure 2. On subsequent orbits, the orbital swath is shifted to the west about 2900 km, and 14 orbits are completed in one day. On the 15th orbit, the subsatellite swath covered by the sensor is shifted 159 km to the west of that swath traced during the first orbit of the previous day. In 18 days, or 251 orbital revolutions, coverage of the Earth is completed. During the next 18-day coverage cycle, the same orbits are traced out again; that is, the 252nd orbit retraces the first orbit. This procedure is repeated for each subsequent 18-day coverage cycle, for the life of the mission.

This repeated coverage requires precise control of the orbital period. Small

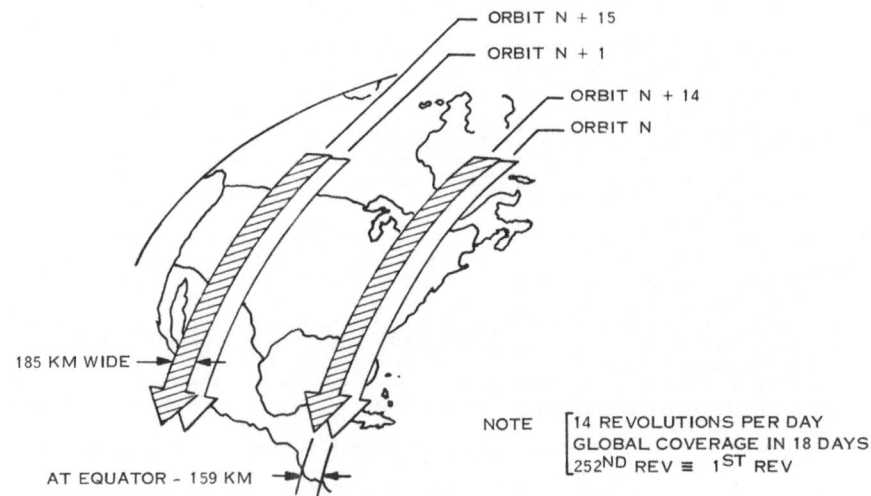

Fig. 2.   ERTS ground coverage patterns.

perturbations to the orbit require that an active orbit adjust system be used on the spacecraft to maintain orbital period control. Figure 3 shows the actual drift in the longitude of the equator crossing from the time that the initial orbital adjustments were made. It can be seen that at the end of about 3 coverage cycles, this drift was beginning to move rapidly away from the nominal position. On September 28, just prior to beginning the 4th cycle, an orbit adjustment was made in order to decrease the orbit period. The subsequent drift in the longitude of the orbit and its expected movement are shown by the dotted line.

3. Spectral Characteristics of Erts

On ERTS, the Multispectral Scanner, referred to as MSS, senses in four discrete spectral bands as shown in Figure 4. Only the first two of these bands are in the visible region of the human eye. The remaining two are in the near IR region of the spectrum.

From the spectral characteristics of water, we can see that it has a decreasing reflectance with increasing wavelength. In the black and white images produced for each of the spectral bands of the MSS, we can see that water becomes progressively darker as we move from Band 1 to Band 4. Vegetation, on the other hand, has an increasing reflectivity with wavelength, with a sharp rise in the near IR region of the spectrum. Correspondingly, we can see that vegetation, in general, will increase in brightness in the black and white images as we move from Band 1 to Band 4. Water and healthy green vegetation, shown as examples here, are only two of many target materials which will be sensed by ERTS. Not all materials, however, show such marked differentiation in their responses in the various spectral bands.

The Return Beam Vidicon camera, or RBV, is the second multispectral sensor employed on ERTS. As shown in Figure 5, it senses in three spectral bands, whose

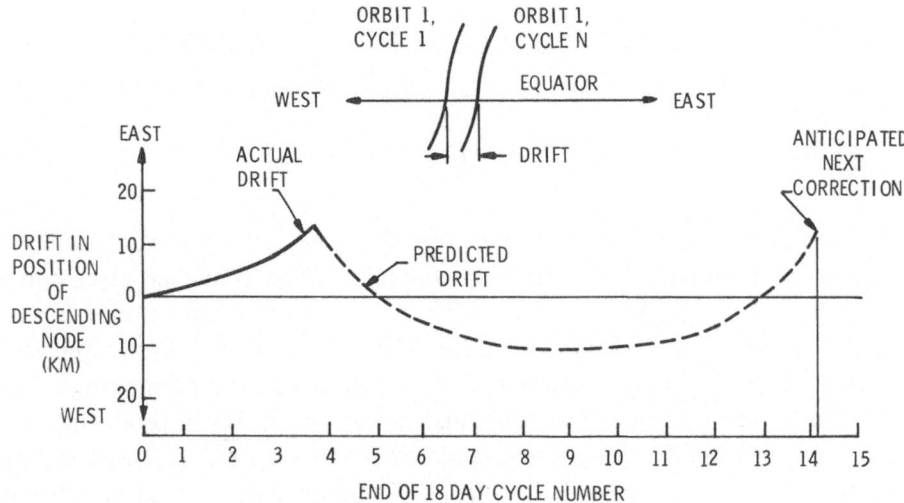

Fig. 3.   ERTS-1 orbital drift.

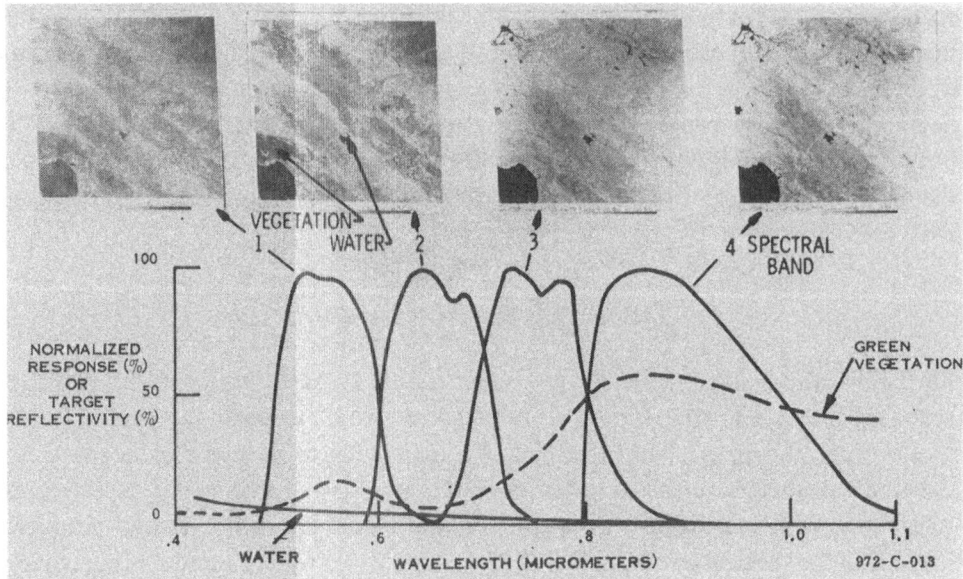

Fig. 4.   ERTS MSS spectral response.

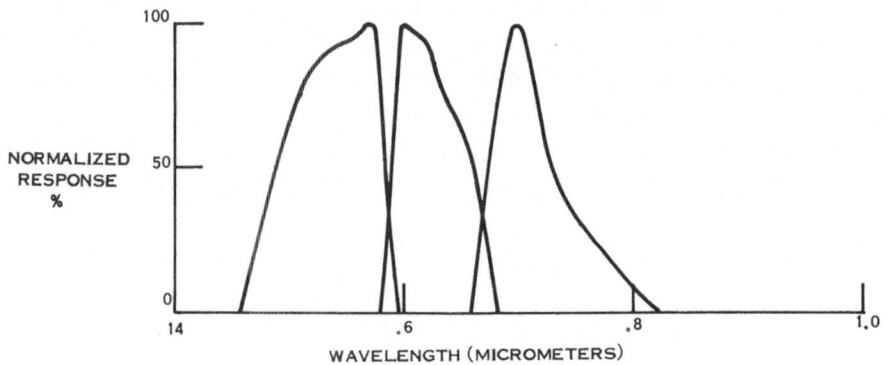

Fig. 5.   ERTS RBV spectral response.

spectral responses are similar to those sensed by the first three bands of the MSS.

In order to combine the information contained in the individual black and white images sensed in the separate spectral regions, false color representation is possible. Figure 6 shows typically how this is done. The black and white transparencies from three of the spectral bands of a given scene are exposed on color film using a different color filter when exposing each individual spectral band. When producing an RBV color image for ERTS, blue, green, and red filters are normally used in Spectral Bands 1, 2, and 3 respectively. For the MSS, these filters are used in either bands 1, 2, and 3, or 1, 2, and 4 respectively. The color produced in a positive film or print

"FALSE" COLOR REPRODUCTION

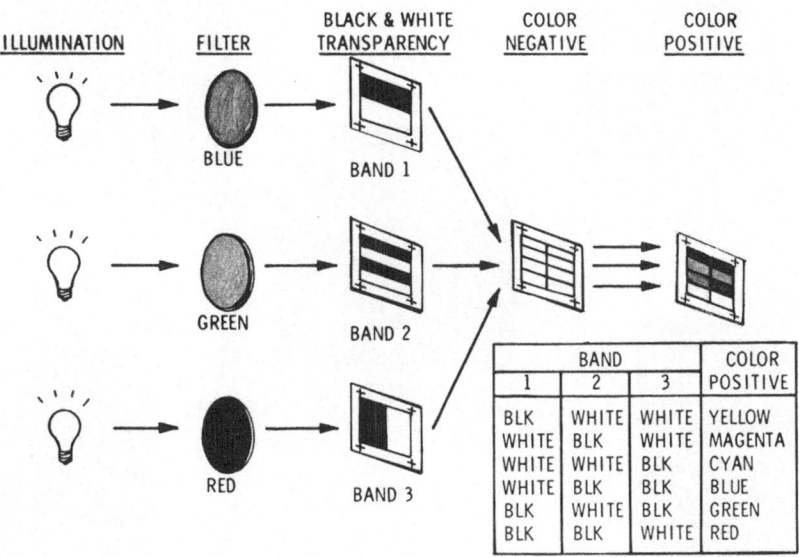

Fig. 6. 'False' color reproduction.

will depend on the relative intensity (grey level) in the various spectral bands. For example, if the black and white transparency is very dense or black in all three bands, the color in the color positive will be printed as black. If the density in the first band is very light, in the second band, very dark, and in the third band, very light, a magenta color will be reproduced in the color positive, and so on. In this manner, we can reproduce in the color positive a composite of much of the information contained in the separate black and white transparencies.

4. Performance of the ERTS System

The quality characteristics of the imagery and digital data tapes produced by ERTS will determine the limits of its utility. Several specific performance characteristics are very important in determining the utility of this data for many applications.

For example, the measure of resolution used to assess ERTS imagery is the response to a specific tri-bar pattern of vertical and horizontal bars, whose length-to-width ratio is 5 to 1. We define resolution to mean the width of a bar that can be detected under specific contrast and mean radiance conditions.

Since specific tri-bar patterns are not inherent in nature, it was necessary to observe some other type of objects in ERTS imagery to assess actual performance. One target type which proved to be very useful was the piers in harbor areas. In Figure 7 we see the harbor area of Boston, in the United States, which contains sets of piers of three sizes, ranging from 50 m to 90 m in width, with approximately equal water spacing

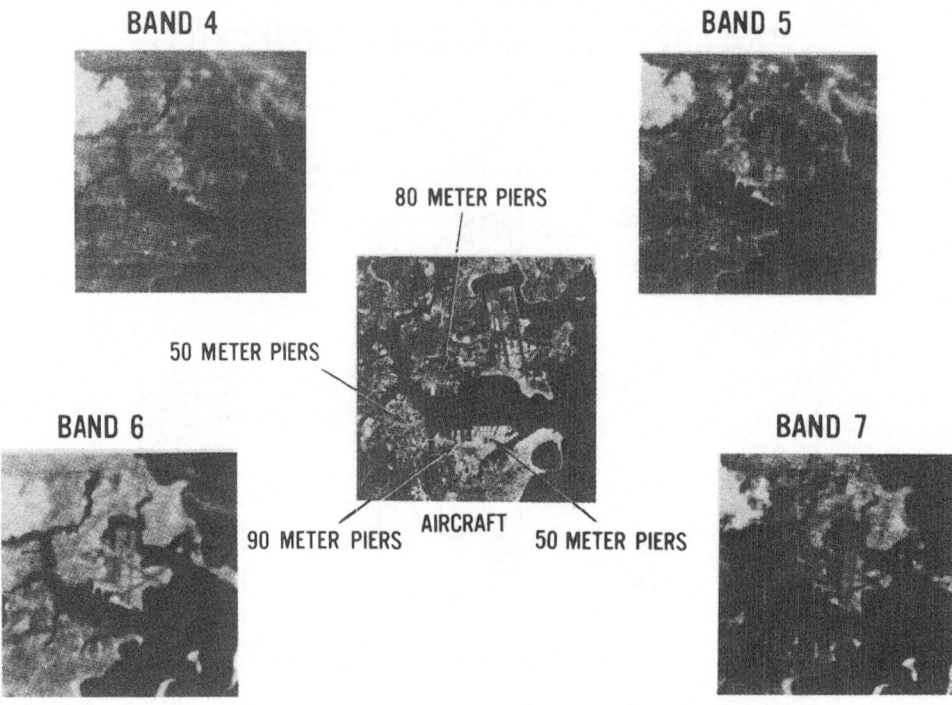

Fig. 7.   ERTS imagery ground resolution evaluation.

in between. Portions of the 4 bands of the multispectral scanner have been enlarged and are shown in the same figure. By careful observation of this type we can examine the ERTS imagery collected to date and evaluate which type targets are visible and which are not in order to compare actual performance with expected performance. This has been done for several type targets, as shown in Figure 8. Agreement for both the RBV and the MSS has been quite good with respect to expected system resolution.

5. Photographic Interpretation and Information Extraction

Development of multispectral analysis techniques for extracting the information content of photographs is a necessary step if this flood of image data is to provide useful information while it is still current. GE, together with other industrial concerns, government agencies and universities in the United States, have been developing these techniques for several years. The GE multispectral information extraction system, GEMS, shown in Figure 9, is one such development which permits scientific investigators to quickly derive useful information from multispectral imagery. In many cases, geological and other Earth surface features are very subtle for conventional photo interpretation techniques. GEMS, with the investigator controlling the machine on an inter-active basis, can detect and enhance these features and extract useful information faster, more effectively and at lower cost.

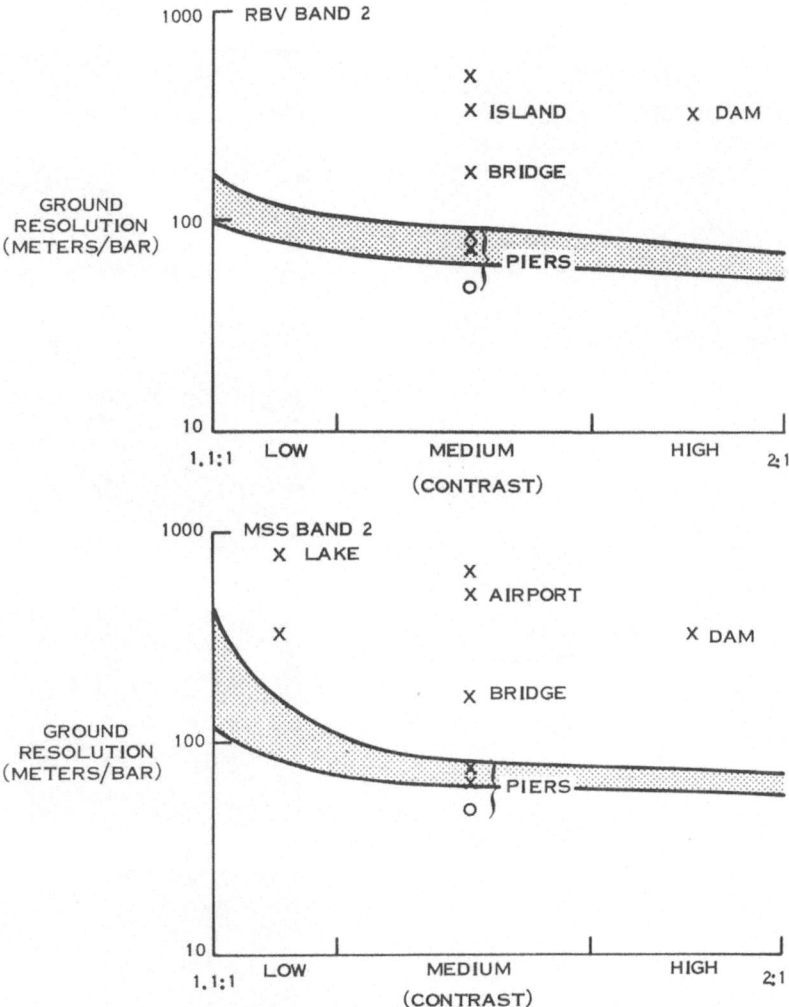

Fig. 8.   Observed vs predicted resolution.

What GEMS and other systems like it do, is to extract the multispectral signature of the material or feature under investigation and define all other areas in a given scene that have the same signature.

In the example shown, GEMS was used to classify the scene into four broad categories of water, IR vegetation, bare soil, and works of man, and to present these as thematic displays.

In summary, many of the technological challenges of the Earth resources technology satellite program have been met and solved. However, to prove the practical utility of multispectral remote sensing, the future challenges that remain are the rapid extraction and application of useful information from this overwhelming source of raw data. Spaceborne data compression, filtering, cloud cover detection, video

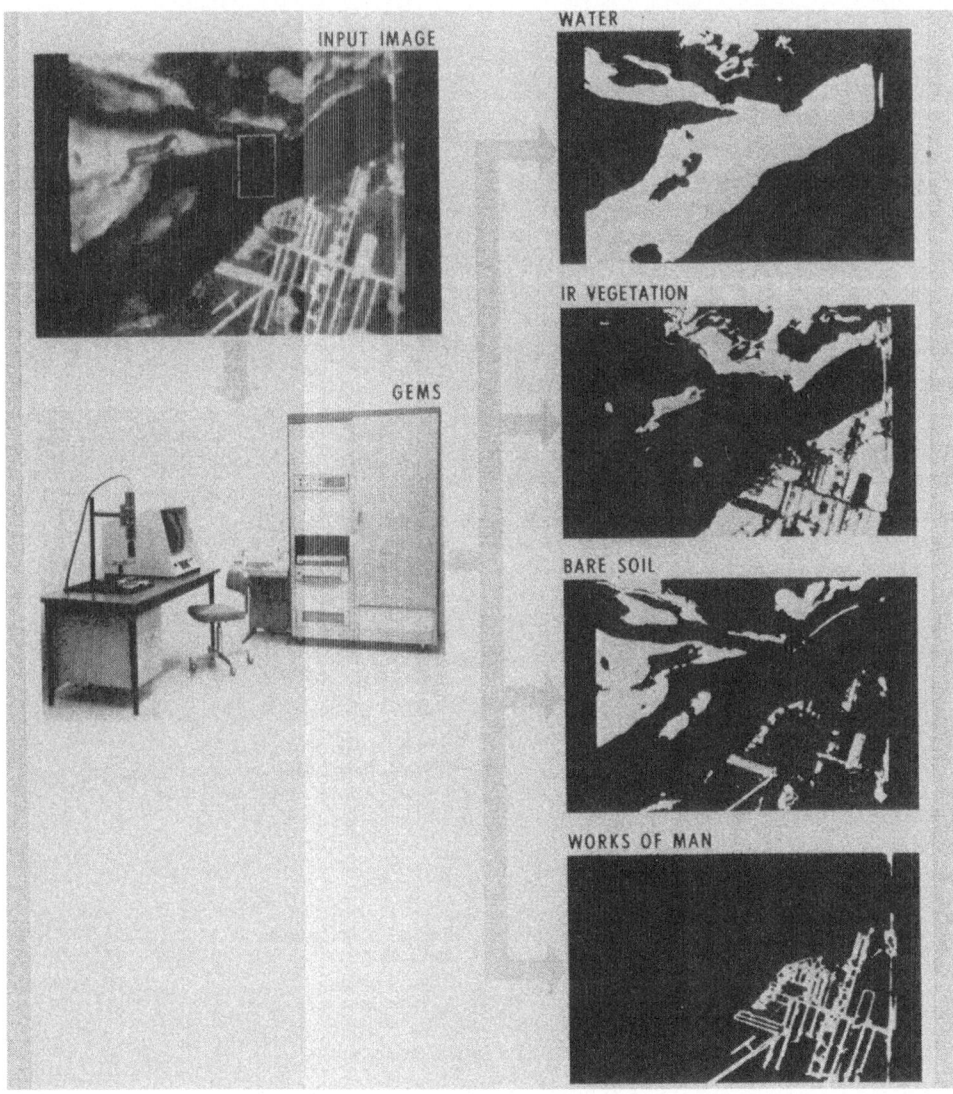

Fig. 9.   Information extraction via GEMS.

pre-processing and other techniques, need further emphasis as a means of minimizing the collection or transmission to Earth of useless data. Full application of data processing and computer technology to the field of remote sensing has not yet been achieved. Much work remains in the application of the single process of converting remotely sensed data to useful statistics.

UTILISATION D'UN SYSTÈME À SATELLITES
POUR LA NAVIGATION ET LE CONTRÔLE MARITIME

A. PINGLIER

Centre National d'Études Spatiales, Division Analyses de Missions, 91220 Brétigny, France

Abstract. A ship must be able to determine its position and may wish to communicate with places on land.

These two requisites of maritime navigation can be satisfied thanks to a satellite system which makes radio-determination and telecommunications with the land possible. Present requirements concern almost exclusively radio-communications, present day techniques making it impossible to ensure service under good conditions (connections are often of a poor quality and waiting time is long).

The needs of a ship for its own use (for navigational purposes) can be satisfied by having recourse to satellites, but also by other means already in existence (optical, radio-electric means, etc.). While it is evident that a ship must know its position, the advantage for some outside element to have this information is beginning to be realised. Radio-determination coupled with a reliable means of communication between ground-station and ships makes an efficient control of maritime traffic possible. A satellite system makes such a system possible.

After an analysis of the fields where such a system could be of use to navigation, the main means of determination not implying satellites are quickly reviewed. The various methods of determination from satellites are then analysed in order to show which type of system is the best. The main characteristics of a satellite system for traffic control, using the determination principle chosen, are then given.

A second chapter attempts to give some idea of the advantages of a satellite system from an economic point of view.

Lastly, the political aspect is mentioned and the difficulties of such an undertaking: international agreement required, problems arising from the necessity of equipping all ships, etc.

1. Intérêt général d'un tel système

Un navire doit être capable de déterminer sa position et souhaite pouvoir communiquer avec la terre.

Ces deux impératifs de la navigation maritime peuvent être satisfaits à l'aide d'un système à satellites qui permet le radio-repérage et les télécommunications avec la terre. Les besoins présents portent presque exclusivement sur les radiocommunications, les techniques présentement utilisées ne permettant pas d'assurer un service dans de bonnes conditions (liaisons souvent de mauvaise qualité, temps d'attente important).

Les besoins de localisation du navire pour son propre usage (pour la navigation) peuvent également être satisfaits par l'emploi de satellites, mais aussi par d'autres moyens déjà existants (moyens optiques, radio-électriques...). S'il est évident que la connaissance, par un navire, de sa position est indispensable, l'intérêt, pour un élément extérieur, de disposer de cette information commence à apparaître.

Cette information va permettre de répondre à des besoins potentiels dans les domaines de la navigation et du contrôle du trafic.

La radiolocalisation associée à un moyen de communication sûr va, en effet, permettre de remplir des fonctions qui ne sont pas, ou mal, remplies à l'heure actuelle. Il convient de placer en premier lieu dans cette catégorie trois grandes fonctions:

– la surveillance et le contrôle du trafic,

L. G. Napolitano et al. (eds.), Astronautical Research 1972, 281–293. All Rights Reserved

- la fonction anti-collision,
- la recherche et le sauvetage (SAR).

1.1. LA SURVEILLANCE ET LE CONTRÔLE DU TRAFIC

Dans certaines zones marines resserrées, au trafic extrêmement dense, les risques de collisions sont importants et il apparait souhaitable de surveiller et de coordonner de façon centralisée le mouvement des divers navires. La Manche et plus particulièrement le Pas-de-Calais constitue l'exemple type d'une telle zone à forte densité de trafic (en moyenne 250 navires franchissent le Pas-de-Calais dans chaque sens en 24 h et il y a environ un accident par mois). Parallèlement à l'augmentation du nombre et du tonnage des navires, les risques et les conséquences d'accidents deviennent de plus en plus graves. Si bien qu'il n'est pas déraisonnable d'envisager pour un avenir, pas trop lointain, la mise en place d'un contrôle dans les zones critiques. Ce contrôle peut s'effectuer de diverses façons, en particulier à l'aide de satellites. Les points saillants d'une étude préliminaire, faite en France sur ce sujet, seront donnés plus loin.

1.2. LA FONCTION ANTI-COLLISION

Le contrôle mentionné plus haut constitue un dispositif anticollision mais ici nous considérons les systèmes qui permettent au navire d'apprécier, par lui-même, le mouvement d'autres navires se trouvant dans son voisinage et de déterminer s'il y a ou non risque de collision. Dans l'affirmative le système doit aider le navigateur à trouver la meilleure manœuvre d'évitement possible.

Dans un système classique la route que vient de suivre et l'extrapolation donnant la route que va vraisemblablement suivre l'autre navire sont déterminées à partir des indications du radar de bord.

– Dans un système à satellites, la radiolocalisation avec identification des navires permet à un navire d'avoir une idée précise de la situation autour de lui. Parallèlement aux informations de localisation et d'identification, le navire reçoit certains paramètres concernant les autres navires (cap et vitesse). Les données de radiolocalisation, les paramètres de marche des autres navires associés aux indications du radar de bord permettent une très bonne estimation des routes que vont suivre les différents navires. D'autre part, l'identité des divers navires en cause étant connue, il sera possible à deux navires, grâce à un canal de communication, d'entrer en contact et de se mettre d'accord sur une manœuvre d'évitement si nécessaire.

1.3. EN MATIÈRE DE SÉCURITÉ MARITIME (SAR)

L'intérêt d'un système alliant une radiolocalisation automatique et précise à des télécommunications sûres est évident. Présentement, en cas desinistre, les opérations de repérage prennent quelquefois plusieurs heures, faute d'une localisation précise. La coordination des moyens de sauvetage se fait plus ou moins bien. Des télécommunications sûres associées à un radiorepérage permettent d'avoir connaissance quasi immédiate du sinistre et de son lieu. Les navires dans le voisinage sont connus et peuvent être à leur tour contactés très rapidement.

Les canaux de télécommunication permettent d'autre part, la diffusion aux navires d'un certain nombre d'informations utiles à la navigation par exemple, un bulletin météorologique spécialisé par zone, l'indication de l'emplacement d'une épave etc. …

L'intérêt de la radiolocalisation se manifeste dans d'autres domaines que ceux mentionnés précédemment:

– la connaissance de la position de ses navires en haute mer permet par exemple à une compagnie de navigation de tenir à jour le tableau de marche de ses navires ('plotting' de la flotte) et peut permettre une meilleure gestion.

– à l'arrivée au voisinage des côtes, la connaissance de la position et des intentions des navires voulant entrer dans un port permet aux autoritiés portuaires d'organiser les arrivées de telle manière que l'écoulement du trafic se fasse de façon satisfaisante.

2. Moyens de localisation d'un bateau

2.1. MOYENS CLASSIQUES

Bref rappel sur ces moyens classiques (c'est-à-dire non spatiaux) qui sont les seuls présentement utilisés.

2.1.1. *Navigation astronomique*

– est la plus ancienne et encore très utilisée
– la précision de localisation se situe entre 2 et 10 m nautique
– est difficile à mettre en œuvre dans les zones où le ciel est souvent couvert
  les équipements ont un prix négligeable.

2.1.2. *Moyens radioélectriques*

(a) radiophares ('Direction Finder')
  – donnent la direction dans laquelle se trouve le navire par rapport à une station fixe de position connue. Deux radiophares permettent à un bateau de déterminer sa position.
  – la précision est de 4 à 10 m nautique.
(b) systèmes hyperboliques
  – Deux stations sol synchronisées émettent au même instant un signal. Le navire mesure la différence des instants de réception (ou le déphasage) des deux signaux qui donne la différence des distances du bateau aux deux stations. Le bateau est sur une hyperbole ayant comme foyers les deux stations. Une mesure avec une troisième station détermine une autre hyperbole et l'intersection de ces deux hyperboles indique la position du navire.
  – Divers systèmes existent sur ce principe. Ils différent par les fréquences émises et la couverture obtenue.
    . le Decca couvre certaines zones côtières; précision $\frac{1}{4}$ à 2 m nautique
    . le Loran A couvre l'Atlantique nord; précision 2 à 5 m nautique
    . l'Omega a une couverture mondiale; précision 1 à 3 m nautique.

Dans tous les systèmes classiques la détermination du point se fait à bord du navire et l'information n'est disponible qu'à bord. Ce sont des systèmes permettant uniquement de remplir des besoins de navigation pure.

2.2. Radiolocalisation par satellite

Les satellites permettent de remplir outre les besoins de navigation pure (la position du bateau n'a besoin d'être connue que de lui seul) des besoins liés à un quelconque système de surveillance dans lequel il est nécessaire que l'organisme qui assure cette surveillance (stations à terre) dispose des informations de position, vitesse... des navires.

Il existe quatre types fondamentaux de localisation par satellite.

(1) Localisation à partir de mesures de distances.

(2) Localisation à partir de mesures de différences de distances.

(3) Localisation à partir de mesures d'angles.

(4) Localisation à partir de mesures de vitesses relatives.

2.2.1. *Localisation à partir de mesures de distances*

Le système permet de déterminer la distance du mobile à deux satellites dont les positions au moment de la mesure sont connues; soit d_1 et d_2.

Le mobile est:

– sur la sphère terrestre,

– sur la sphère de centre S_1 et de rayon d_1,

– sur la sphère de centre S_2 et de rayon d_2.

Il est à l'intersection de ces trois sphères. Il y a deux points d'intersection, la connaissance approximative de la position permet de lever l'ambiguité.

Méthode de mesure de la distance:

La distance est déterminée à partir de la mesure d'un temps de propagation.

– Une première possibilité consiste à envoyer d'une station T vers chacun des deux satellites un signal qu'ils renvoient vers le mobile. Une horloge très précise, à bord du bateau permet de mesurer le temps de propagation sol-satellite-bateau c'est-à-dire déterminer $r_1 + d_1$ et $r_2 + d_2$. r_1 et r_2 étant connus on en déduit les deux distances cherchées. Concrètement une mesure de distance à 50 m près nécessite une précision de calage entre l'horloge sol et l'horloge de bord de 0.2 μs environ ce qui est technologiquement impossible. Cette version du système ne peut donc être concrètement utilisée.

– Deuxième possibilité: la station émet en direction de l'un des deux satellites (ex S_1) qui réémet le signal vers le bateau. Le navire réémet à son tour en direction des deux satellites S_1 et S_2 qui eux-mêmes renvoient le signal vers la station terrestre. Le temps de propagation station-sol et retour via les deux trajets permet de déterminer:

et $\qquad \left. \begin{matrix} r_1 + d_1 + d_1 + r_1 \\ r_1 + d_1 + d_2 + r_2 \end{matrix} \right\}$ d'où d_1 et d_2.

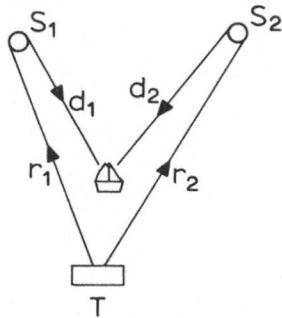

Fig. 1.

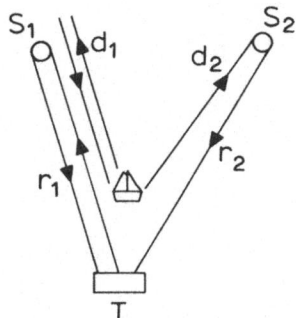

Fig. 2.

Dans ce mode de fonctionnement le navire réémet le signal reçu. Le système est dit actif. La détermination de la position se fait à terre. La mesure du temps aller retour se fait à partir d'une seule horloge et il n'y a de ce fait aucun problème de calage.

2.2.2. *Localisation à partir de mesures de différences de distances*

Le système se présente de la même façon mais au lieu de mesurer les distances, l'on mesure les différences des longueurs des trajets. Le mobile se trouve sur un hyperboloïde dont les foyers sont les satellites. Comme précédemment deux versions peuvent être envisagées.

– Un système passif dans lequel le navire ne fait que recevoir les signaux mais ne les réémet pas.

La différence des temps de propagation des deux trajets est mesurée à bord du navire d'où $(r_1 + d_1) - (r_2 + d_2)$ et finalement $d_1 - d_2$.

Ce système ne souffre pas des difficultés techniques signalées par le dispositif passif comportant deux mesures de distance. L'information cherchée ici étant le décalage en temps à la réception de deux signaux émis au même instant, il n'est pas nécessaire de connaître l'instant d'émission.

– Un système actif dans lequel le signal reçu par le navire est, comme dans le cas de mesures distance, renvoyé vers les deux satellites.

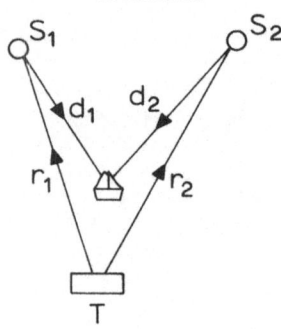

Fig. 3.

L'on mesure

$$2(d_1+r_1)-[(d_1+r_1)+(d_2+r)]$$

soit

$$(d_1+r_1)-(d_2+r_2) \quad \text{d'où} \quad d_1-d_2.$$

2.2.3. *Localisation à partir de mesures d'angles*

Les mesures d'angles peuvent être faites de plusieurs façons.

2.2.3.1. *Le navire est équipé d'un radiosextant.* Le satellite se présente comme une source radioélectrique de position bien connue. On mesure l'angle de la droite satellite-bateau avec une direction de référence.

La qualité de la localisation est fonction de la précision de la mesure d'angle. Une bonne précision implique une antenne de radiosextant de grand gain, c'est-à-dire de dimensions importantes (avec en plus un problème de stabilisation). Le système n'est pas retenu pour cette raison.

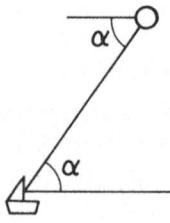

Fig. 4.

2.2.3.2. *Interféromètre.* L'interféromètre est constitué de deux antennes A_1 et A_2 séparées par une distance l_0 (base). La mesure du déphasage entre les signaux reçus en A_1 et A_2 et provenant d'une même source M, permet la détermination de l'angle α (angle entre la base et la direction de la source).

Un double interféromètre constitué par deux bases (en général perpendiculaires) permet de connaître la direction de la source.

Les deux interféromètres peuvent être montés sur un satellite géostationnaire,

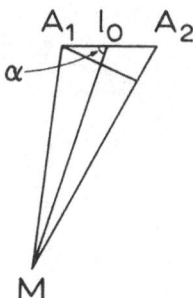

Fig. 5.

les antennes étant fixées au bout de bras. La source, dont la direction est déterminée dans un repère lié au satellite, peut être localisée après passage en coordonnées terrestres (En fait le signal venant de la source c'est-à-dire du mobile est déclenché par le satellite lui-même commandé par une station à terre).

Pour avoir une précision de localisation de 1 m nautique, il faut une base de 80 à 100 m pour des émissions en bande L. Cette longueur importante rend, dans l'état présent de la technologie le système très difficile à installer à bord de satellites (Problème de stabilité thermique des bras...).

D'autre part, il faut connaître de façon très précise l'attitude du satellite.

Ce système est en théorie séduisant puisque la localisation se fait à partir d'un seul satellite mais les difficultés de réalisation technique font qu'il ne peut être présentement retenu.

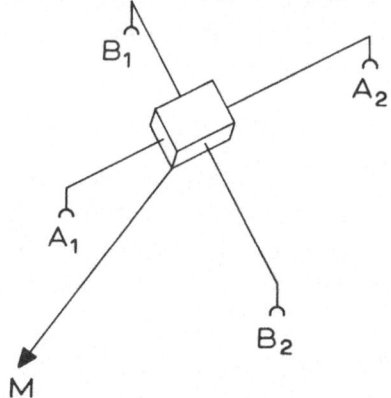

Fig. 6.

2.2.4. Localisation à partir de mesures de vitesses relatives

Localisation à partir de mesures de vitesse relative satellite-mobile. On utilise des satellites à défilement en orbite basse (ex. 1000 km), c'est-à-dire à forte vitesse relative par rapport à la terre. On peut considérer dans ces conditions la vitesse des mobiles comme pratiquement sans influence.

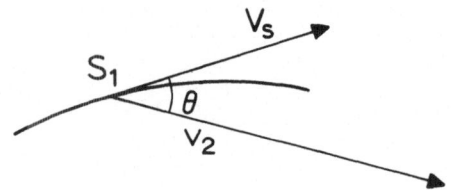

La vitesse relative $v_2 = v_s \cos\theta$.

v_2 est déterminé à partir de la mesure du saut Doppler Si la source émet une fréquence f_0 le récepteur reçoit une fréquence $f_0 + \Delta f$

Δf représente le saut Doppler

$$\Delta f = \frac{v_2}{c} f_0 \, (c = \text{vitesse de la lumière}).$$

La fréquence reçue est

$$f_r = f_0 \left(1 + \frac{v_2}{c}\right).$$

La connaissance de f_0 et f_r permet de déterminer v_r et donc $\cos\theta$ (v_s est la vitesse du satellite qui est déterminée par une orbitographie fine).

A partir de la valeur de θ et de la position du satellite au moment de la mesure, l'on détermine un cône, de sommet S_1 et de demi-angle θ, sur lequel se trouve le mobile. Une autre mesure lorsque le satellite se trouve en S_2 permet la détermination d'un second cône. L'intersection de ces deux cônes avec la sphère terrestre permet de localiser le mobile.

2.2.5. *Causes des erreurs de radiolocalisation*

Tous les systèmes présentent des erreurs ayant la même origine mais dont l'influence varie d'un système à l'autre. Les principales causes d'erreurs sont:

(1) *Erreur ionosphérique*. La vitesse de propagation de l'onde radioélectrique se trouve affectée par la traversée de l'ionosphère d'où une erreur sur la mesure distance et doppler.

(2) *Erreur due aux trajets multiples*. A l'onde directe satellite mobile (ou vice versa) s'ajoute une onde secondaire réfléchie par la mer. Cette onde agit comme élément parasite qui altère la mesure.

(3) *Erreur due aux appareils de mesure*.

(4) *Erreur due à une mauvaise connaissance de la position des satellites*. La détermination de la position du mobile se fait à partir de la position des satellites. Une erreur sur la position de ces derniers dans un repère terrestre introduit une erreur de localisation du mobile.

(5) *Erreur 'géométrique'*. A une erreur donnée sur la connaissance de la grandeur de base (ex. la distance satellite mobile) correspond une erreur de localisation différ-

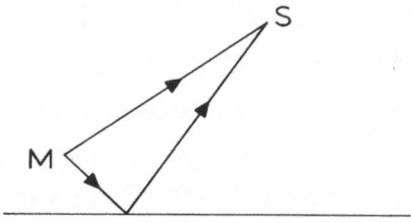

Fig. 8.

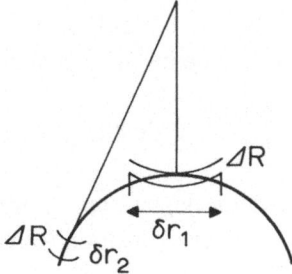

Fig. 9.

ente suivant la position relative du satellite et du mobile par rapport à la terre. Une même erreur sur la distance entraîne une erreur de localisation δr_1. Si le mobile est sous le satellite et δr_2 si le mobile est loin de la trace $\delta r_2 \leqslant \delta r_1$.

2.2.6. *Type de système retenu*

Le système retenu dépend de la mission qui lui est impartie.

 – Les satellites maritimes doivent rendre possible l'établissement de télécommunications avec la terre ainsi que la radiolocalisation aux fins d'aide à la navigation et/ou de contrôle du trafic, ceci en *permanence*. Cette notion de permanence implique qu'un satellite ou groupe de satellites soit toujours en vue ce qui élimine les satellites à défilement à orbite basse, c'est-à-dire une localisation par mesure Doppler. Pour assurer une couverture permanente un trop grand nombre de satellites seraient nécessaires.

 – Les difficultés techniques de mise au point nous ont fait éliminer les systèmes permettant des mesures d'angles (interféromètre embarqué).

 – Le choix doit donc se faire entre le système hyperbolique et sphérique (à partir de mesure de distances).

Le système hyperbolique nécessite l'emploi de trois satellites (non stationnaires) pour effectuer une localisation. Le système sphérique n'en demande que deux. Ces satellites peuvent être géostationnaires.

En définitive le système retenu pour la radiolocalisation utilisera une mesure distance à partir de deux satellites géostationnaires.

Une couverture globale des océans et des côtes à la fois pour les télécommunications et la radiolocalisation (limitée à 70° de latitude nord et sud) peut être obtenue à l'aide de 6 à 8 satellites géostationnaires.

2.2.7. *Exemple de caractéristiques d'un système de satellites pour le contrôle maritime*

L'étude faite en France au début de 1972 portait sur le problème de l'aide à la navigation et du contrôle du trafic le long des côtes européennes de l'Atlantique (mer du Nord et Manche comprises) ainsi que sur la Méditerrannée occidentale.

2.2.7.1. *Précision de la localisation.* L'erreur est fonction de l'écartement entre les satellites. Si on rapproche les deux satellites la zone de couverture (zone de visibilité commune aux deux satellites) augmente mais la précision diminue. Un écartement de 35 à 40° est souhaitable.

L'écartement choisi dans l'étude est de 40° (avec un satellite à 20° W et l'autre à 20° E).

D'autre part, l'erreur de position dépend beaucoup des incertitudes sur la mesure de distance; par contre les erreurs sur la position (en latitude et longitude) du satellite sont moins critiques.

Si   l'erreur sur la distance σ_d 10 m
            les erreurs sur la position
            du satellite $\sigma_\theta = \sigma_\varphi =$ 50 m
            l'erreur sur l'altitude
            du satellite $\sigma_R =$ 25 m

dans la zone considérée l'ellipse d'erreur est telle que:

$a(\frac{1}{2}$ axe de l'ellipse) $\simeq 150$ m
$b(\frac{1}{2}$ petit axe de l'ellipse) $\simeq$ 80 m
α varie entre 10 et 15°.

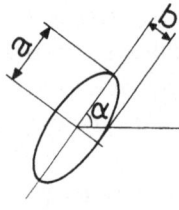

Fig. 10.

Les précisions indiquées pour la mesure distance sont obtenues avec *un satellite*
 – ayant une antenne bande L (1.5–1.6 GHz) de 25 dB de gain (angle d'ouverture 7°)
 – et qui émet une puissance de 55 W par canal de contrôle (qui comprend la radiolocalisation et une voie phonie).
Le navire
 – est équipé d'une antenne de 1 dB de gain
 – la puissance émise par canal de contrôle est de 27 W.

2.2.7.2. Organisation du contrôle et résultats. Différents types de contrôle peuvent être envisagés. L'étude a retenu un système à voies paralléles et à vitesse imposée. Les navires doivent suivre l'axe de la voie avec une vitesse déterminée et constante. Lorsqu'ils s'écartent de l'axe et dépassent un certain seuil, le centre de contrôle le leur signale, de même si la distance avec le bateau qui les précède ou les suit dans la même voie n'est plus correcte. Le problème est, compte tenu de la précision de localisation donnée et des possibilités de tenue de route et de vitesse des navires, de déterminer la largeur des voies et la cadence de localisation des navires. Ces deux grandeurs sont, bien entendu, dépendantes l'une de l'autre: il ne faut pas qu'un navire, dans l'axe de la voie au moment d'une localisation, puisse être sorti de cette voie à la localisation suivante.

Compte tenu des hypothèses sur la tenue de route des navires caractérisés par σ_y (y est l'écart latéral d'un navire par rapport à l'axe de la voie sur lequel il essaie de naviguer) et la loi de distribution de ces écarts, on peut déterminer en fonction de la largeur de la voie, du temps entre localisations et pour un trafic donné à écouler le risque de collision.

Dans le Pas-de-Calais avec 250 navires à écouler dans chaque sens en 24 h l'établissement d'un contrôle du type indiqué avec deux voies dans chaque sens de *deux mille nautique* de large chacune, un temps entre localisations de 10 min et un σ_y de 0.25 mille nautique, le nombre des collisions pourrait être réduit à une tous les 8 ans.

3. Aspects économiques

Nous avons mentionné les domaines dans lesquels un système par satellite pourrait apporter des améliorations à la situation présente.

3.1. RÉDUCTION DES RISQUES DE COLLISIONS

Dans cette catégorie, on place les collisions avec d'autres navires ainsi que les échouages.

D'après une étude norvégienne, entre 1964 et 1968, 53% des pertes de navires ont comme origine une collision (11% par collision entre navires et 42% par échouage). Pour la flotte norvégienne ceci représente une perte de 3 à 4 M $ par an.

Il est difficile d'apprécier exactement la réduction des collisions dues à l'utilisation d'un système par satellites. Si on suppose une diminution de 50% des risques, l'économie sur le matériel perdu serait pour la flotte norvégienne de 1.5 à 2 M $ par an, soit rapporté à la flotte mondiale environ 20 M $ par an.

Cette économie se répercutera, avec un certain retard, sur les primes d'assurances payées par les armateurs. La perte d'un bateau réduit les capacités de transport d'un armateur, d'où un manque à gagner qui peut être important.

Enfin un accident de navigation peut, non seulement entraîner la perte d'un bateau, mais aussi causer un dommage considérable à l'environnement (cas des pétroliers par exemple). Il est très difficile de chiffrer le montant des frais et pertes occasionnés mais on peut penser qu'ils seront de plus en plus élevés avec la croissance du tonnage des navires.

3.2. RECHERCHE ET SAUVETAGE

En ce domaine, il est difficile d'apprécier les gains en termes économiques. La motivation première d'un système de recherche efficace est de sauver davantage de vies humaines. On peut, cependant, dire que les opérations de recherche avec un tel système seront plus courtes et demanderont moins de moyens donc seront plus économiques.

– Le fait, pour le navire, de pouvoir faire le point souvent lui permet de suivre rigoureusement la route la plus courte d'où gain de combustible et gain de temps qui peut se traduire par une économie de fonctionnement non négligeable. L'allongement du trajet peut aller jusqu'à 7% lorsque les conditions météorologiques sont mauvaises et que le bateau reste très longtemps sans faire le point. Par beau temps, il est de l'ordre de 0.5 à 1%. En définitive les gains obtenus dépendent de la route suivie et du bateau. Le chiffre de 16000 $ d'économie par an pour un bateau de 75000 tjb a été avancé. Il est à remarquer que ce type de navigation précise peut être effectué avec les moyens de radiolocalisation classiques comme l'Oméga.

– Dans le même ordre d'idées, une connaissance des conditions météorologiques peut permettre d'éviter les zones de mauvais temps et tout en améliorant la sécurité faire gagner du temps et du combustible (un bateau par grosse mer progresse beaucoup moins vite que par mer calme). Les norvégiens estiment que les gains apportés par ce type d'aide (weather routing) peut être de l'ordre de 1 M $ par an pour leur flotte.

En conclusion, indiquons que la navigation et le contrôle par satellite apporteront des améliorations sensibles aux conditions d'exploitation. Il est difficile de les chiffrer précisément en termes monétaires. Cependant, il ne semble pas qu'un rapport efficacité-coût acceptable puisse être assuré par un système à satellites axé sur la navigation et le contrôle. Il sera indispensable d'associer les fonctions télécommunications et aide à la navigation pour rendre un tel système rentable.

4. Aspect politique

Tout système d'aide à la navigation et surtout de contrôle doit, pour être efficace, s'appliquer à l'ensemble des bateaux. Ceci implique au premier chef une entente internationale entre tous les pays pour créer une organisation mondiale capable de mettre concrètement en place et de gérer un système à vocation planétaire. Ce problème, on s'en doute, n'est pas simple. L'OMCI organisme spécialisé des Nations Unies sur la navigation maritime a proposé aux états membres d'étudier le dossier. Des réunions d'experts ont lieu périodiquement et une réunion intergouvermentale est prévue en 1974.

La première génération de satellites maritimes mettra sans doute l'accent sur les télécommunications avec vraisemblablement une évaluation des possibilités en ce qui concerne l'aide à la navigation.

On peut penser que la deuxième génération permettra l'exploitation opérationnelle des moyens d'aide à la navigation (y compris le contrôle dans certaines zones). D'ici là une solution devra être trouvée au problème de l'équipement de tous les bateaux, jusqu'à un tonnage très faible (100 tjb par ex.), d'un matériel de bord permettant la radiolocalisation et les communications associées. Cet équipement est estimé à 10 000 $ pièce; des modes de financement convenables devront être trouvés pour permettre à tous les bateaux de s'équiper.

Il reste donc un nombre considérable de problèmes à régler avant d'arriver à un système d'aide à la navigation par satellite vraiment opérationnel. Dix ans au moins seront nécessaires mais les possibilités immenses d'un système à satellites fait que l'on peut être optimiste quant à sa mise en place à plus ou moins brève échéance.

MATERIALS PROCESSING IN ZERO GRAVITY

(An Overview)

HANS F. WUENSCHER

Assistant Director, Process Engineering Laboratory, NASA-Marshall Space Flight Center, Huntsville, Ala., U.S.A.

1. Introduction

While the exploration of space is under continuous development, the exploitation of the orbital flight capabilities is already successfully underway in unique world-wide communications, weather and Earth observation applications. Considering the accomplishments of space technology over the last decade, turning now toward realization of low cost space transportation systems, the time has come to align our conventional thinking about manufacturing towards application of the unique space environments for the achievement of product characteristics superior to, or not attainable on Earth.

The Earth bound environment determines but also limits many processes, especially the ever present, large gravitational action of the Earth which can be controlled only for a few seconds during free fall, too short for practical utilization. Space environment offers total gravity control and in addition unequaled vacuum, temperature, pressure and radiation characteristics. The improvement or even uniqueness of the product might easily pay for the effort to relocate the critical phases of a manufacturing process into space.

The fundamental question is whether there is any other way to get the benefits of space environment. In fact, the search for matchless space processes might even yield a few new terrestrial concepts, using space conditions as transient effects as far as they can be reproduced on Earth. But the increasingly easier access to space will provide the extraterrestrial environment as an unrestricted natural resource to be occupied like a newly discovered continent.

While other fields of space utilization, such as astronomy, meteorology, communications and Earth resources survey, consist essentially in services, manufacturing in space is aimed at products and has, therefore, a considerable growth potential.

2. Unique Space Manufacturing Processes

The most apparent effect of the weightless environment upon matter is the absence of relative mass accelerations. This not only eliminates the need for support of solid matter, but also precludes any relative motion in fluids, due to differences in density, resulting in the stability of liquid-solid, liquid-liquid or liquid gas mixtures and the absence of thermal convection.

Liquid-solid mixtures find primary application in the casting of composites [1, 3–5]. On Earth, the liquid-matrix preparation of composites is limited to liquids of high

L. G. Napolitano et al. (eds.), Astronautical Research 1972, 295–307. All Rights Reserved
Copyright © 1973 by D. Reidel Publishing Company, Dordrecht-Holland

viscosity like polymers and glasses. Metal matrix composites are exclusively produced in solid state in view of the extremely low viscosity of molten metals. Under zero gravity conditions, preparation in liquid matrix state can be applied in various combinations. Fine particle dispersion may be used as nucleation sites during solidification, resulting in fine grain castings. Dispersed particles also may serve as nucleation sites for the formation of gas bubbles and foams.

In alloying, the liquid-liquid mixture stability in space will prevent segregation between elements of different density and permit preparation of supersaturated alloys including combinations which exhibit liquid phase immiscibility.

Zero-g effects on supercooling, solidification, crystal formation and single crystal growth should improve uniformity and perfectness because of elimination of thermal convection at the solid-liquid interface and nonuniform concentration gradients of the solute.

The absence of gravity has an influence on chemical reactions in liquids [6]. In cases where the chemical energy is comparable to the gravity induced hydrostatic pressure gradient, the small change due to lack of gravity energy in space can serve as trigger for chemical reaction.

Probably the most typical of all space processes is the containerless free suspension of materials. On Earth we help ourselves with levitation melting, which is limited to very small amounts of metals at high temperatures.

It is not suitable for low melting metals, glasses, ceramics and liquid chemical compounds. In space, levitation is the natural condition for all materials and only small forces for positioning and handling are needed for processing, as shown in Figure 1.

Processes which are expected to show drastic changes in space environment

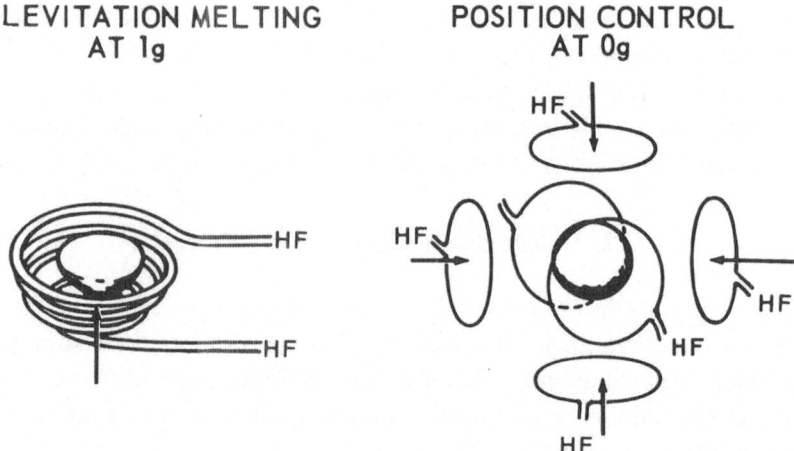

LEVITATION MELTING AT 1g   POSITION CONTROL AT 0g

Fig. 1.   On Earth small quantities of high temperature liquid metals (L) can be levitated by high frequency (HF) electromagnetic field forces as shown on the left side. On the right side, very small electromagnetic fields are sufficient to position the weightless liquid materials during processing. HF = High Frequency Coils. L = Liquid Material.

because of the absence of Earth gravitational action can be classified by the following two groups or combinations thereof:

– Group A: *Buoyancy and Thermal Convection Sensitive Processes*
– Group B: *Molecular Forces like Cohesion and Adhesion remain as the relatively strongest and hence process controlling factors*

Examples for typical space manufacturing processes are listed in Table I.

TABLE I

Space manufacturing unique to weightless environment

Group A *Buoyancy and thermal convection free processes*	Group B *Cohesion and adhesion controlled processes*
1. Free and captive suspension	1. Surface tension casting
2. Mixing, homogenization	2. Surface tension drawing
3. Separation and purification	3. Adhesion casting
4. Alloying and supersaturation	4. Blow casting
5. Composite casting	5. Controlled density casting
6. Solidification, crystal growth, supercooling of liquid phase	
7. Decomposition	
8. Polymerization	
9. Biological processing	

3. Processes Development

Space processes utilize the weightless environment by applying liquid state phenomena in various combinations. Liquid state theory is not yet developed to the extent of predicting fluid behavior during critical phases of zero-g processing. The use of energy scaling laws is limited as g approaches zero. Therefore, the validation of postulated space processes and the recognition of the real problems are only possible through experimentation in real zero-g.

This is the large obstacle in doing the 'Laboratory Homework'. Simulation of zero-g in neutral buoyancy (suspension in water or other liquids) and low friction facilities (air bearings) is only useful for demonstration of operational aspects, training and some mechanical testing. For process development only the events during the time limited 'free fall' period are valid. The following means for extended zero-g testing are available:

– Drop Tower (300-ft drop delivers 4 s zero-g).
– Aircraft Zero-g Maneuver (KC-135 Research Aircraft delivers 10 to 20 s).
– Sounding Rocket (Aerobee delivers 4 to 10 min).
– Suborbital Rocket (recoverable pod up to 30 min).

A number of concepts for zero-g deployment of liquid metal were tested during the 4 s drop tower fall. A test package suitable for free flight use in the drop tower cabin or in the KC-135 aircraft cabin and which can accommodate a wide range of experiments is shown in Figure 2. Being readied for a drop is the electromagnetic positioning

Fig. 2.   A test package for free flight use in the drop tower or in the KC-135 aircraft cabin. This package
is being readied for zero-g testing of an electromagnetic position system for free floating material, visible
in the upper front corner.

system, visible in the upper front corner. This first prototype employing six positioning
coils is in development [7]. The development goal here is to provide positioning of
free floating liquid materials for processing in the low gravity gradient field of an
orbital laboratory. The capability of rotating and pulsing of the containerless liquid
for processes like shaping, centering of the bubble, degassing, blending and HF
temperature control is planned to be provided as well. Facilities for homogenization
of premelted materials by ultrasonic agitation and rapid chilling of small specimen
by water quench are in development in order to exploit the few seconds of drop tower
time for exploring the fasibility of combined processes.

For use with sounding rockets, a melting and water quench facility for samples of
18 mm diam and 50 mm long has been recently flown on an Aerobee 170. The zero-g
coast phase lasted 240 s. A prebuilt metal foam sample of low melting indium bismuth
alloy for investigation of the stability of metal gas dispersions in liquid state at zero-g
was processed. Figure 3 shows the cross section through a segment of the flight
sample. The gas bubbles were maintained during liquefaction and solidification
phases. Similar processed samples on the ground show complete segregation.

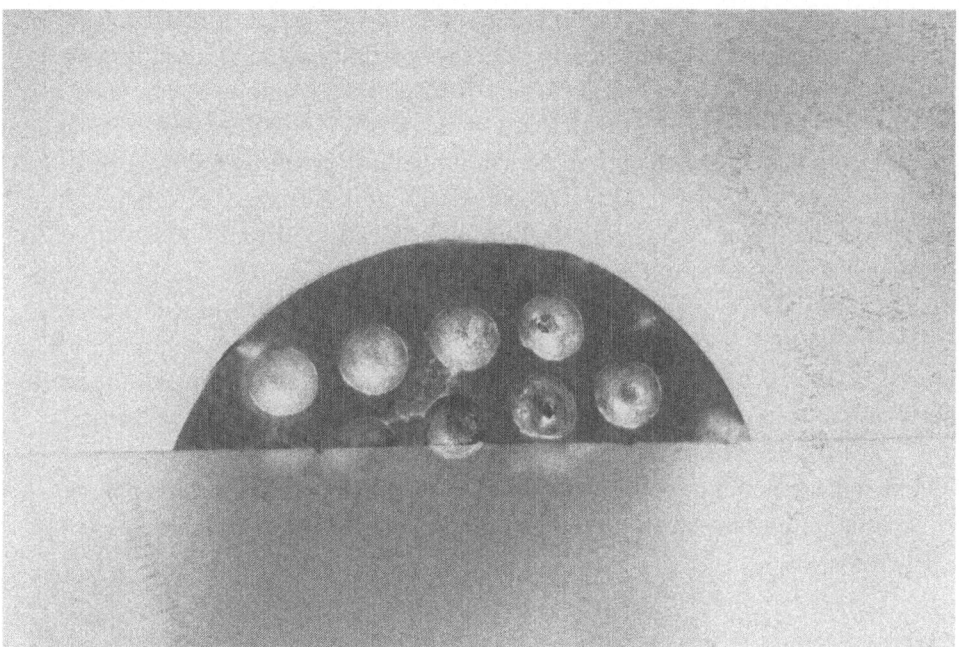

Fig. 3. Cut through a segment of a metal sample with 'built in' gas inclusions after being melted and solidified during the zero-g coast phase of a Sounding Rocket flight. Gas bubbles have remained in place, while Earth processed samples show complete segregation. Magnification 6 × .

A program for process development and verification by making extensive use of extended zero-g testing is in progress with the primary objective of obtaining preliminary data to guide the planning of future space experiments.

4. Space Experiments During the Apollo 14 Mission

During the Apollo 14 lunar mission, a number of 'Zero Gravity Inflight Demonstrations' during the transearth coasting phase have been processed by the Astronauts Shepard, Mitchell and Roosa. Two process groups suitable for space manufacturing application were used for the demonstrations. One group concerned the chemical separation process by electrophoresis [9]. The other consisted of a number of test samples concerning composite casting [10].

Electrophoresis uses the weightless environment to improve the process reducing sedimentation and thermal convection mixing. A sample of watery solution of purple organic dye was separated into its red and blue components. On Earth the boundaries are highly irregular due to a combination of electroosmosis, thermal convection and sample density. As a result of the space processing, a clear separation line between the components indicating that no lateral motions of the fluid have existed was photographically recorded during the process.

The composite casting demonstration was designed to survey the phenomena that occur when a variety of metallic and nonmetallic compositions alone and with solid particles and/or gas are melted, mixed and solidified in weightless environment. In order to keep the power, heating, cooling and safety requirements compatible with the Apollo spacecraft, a low melting metal matrix, indium-bismuth eutectic alloy, and paraffin and sodium acetate for transparent samples were chosen. For reasons of equipment simplicity, the mixing process was manually done by the astronauts by shaking the heater with the molten sample inside. Figure 4 shows the sample capsule $\frac{3}{4}''$ diam, 3″ long) before being loaded into the heater. Equilibrium of the liquefied composite was reached after the heater was put stationary onto the heat sink of the experiment box which was attached to the spacecraft. The cool-down proceeded through the bottom where the heat sink pin was in springloaded contact. In total, 11 capsules were processed. Examination of the samples has shown the following differences between materials processed in space and those processed on Earth.

In the space processed samples the evaluation revealed:

(1) More homogeneous distribution and stability during the solidification phase of both wetting and non-wetting particles and fibers in a metal matrix.

Fig. 4. The Aluminium capsule containing a processing sample for the Apollo 14 Composite Casting Experiment before being slipped into the electric heater for melting and shaking by the astronaut in order to achieve mixing of the sample contents.

(2) More homogeneous distribution and stability during the solidification phase of gas bubbles in a fiber reinforced metal matrix.

(3) Dispersions of immiscible liquid materials achieved stability during the solidification phase.

Three specimens are shown demonstrating some background for these conclusions:

Figures 5a and 5b are longitudinal cross sections of identically processed samples containing metal matrix (InBi) with 6% copper fibers*, which were surface treated to achieve wetting. The samples in Figure 5a was processed on the ground while the sample in Figure 5b was processed in the Apollo capsule. The segregation of the copper fibers, which are only 10% more dense than the matrix, has clearly occurred in Figure 5a, while a fairly even distribution has been maintained in Figure 5b.

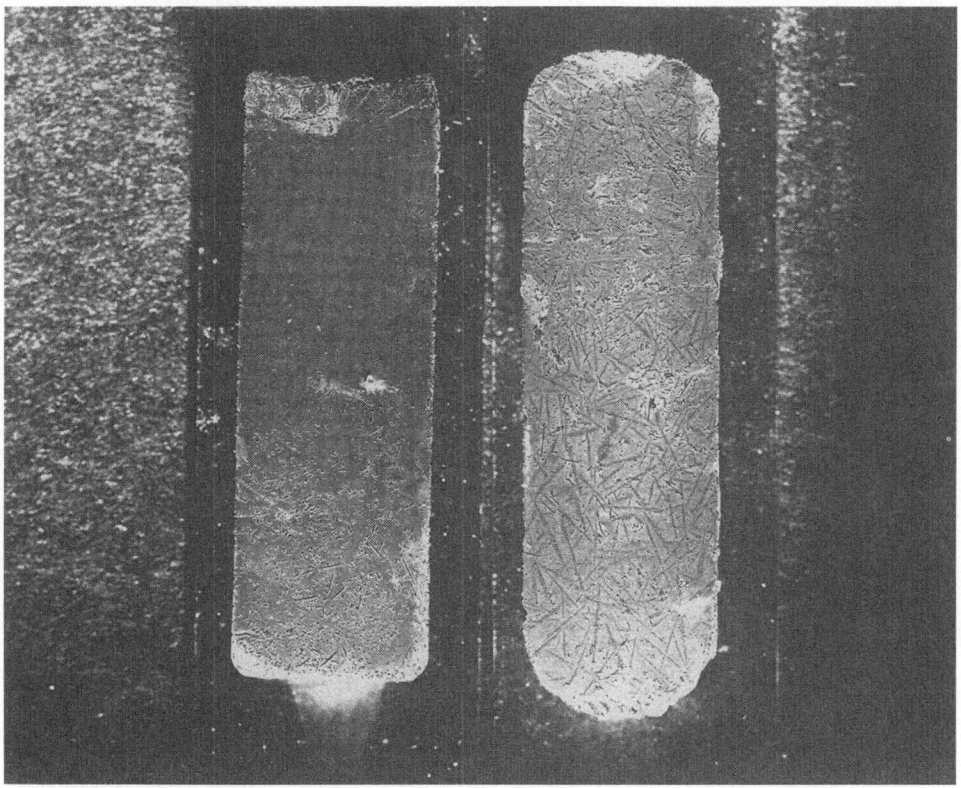

Fig. 5. (a) Longitudinal cross-section through an Apollo 14 composite metal sample containing 6% copper fibers after melting, shaking and solidification on Earth. The gravity-induced segretion of the copper fibers is clearly visible. (b) A sample containing 6% copper fibers, similarly processed as (a) by the astronauts in the weightless environment during flyback from the Moon. The copper fibers maintained distribution throughout the sample.

* % by volume.

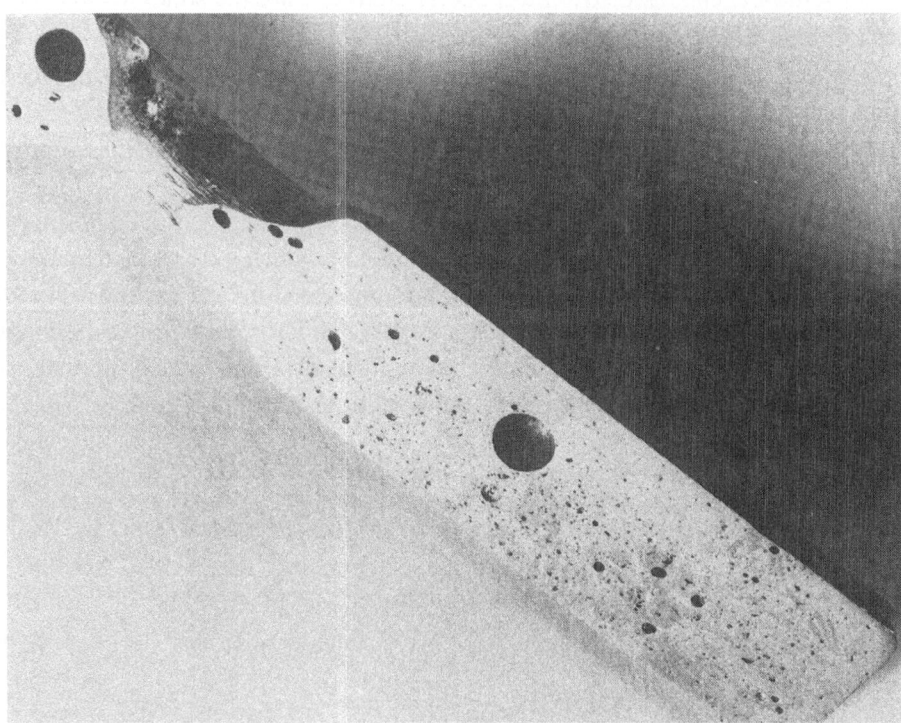

(b)

(a)

Fig. 6. (a) A ground processed sample containing 4% SiC whiskers and argon bubbles shows segregation at the bottom. (b) The flight processed sample contains on signs of segregation and has twice the pore density of the groundsample.

In Figures 6a and 6b a metal foam (InBi) containing 4% of SiC whiskers* was used to demonstrate that mixture stability in metal-gas bubble dispersions exist. The ground processed sample shows segregation at the bottom; the flight processed sample shows no gravity segregation and twice the pore density of the ground sample.

As a model for immiscible metal systems, a sample of 50% paraffin and 50% sodium acetate was used. Figure 7a shows that the ground processed sample is completely segregated. The flight sample (Figure 7b) shows that dispersion was maintained during solidification in many different types and sizes down to below one micron, which was essentially caused by the marginal mixing procedure. Such dispersions have been observed during agitation of immiscible liquid systems but never have been 'frozen in' before. With melting and solidification equipment suitable for metal systems, it is anticipated that zero-g environment can form an intermetallic compound which cannot be presently made.

5. Experiments on Skylab (1973)

The Skylab mission carries a major Materials Processing Facility M512 [11], serving as the basic apparatus and a common spacecraft interface for a group of space processing experiments, Figure 8. It provides for three power sources. The main system consists of a battery powered electron beam gun of 2 kW power at 20 kV for 10 min continuous operation. Onboard power of 150 W is available for direct heating of experiments and temperature control purposes like for sustaining slow cooling rates. Exothermic heating devices were developed for brazing and can be utilized for sample melting as well.

The processing takes place in a spherical chamber, which has a large door with observation window. The chamber can be vented through the wall of the orbital laboratory to provide for evacuation in addition to the weightless environment in orbit.

The following space processing experiments are to be performed in the M512 Experiment Facility:

5.1. M551 METALS MELTING EXPERIMENT

This experiment will examine the molten flow and solidification characteristics of three metal alloys: stainless steel, thoria dispersed nickel, and a structural aluminum alloy. Samples of the first two metals will be disks graduated in thickness from $\frac{1}{64}''$ to $\frac{1}{4}''$, and the aluminum sample will consists of a disk to be joined to an annular ring by welding along the abutting edges. All of the samples will be melted along a circular line by rotation under the electron beam in the M512 facility's vacuum chamber, using a special rotation fixture designed for this purpose. The melting operatio ns will be recorded on motion picture film, and the samples and film will be returned to Earth for analysis and evaluation.

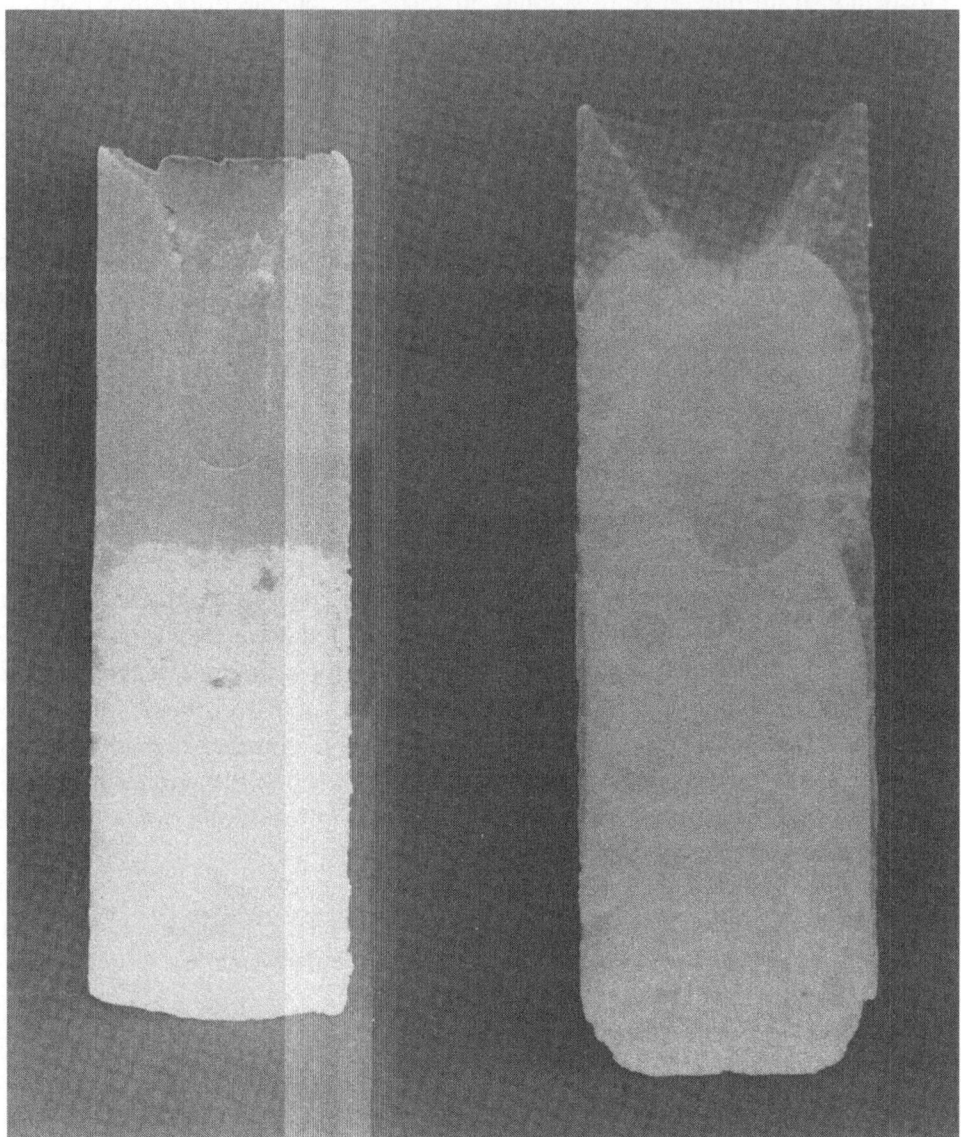

Fig. 7.   (a) A sample containing paraffin and sodium acetate used as a demonstration model for immiscible metal systems shows complete segregation after processing on the ground. (b) A similarly processed sample as shown in (a) processed in space shows dispersion patterns to sizes down to below one micron.

5.2. M552 EXOTHERMIC BRAZING EXPERIMENT

A technique for joining stainless steel tubes will be tested in this experiment, and the flow and solidification behavior of weightless molten braze alloys will be studied. The joining technique will use a solid mixture that produces heat by exothermic chemical reaction to braze sleeves over $\frac{3}{4}''$ diam tubes, using a copper-silver-lithium braze alloy.

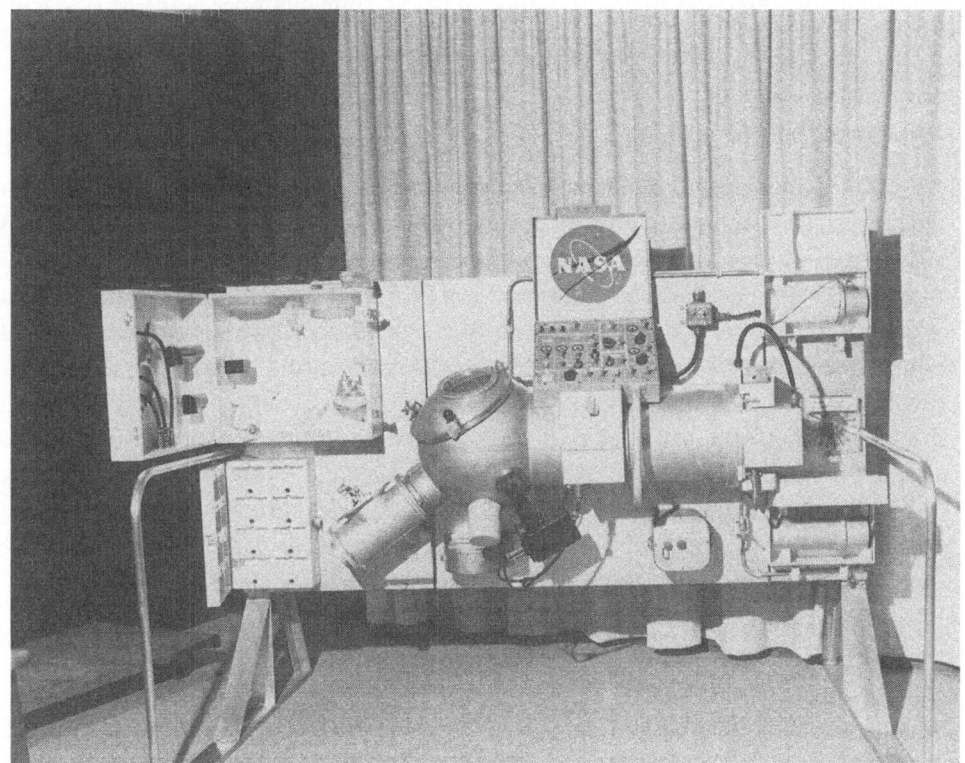

Fig. 8.   M-512 Skylab Materials Processing Facility.

A package containing four such assemblies, each comprising a tube with sleeve and preformed braze alloy surrounded by exothermic material, will be mounted in the M512 facility's vacuum chamber. The exothermic reaction in the four assemblies will be initiated in sequence and the whole package will be returned to Earth for analysis.

5.3. M553 SPHERE FORMING EXPERIMENT

Containerless melting and solidification will be studied on three metals: pure nickel, a nickel-tin alloy, and Star-J Stellite. Beads of these metals about $\frac{1}{4}''$ in diameter will be mounted on this wire supports around the periphery of a wheel-shaped sample holder and rotated sequentially into the electron beam path by an indexing motor. A few of the samples will be retained on their supports after being melted by the electron beam, but most will be released and allowed to float freely in the vacuum chamber while they solidify by radiation cooling. The melting and solidification process will be filmed, and the samples and film will be returned to Earth for study.

5.4. M554 COMPOSITE CASTING EXPERIMENT

Three samples of aluminum-copper eutectic alloy will be melted and directionally solidified simultaneously to investigate the effects of convectionless heat and mass transfer on metal solidification. The experiment will be performed in a furnace module containing the three specimens, mounted in the vacuum chamber of the M512 facility and controlled from the facility's control panel. After solidifications, the samples will be removed from the furnace and returned to Earth for analysis.

5.5. M555 GALLIUM ARSENIDE CRYSTAL GROWTH EXPERIMENT

Layers of the compound semiconductor gallium arsenide (GaAs) will be deposited on singlecrystal GaAs seed plates by solution transport from a polycrystalline GaAs source to the seeds through a temperature gradient in a column of molten gallium metal. The experiment will be performed in a furnace module containing three quartz growth ampoules and mounted in the M512 facility's vacuum chamber. Thermal control for the furnace will be passive, since the temperature controller on the facility control panel is preempted by the Composite Casting experiment. After the crystal growth run, the furnace will be returned to Earth unopened for removal and evaluation of the samples.

Furthermore, a multipurpose furnace for additional experiments is presently under development. New proposals for processing experiments for later flight opportunities are under study and a widening of the already considerable field is very probable.

6. Conclusion

In conclusion, gravity plays a very often hidden and indirect role in our daily processing. It might at first appear simple to derive processes which take unique advantage of the absence of the dominating Earth gravitational action during space flight. But be on guard; it is a change completely abstract to our experience.

The time has now come to use the achievements of space technology for removing the Earth gravity action in a large number of processes, which may start a new phase of technical civilization. From an evolutionary point of view, the most powerful assurance of a high potential of space manufacuring lies in the effective utilization of space for life on Earth, as this is the ultimate purpose of space endeavors. It is an area of great promise and even greater challenge.

References

[1] Wuenscher, H. F., *Method of Making Foamed Materials in Zero-g*, Patent 3, 592, 628, original date July 66.
[2] Wuenscher, H. F., *Space Manufacturing Machine*, Patent 3, 534, 926, original date July 66.
[3] Steurer, W. H. and Gorham, D. J., *Processes for Space Manufacturing*, General Dynamics/Convair, Report June 1970.
[4] *Space Processing and Manufacturing*, NASA/MSFC ME-69-1, 32 Presentations, Oct. 1969.
[5] Steurer, W. H., *Selected Examples for Space Manufacturing Processes, Facilities and Experiments*, General Dynamics/Convair, Seventh Space Congress, April 1970.
[6] Kober, C. L., *Chemical and Biochemical Space Manufacturing*, Martin-Denver, Seventh Space Congress, April 1970.
[7] Frost, R. T., *Techniques and Examples for Zero-g Melting and Solidification Processes*, General Electric, Seventh Space Congress, April 1970.

[8] Wuenscher, H. F., *Unique Manufacturing Processes in Space Environment*, NASA/MSFC ME-70-1, April 1970.

[9] McKannan, E. C. and McCreight, L. R., *Electrophoresis Separation in Space – Apollo 14*, NASA TM X-64611, August 1971.

[10] Yates, I. C., *Apollo 14 Composite Casting Demonstration*, NASA PT-71-1, July 1971, and TMX-64641, March 1972.

[11] Parks, P. G., *et al.*, *Experiment Summary/Materials Processing in Space M512*, NASA Form 1347, June 1971.

SKYLAB STUDENT PROJECT

KENNETH S. KLEINKNECHT (MANAGER)

Skylab Program, Manned Spacecraft Center, National Aeronautics and Space Administration,
Houston, Tex., U.S.A.

and

JAMES E. POWERS, JR.

Skylab Program Office, National Aeronautics and Space Administration, Washington, D.C., U.S.A.

Abstract. As part of the Skylab program, in which the first United States experimental space station will be launched and operated, the National Aeronautics and Space Administration (NASA) has offered to students of the United States the opportunity to participate as scientific investigators. This part of the Skylab program is called the Skylab student project and was conceived to stimulate interest in science and technology by directly involving students in a major research program. The student project is under the joint sponsorship of the NASA, an agency of the United States Government, and the National Science Teachers Association (NSTA), a non-Government professional organization of science teachers.

The NSTA conducted the national competition. Over 4000 students, ranging in age from 11 to 19 yr, submitted proposals from which 301 regional winners and, then 25 national winners were selected. These selections were based solely on their scientific and technical merit and were chosen because of the creative ability displayed by the student. The NASA evaluated each of the 25 winning proposals to determine which experiments could actually be performed with minimum impact to the Skylab program. As a result of this evaluation, 19 of the 25 are planned to be accomplished in flight.

The 19 student experiments that will be performed fall into three categories. First, for two experiments, we can provide data that will be gathered for other Skylab experiments. Second, there are six proposals for which we can satisfy the data requirements by modifying the way we plan to use existing Skylab equipment. For experiments in these two categories, the student will be associated with a Skylab principal investigator who will assist the student in developing his experiment and obtaining and evaluating the data he will require. Third, the remaining eleven experiments require new equipment.

Ways to provide broad classroom participation are being planned. Included will be the provision to teachers of information packages describing the experiments and suggesting ways to conduct associated experiments in the classroom.

The response to the Skylab student project has been enthusiastic and it is foreseen that we will achieve our objective of stimulating interest in science and technology, both among students and among teachers.

1. Introduction

In October of 1971, as a part of its overall educational program, the National Aeronautics and Space Administration (NASA) recognized the opportunity to involve students in the Skylab program by direct participation as investigators; thus, the student project was conceived to stimulate interest in science and technology by directly relating students to a research program. The Skylab student project was open to students in all secondary schools of the United States and to United States students in foreign countries.

2. Skylab Program Description

The Skylab program capitalizes on the capabilities and resources developed in the Apollo program in order to accomplish scientific, technological, biomedical, and

L. G. Napolitano et al. (eds.), Astronautical Research 1972, 309–317. All Rights Reserved

Earth-application investigations in space. It will also develop the techniques for man to operate in space for increasingly longer periods of time, thus building on the foundation for future major steps in manned exploration of the universe. Skylab is a manned space station that will orbit the Earth between 50 deg north and 50 deg south latitude (Figure 1) at an altitude of 435 km. It will fly over 75% of the Earth's surface, 90% of the world's population, and 80% of the food-producing areas of the world.

The first Skylab mission, scheduled in the first half of 1973, will consist of two launches, from the NASA John F. Kennedy Space Center, located at Cape Kennedy, Florida, approximately one day apart. The first launch will be that of the Skylab Saturn workshop atop a Saturn V launch vehicle. One day later, a modified Apollo command and service module (CSM), with a crew of three, is launched by a Saturn IB. The CSM will rendezvous and dock with the Skylab, the crew will activate the laboratory, and will perform experiments for up to 28 days.

The second mission, a revisit by another CSM, will be launched approximately 90 days after the first mission. This mission is planned for a duration of up to 56 days, during which time its crew of three will continue to perform assigned experiments.

The third mission will also be a revisit, launched approximately 90 days after launch of mission 2. It, too, is planned for a duration of up to 56 days, during which time the experiment activities will be completed.

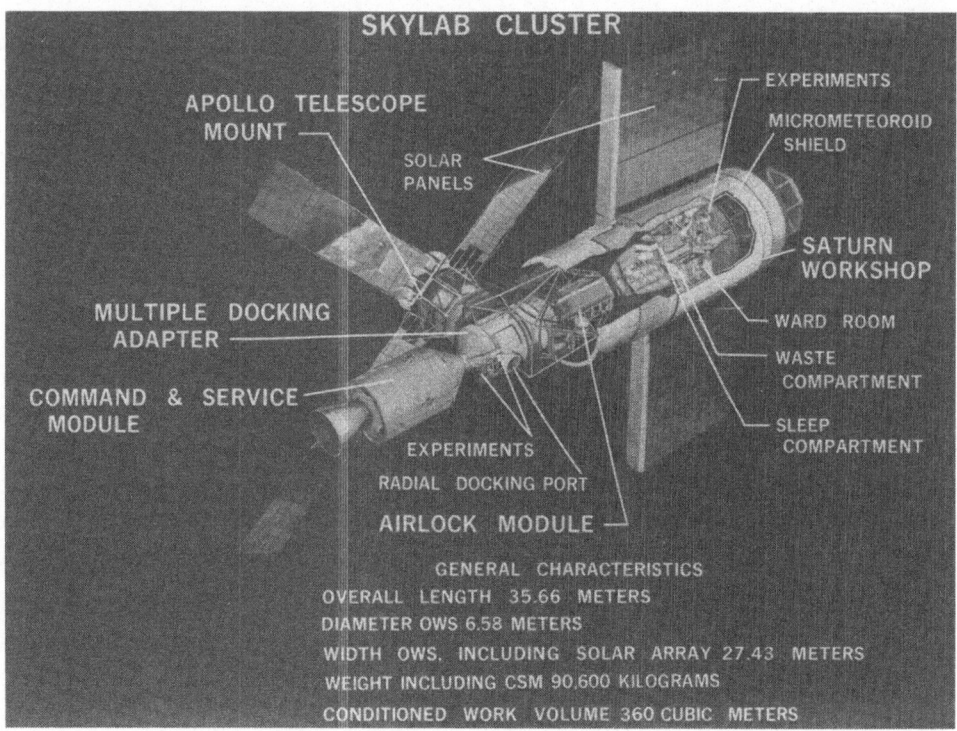

Fig. 1.

The on-orbit configuration of Skylab, commonly referred to as the 'cluster' (Figure 2), will be 35.66 m in length. The diameter of the orbital workshop will be 6.58 m; the total width of the orbital workshop, including the solar arrays that extend to the side for the purpose of generating electrical power, will be 27.43 m. The cluster, including the CSM, will weigh 90 600 kg. The volume of the cluster in which the crew works is 360 m^3.

The Skylab experiments fall into four categories of biomedical science, physical science, Earth applications, and space applications.

The biomedical-science experiments will increase man's knowledge of the functions of living organisms by making observations under conditions different from those on Earth, in order to determine the influence of the Earth's one-gravity environment upon these physiological functions.

The physical-science experiments will increase man's knowledge of the sun which will help him to understand its importance to Earth and to man's existence. With data taken from outside the Earth's atmospheric filter, the investigators will evaluate the lighting, radiation, and particle environment of near-Earth space and the radiations emanating from the Milky Way and remote regions of the universe.

Earth-applications experiments will develop techniques for observing Earth phenomena from space, in the areas of agriculture, forestry, geology, geography, hydrology, air and water pollution, land use, and meteorology as well as the influence of man upon these ecological elements.

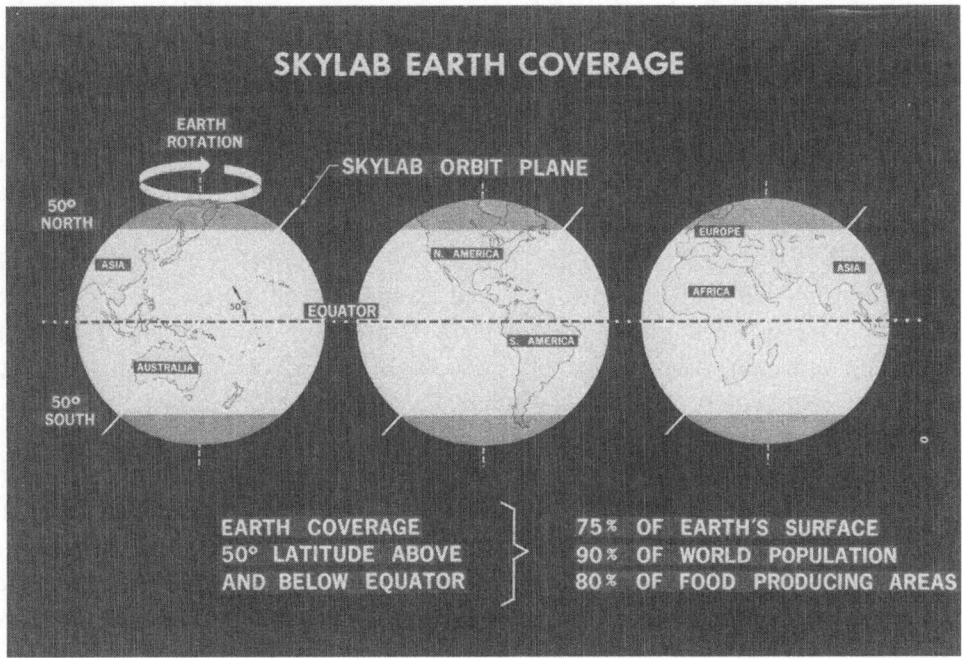

Fig. 2.

The space-applications experiments will develop improved techniques for space operations in the areas of crew habitability, crew-vehicle interrelationships, and space-vehicle structures and materials. The data obtained will be used to evaluate various equipments necessary for successful long-period habitation of the unique environment of space.

3. Skylab Student Project

The Skylab student project is under the joint sponsorship of the NASA, an agency of the United States Government, and the NSTA, a non-Government professional organization of science teachers.

The NASA established general guidelines and constraints that would not limit the student's creativity, but would still assure simplicity and compatibility with the Skylab program. Any equipment needed to perform the selected experiments was to be furnished by the NASA, and each student proposal had to be certified by a teacher or advisor that it was the independent work of a student or a team of students.

A preliminary indication of the success of the objective to stimulate interest in science and technology was evidenced by the fact that over 9000 science teachers requested entry kits for over 87 000 students. The requests came from every state in the United States as well as from United States students residing in foreign countries. Over 4000 students, ranging in age from 11 to 19 yr, submitted proposals from which 301 regional winners and, then, 25 national winners were selected. The initial selection process, performed for the NSTA in each of their 12 regions, was conducted by members of the teaching and industrial-research communities, with no knowledge of the student's name, sex, background, or where he lived. These experiments were selected solely on their scientific and technical merit.

A second selection process was conducted by the National Science Teachers Association in late March of 1972, in Washington, D.C., to identify the 25 national winners; these 25 were to be the candidate experiments to fly in Skylab. These finalists, with their teacher or sponsor, will be rewarded by a trip to the Kennedy Space Center, where they will participate in a Skylab educational conference, and will witness the initial Skylab launches. Each of the students and teachers also has received a medallion, as has their school. The quality of the proposals received was outstanding and there were many original, imaginative ideas. Because of this general high quality, after the selection of the 25 national winners, 22 additional proposals were recognized as worthy of special mention.

After the selection of the national winners, it was necessary for the NASA to determine which experiments could be accommodated on Skylab. Teams of scientists and engineers analyzed each experiment to determine the feasibility of performing it in Skylab. Preliminary design preparation, estimated crew time required, and analysis of the viewing opportunities for earth and astronomical observations were accomplished. Hardware stowage provisions and the effect the experiment might have on the integrity of the Skylab missions and on crew safety were also analyzed. The necessity to keep any new equipment simple and basically self-contained had to be the paramount consideration, in view of the fact that nearly all other Skylab flight equipment had been manufactured and was already undergoing testing with the major Skylab flight modules.

The preliminary results of these design and analytical activities were presented NASA management in early May of 1972. In these reviews, the students participated, together with NASA scientists and engineers, in order to determine what problems existed for flight of each experiment as well as the overall group. We were very pleased when it became apparent that we would be able to accommodate many more than the half-dozen or so that we had originally anticipated. The students were very impressive, both in the knowledge they displayed as well as the poise with which they conducted themselves in what could have been an overwhelming environment; also, the attitude and enthusiasm of the NASA participants were outstanding. We were able to select 19 of the 25 experiments as the Skylab student project flight complement. The remaining six proposals were not selected to be flown because either their performance requirements were inconsistent with the Skylab environment or the equipment required entailed development programs that were incompatible with the Skylab schedule.

Several Skylab principal investigators have volunteered to assist the students throughout the Skylab program and very meaningful relationships are anticipated. In addition, arrangements have been made to keep the six students involved whose experiments will not be performed on Skylab.

The 19 investigations that we are planning to perform fall into three basic categories: (1) For two experiments, we can provide data that we will obtain through our other Skylab experiments; (2) Six experiments require data only and no new equipment, but for these, some change will be required in the way we plan to use the existing Skylab equipment; and, (3) Eleven experiments require new equipment. As noted earlier, for all of these, the equipment will be as simple as possible.

In the first category, there are two investigations that can use data that already were to be made available; these are as follows:

'Photography of Libration Clouds'
Proposed by Alison Hopfield of Princeton, New Jersey

This experiment will use the Skylab white-light coronagraph to obtain information on two regions in the moon's orbit. At two points in the orbit of the moon, ahead of and following the Moon in its path, a condition of gravitational equilibrium is conducive to the collection of particles.

'Possible Confirmation of Objects Within Mercury's Orbit'
Proposed by Daniel C. Boschler of Silverton, Oregon

The aim of this experiment is to detect planetary bodies whose orbits lie within the orbit of Mercury. It is theorized that an additional planet, tentatively named 'Vulcan', may exist in the solar system at an orbital radius about $\frac{1}{10}$ of Earth's orbital radius.

The six investigations in the second group of student experiments require data that can be obtained by existing Skylab scientific instruments; these are:

'Earth's Absorption of Radiant Heat'
Proposed by Joe B. Zmolek of Oshkosh, Wisconsin

The objective is to derive information on the attenuation of heat energy in the Earth's atmosphere.

'Space Observation and Prediction of Volcanic Eruptions'
Proposed by Troy A. Crites of Kent, Washington

The aim is to analyze infrared surveys of known volcanoes obtained by the baseline Skylab Earth-sensing experiment equipment. The data will be compared to ground-based data to determine whether remote sensing can detect increased thermal radiation which precedes an imminent eruption.

'Spectroscopy of Selected Quasars'
Proposed by John C. Hamilton of Aiea, Hawaii

Selected photographs obtained by the ultraviolet stellar telescope will be analyzed.

'X-Ray Content in Association with Stellar Spectral Classes'
Proposed by Joe W. Reihs of Baton Rouge, Louisiana

Observations will be made, using the X-ray telescopes, of celestial regions in X-ray wavelengths in an attempt to relate X-ray emissions to other spectral characteristics of stars observed.

'X-Ray Emission from the Planet Jupiter'
Proposed by Jeanne L. Leventhal of Berkeley, California

The intent of this experiment is to attempt to detect X-rays emitting from Jupiter.

'A Search for Pulsars in Ultraviolet Wavelengths'
Proposed by Neal W. Shannon of Atlanta, Georgia

The objective is to make ultraviolet observations of selected celestial regions in an attempt to relate ultraviolet emissions with known radio emitting pulsars and with emissions from the pulsar in the Crab Nebula.

There are eleven experiments in the third category; these are:

'Web Formation in Zero Gravity'
Proposed by Judith S. Miles of Lexington, Massachusetts

The student will observe the web-building process and the detailed structure of the web of the common cross spider (Araneus Diadematus) in an Earth environment and in a Skylab environment.

'Earth Orbital Neutron Analysis'
Proposed by Terry C. Quist of San Antonio, Texas

Approximately 16 detectors are mounted in several locations inside the Skylab to record impacts of high-energy neutrons.

'An In-Vitro Study of Selected Isolated Immune Phenomena'
Proposed by Todd A. Meister of Jackson Heights, New York

This experiment is structured to determine if the absence of gravity affects representative life processes. It contains two parts:

In Part 1, 'Chemotaxis', guinea pig macrophage and casein will be injected into each side of a two-compartment vessel.

In Part 2, 'Antigenicity', the effect of the Skylab environment on the interaction between antibodies and antigens is studied by injecting antibodies of varying concentrations into an antigen-impregnated gel.

'Cytoplasmic Streaming in Zero Gravity'
Proposed by Cheryl A. Peltz of Littleton, Colorado

The purpose of this experiment is to perform microscopic observation of leaf cells of plants in zero gravity to determine if there is any difference between the motion of the intracellular cytoplasm and the cytoplasmic motion of similar leaf cells on Earth.

'Plant Growth in Zero Gravity'
Proposed by Joel G. Wordekemper of West Point, Nebraska

and

'Phototropic Orientation of an Embryo Plant in Zero Gravity'
Proposed by Donald W. Schlack of Downey, California

It has been possible to combine and package these two experiments into a single joint experiment which has the following objectives:

(1) To determine the differences in root and stem growth and orientation of rice seeds in specimens grown in zero gravity and on Earth under similar environmental conditions.

(2) To determine whether light can be used as a substitute for gravity in causing the roots and stems of rice seeds to grow in the appropriate direction in zero gravity as well as to determine the minimum light level required.

'Behavior of Bacteria and Bacterial Spores in the Skylab
and Space Environments'
Proposed by Robert L. Staehle of Rochester, New York

In this experiment, colonies of various species of bacteria will be studied in the Skylab zero-gravity environment to determine if this environment induces variations in survival, growth, and mutations of the spores which are different from those observed in identical colonies on Earth.

'Capillary Action Studies in a State of Free Fall'
Proposed by Roger G. Johnston of St. Paul, Minnesota

The aim of this experiment is to determine if the zero-gravity environment induces changes in the characteristics of capillary and wicking action from the familiar Earth-gravity characteristics.

'Wave Motion Through a Liquid in Zero Gravity'
Proposed by W. Brian Dunlap of Youngstown, Ohio

The objective of this experiment is to observe the motion of a gas bubble surrounded by a fluid when excited by a calibrated oscillation.

'Zero-Gravity Mass Measurement'
Proposed by Vincent W. Converse of Rockford, Illinois

The purpose of this experiment is to develop a simple mechanical system which demonstrates the principals of mass measurement in a zero-gravity environment.

'A Quantitative Measure of Motor-Sensory Performance
During Prolonged Inflight Zero Gravity'
Proposed by Kathy L. Jackson of Houston, Texas

This experiment uses a standard eye-hand coordination test apparatus to measure motor-sensory skill of crew members before, during, and after Skylab flights.

The six experiment proposals selected as national winners, but which we cannot perform on Skylab are:

'Effects of Intermittent Long-Duration Exposure to Zero and Artificial Gravity'
  Proposed by Keith L. Stein of Westbury, New York

'Chicken Embryology in Zero Gravity'
  Proposed by Kent M. Brandt of Grand Blanc, Michigan

'Effect of Zero Gravity on the Colloidal State of Matter'
  Proposed by Keith D. McGee of Garland, Texas

'Testing Flow Properties of Powdered Solids in Zero Gravity'
  Proposed by Kirk M. Sherhart of Berkeley, Michigan

'Brownian Motion and Dissolution of a Salt in Zero Gravity'
  Proposed by Gregory A. Merkel of Springfield, Massachusetts

'Universal Gravitational Constant Determination in Space'
  Proposed by James E. Healy of Bayport, New York

4. Conclusion

We are enthused with the response to the Skylab student project and foresee that we will achieve our goal of stimulating interest in science and technology, both among students and among teachers. In addition to the direct involvement of students in the Skylab activities, we are, through our educational program channels, providing the Nation's science instructors with teacher-information packages concerning the Skylab student experiments.

The Skylab student project is the first venture that involves this type of close relationship between space research and classroom science and technology. We are confident that it will prove extremely worthwhile and look forward to continued involvement of students and their teachers in our space flight programs.

SAFETY IN YOUTH ROCKET EXPERIMENTS (SYRE)

(Summary of Session)

A. INGEMAR SKOOG (EDITOR)

Chairman of the 4th SYRE Session, Dornier System, G.F.R.

and

GEORGE S. JAMES (CO-EDITOR)

Chairman of the SYRE Study Group, Rocket Research Institute, Inc., U.S.A.

The 4th SYRE session was mainly dedicated to the presentation of rocket safety regulations in Europe and the United States of America and the results of the First European Youth and Space Conference.

The review of international youth rocket activities was continued with a presentation of the zinc-sulphur rocket propellant project by the group NERO in the Netherlands.

Model rocket engine manufacturing and the use of model rocketry in high school aerospace curricula in the U.S.A. were also topics on the program.

One of the main tasks of the SYRE Study Group is the elaboration of standardized safety regulations for international use in youth rocket activities. A step towards these new regulations is the reviewing of existing national regulations. The first result of the SYRE Study Group work was presented in the two survey-papers on safety regulations in Europe and U.S.A. Due to their importance for the future work of SYRE, these two papers will be partly reproduced, together with a short review of the results of the First European Youth and Space Conference.

1. Preliminary Survey of European Rocket Safety Regulations

(A. Ingemar Skoog, *Dornier System, G.F.R.*)

Safety regulations for youth rocket activities in Europe is a very diverse and complex system of laws and regulations. Every country has its own set of regulations and they vary from complete prohibition to no restrictions at all. Only model rocketry has a standardized safety procedure due to its character of sport activity.

The existing regulations are mainly of three different kinds (Figure 1):

 – Governmental laws including regulations for Handling of Explosives, Propellant Manufacturing and Air Traffic.

 – Regulations, due to the use of particular facilities, like Military Safety Regulations, Police Permission and Insurance Requirements.

 – Safety regulations established by a central agency or by the rocket group itself.

1.1. TYPES OF REGULATIONS

In a few countries like France, Italy and Spain *Ministerial Decrees* have been issued

L. G. Napolitano et al. (eds.), Astronautical Research 1972, 295–307. All Rights Reserved
Copyright © 1973 by D. Reidel Publishing Company, Dordrecht-Holland

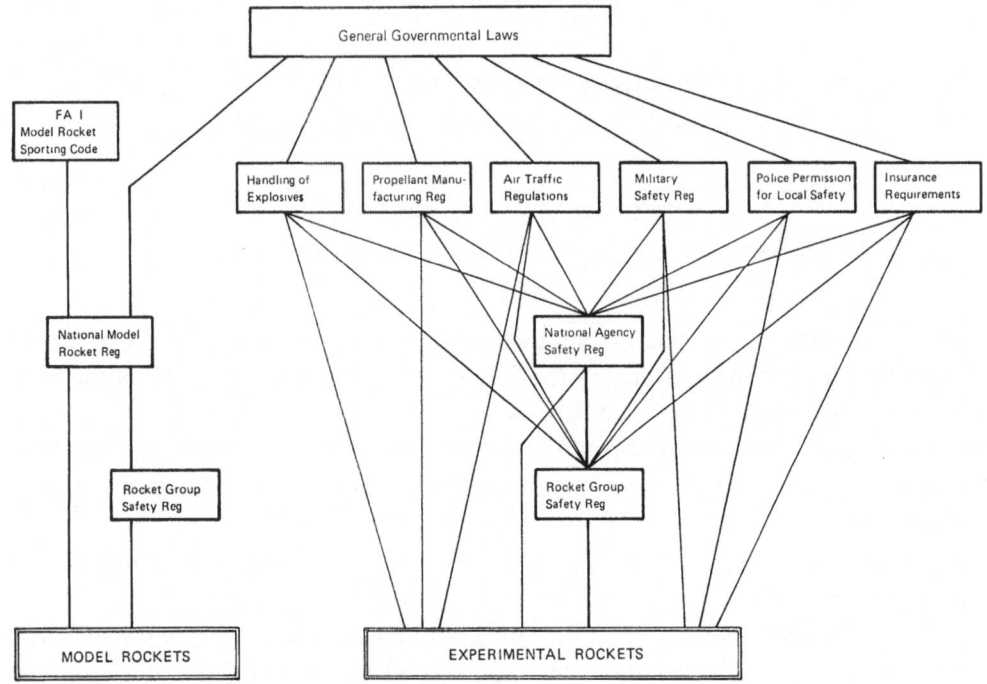

Fig. 1.   Regulations applicable to safety in youth rocket activities.

in order to regulate the youth rocket activities. These decrees appoint a national agency or the military to be responsible for the activities. The responsibility for safety procedures is given to the authority to be in charge of the youth rocket activities.

Regulations for Handling of Explosives contain information on handling, storage, transport and trade with various types of explosives. Generally a certain minimum age is given for those to handle explosives. Many of the components used for rocket propellants are contained in these regulations. To this kind of regulations can also be counted regulations for fireworks, which due to their character are directly applicable to rockets.

Propellant Manufacturing Regulations are very similar to those for Handling of Explosives and contain advice on how to handle various chemical components. Age limitations are generally included.

In order to provide maximum safety for air traffic many countries have issued *Air Traffic Regulations* concerning the launching of rockets. These regulations generally require the use of areas outside any controlled airspace and sometimes radio or telephone contact with the next air traffic control. In some countries permissions for rocket launchings have to be applied for by the civil air traffic board. ·

Military Safety Regulations are very frequently used, due to the fact that military ranges are the only possible launch areas in many countries. In some countries the use of military facilities are stated in governmental decrees (e.g. Italy and Spain).

In cases where rockets are to be launched outside military ranges a *Police Permis-*

sion for Local Safety is often required. This is to ensure that only those areas are used, where no risk for man or property exist.

Insurance Requirements exist in some countries and are voluntarily used in some others.

Model rocket activities are generally excluded from the above mentioned regulations if those activities follow the *FAI Model Rocket Sporting Code* or a national model rocket regulation based on the FAI regulations. The FAI Model Rocket Sporting Code gives specifications on the rocket, the model rocket engine, the launching and the competition with model rockets. A maximum rocket weight of 500 grams with a maximum propellant weight of 125 grams is stated. The engine must not produce a total impulse of more than 100 Ns and must have a thrust duration longer than 0,050 s. No substantial metal parts shall be used.

The above mentioned regulations are either used directly in their original form or they are incorporated in National Agency Safety Regulations and/or Rocket Group Safety Regulations (Figure 1).

1.2. EXPERIMENTAL ROCKETS

Most European countries do have some kind of regulations which applies to experimental youth rocket activities (Figure 2). Military safety regulations are used due to the fact that in practically all countries military ranges are either requested or the only available. Frequently used are also regulations for Handling of Explosives and Air Traffic. Known regulations are listed in Table I.

Austria: Regulations for Handling of Explosives and Air Traffic are the two general ones. Military Safety Regulations are used as military ranges are the only available launching areas. As a result of the second world war peace-treaty certain restrictions for manufacturing of rockets exist.

Belgium: Handling of Explosives and Air Traffic Regulations govern the activities in Belgium. Rocket launchings have been performed under supervision of the Royal Military Academy, which has applied army safety regulations.

Denmark: Handling of Explosives, Propellant Manufacturing Regulations and the Use of Fireworks have to be followed. Military ranges have been used.

German Federal Republic: Regulations for Handling of Explosives and the Use of Fireworks exist. Rockets with a diameter over 0,25 m are not allowed to be manufactured. Military Safety Regulations are used in most cases.

France: In a ministerial decree the responsibility for youth rocket activities are given to Centre National d'Études Spatiales (CNES), which is supervising the youth organization Association Nationale des Clubs Aerospatiaux (ANCS). CNES provides commercially manufactured rocket engines and is in charge of the rocket handling during launch campaigns. Military ranges are used and military assistance for e.g. ground support and radar tracking is given. In the instructions of ANCS an insurance for each launching is requested. Private construction of rockets is forbidden.

Holland: The Dutch rocket club NERO uses Military Safety Regulations, and

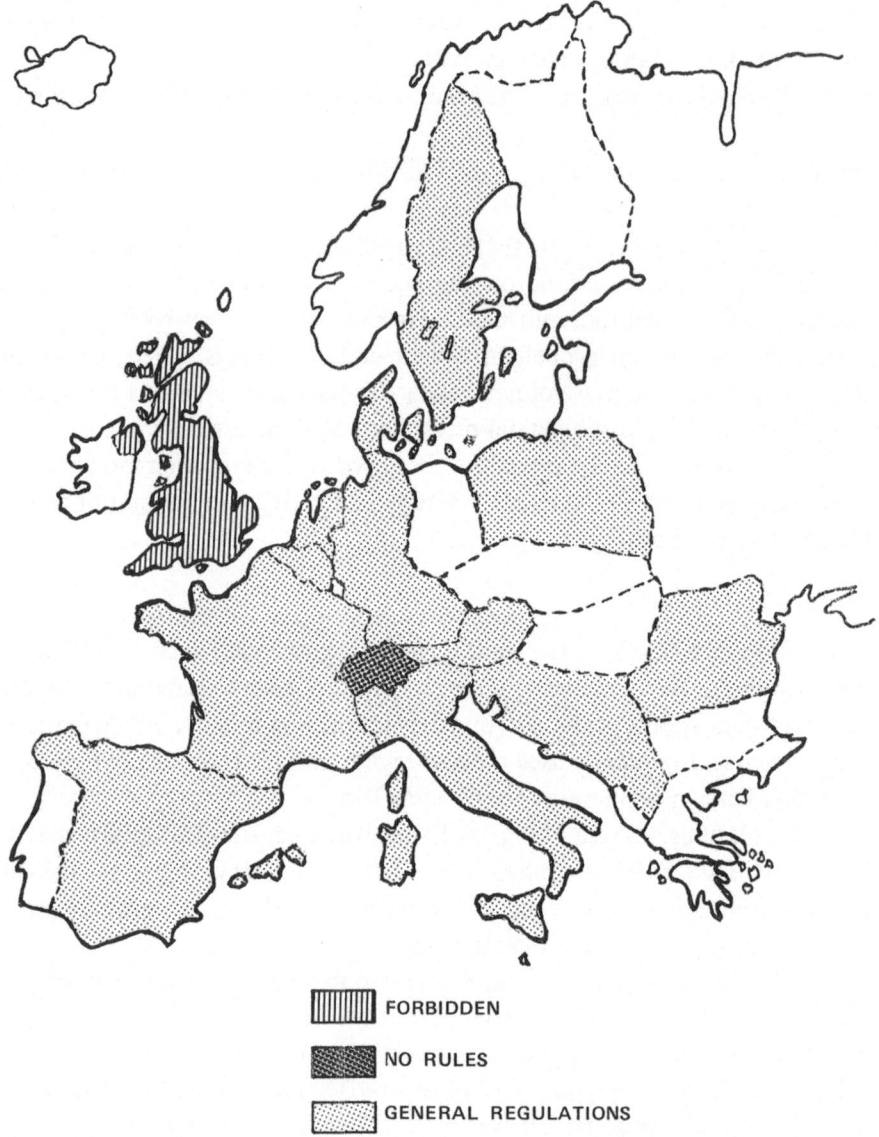

Fig. 2.   Regulations for experimental rockety in Europe.

Fig. 2.   Regulations for experimental rockety in Europe.

it has also been able to get a voluntary insurance for its launchings.

Italy: Only military ranges are to be used and the launchings shall be controlled by the Air Force according to a general decree from the Ministry of Interior. Furthermore an insurance has to be paid for and military supervisors will assist.

Poland: Air Traffic Regulations form the basis for rocket activities in this country.

Rumania: Air Traffic Regulations concerning rocketry exist and Military Safety Regulations are used.

TABLE I

Safety Regulations in Europe

Country	Model Rockets	Experimental Rockets
Albania		
Austria	FAI	Handling of Explosives, Air Traffic Regulations, Military Safety Regulations.
Belgium	FAI	Handling of Explosives, Air Traffic Regulations, Military Safety Regulations.
Bulgaria	FAI	
Czechoslovakia	FAI	
Denmark		Handling of Explosives, Propellant Manufacturing Regulations, Use of Fireworks, Military Safety Regulations.
Federal Republic of Germany	FAI + restrictions	Handling of Explosives, Use of Fireworks, Military Safety Regulations.
Finland		
France	FAI	National Agency, Military Safety Regulations, Insurance Requirement, Handling of Explosives.
German Democratic Republic	FAI	
Greece		
Holland		Military Safety Regulations.
Hungary	FAI	
Iceland		
Ireland		
Italy	FAI	Military Safety Regulations, Insurance Requirement.
Luxemburg		
Norway	FAI	
Poland	FAI	Air Traffic Regulations.
Portugal		
Rumania	FAI	Air Traffic Regulations, Military Safety Regulations.
Russia	FAI	
Switzerland	FAI	No restrictions!
Spain	FAI	Central Agency, Military Safety Regulations, Air Traffic Regulations.
Sweden	FAI	Handling of Explosives, Air Traffic Regulations, Police Permission for Local Safety.
United Kingdom	FAI (to join)	Forbidden!
Yugoslavia	FAI	Air Traffic Regulations, Military Safety Regulations, Police Permission for Local Safety.

Switzerland: No restrictions. An old black powder law from 1874 is not applicable. The Swiss Association for Rocketry appoints a member to be responsible for safety during rocket launchings.

Spain: The Government has prescribed rules by means of which all experimental rocketry has to be made under the authority of the National Institute of Aerospacial (INTA), which controls the rocket engine from a safety point of view. The launchings take place at military ranges with military ground equipment. Attention must be paid to the Air Traffic Regulations.

Sweden: The Handling of Explosives is regulated in a Royal decree (KF of June 10, 1949). The launching of rockets is regulated in the Aeronautical Information

Circular (AIC) B 108/1967 of January 26, 1967 issued by the Civil Air Traffic Board, which gives permission for launches. The local police authority has to be informed too. In some cases military ranges with military supervisors have been used.

United Kingdom: Launchings of rockets are forbidden.

Yugoslavia: Air Traffic Regulations have to be followed. Military ranges are used, but Police Permission must be given.

1.3. MODEL ROCKETS

Unlike experimental rocketry, standardized regulations exist for model rocketry, as almost all European countries have through the national aeroclubs adopted the FAI Model Rocket Sporting Code (Figure 3). The code (latest issue approved by CIAM, as official as of January 1, 1970) contains all necessary information like:

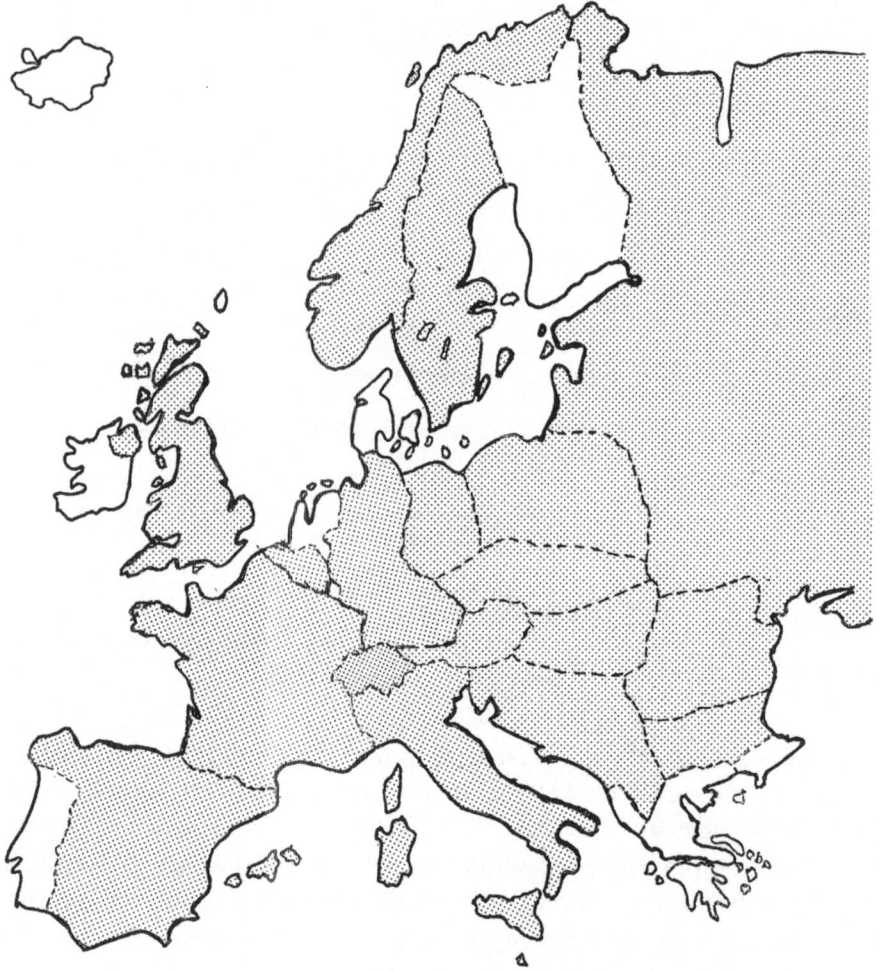

Fig. 3.   FAI model rocket sporting code in Europe.

- Model Rocket Specifications,
- Model Rocket Engine Standards,
- Launching,
- Competition Regulations.

This code requires that all models are approved by the Range Safety Officer before flight.

The code is a necessity because of the nature of sport activity with world championships, which must be based on one international standard regulation.

In most countries model rocketry is excluded from more severe regulations as long as the FAI-regulations are followed.

In a few countries the national model rocket regulations are somewhat more restricted than the FAI Model Rocket Sporting Code. In the German Federal Republic police permission must be given, and propellant loads of more than 20 g are only to be ignited by a professional fireworker. The Swedish AIC-regulations limits the maximum propellant mass to 100 g.

1.4. Conclusions

In the last years an increasing cooperation between various rocket groups in different European countries has taken place. This requires more uniform safety regulations for all participating countries. It is likely that existing regulations will be modified to give a simular recommendation for the safety in experimental rocket activities like the FAI code for model rocketry.

2. Preliminary Survey of Rocket Safety Regulations in the Fifty States of the United States of America

(Edward A. Quarterman, *UNS-RET*, *U.S.A.*)

Data for this survey were generated by requesting of each of the Governors of the fifty States a copy of the safety regulations under which 'Youth Rocket Experiments' were undertaken.

A resume of the data is shown in convenient categories in Table II. Category 'A' shows those states not having specific statutory regulations. It is to be understood that the absence of such regulations does not indicate that 'Youth Rocket Experiments' may be undertaken within these states in a careless and dangerous manner. All of the states have statutes which provide for the protection of life and property. Governor Richard B. Ogilvie, Governor of the State of Illionis in his letter of reply states that 'there are no safety regulations specifically regulating rocket experiments'. However, he cites two statutes which would be generally applicable. One of these is an 'Act to Regulate Storage, Sale and Use of Volatile Combustibles' (State Fire Marshal), and the other is a statute, 'Uniform Hazardous Substances Act' (Public Health). The Governor further points out, that "it is expected that the State rule for model rocket propellant devices will be the same as the rules under the Federal Hazardous Substances Labeling Act."

TABLE II

Rocket Safety Regulations in the United States of America

The respective states may be grouped into the following typical categories with respect to 'Rocket Safety Regulations'.

(A) Those states not having specific statutory regulations:
Alaska, Arizona, Georgia, Illinois, Iowa, Mississippi, Nebraska, Nevada, New Hampshire, New York, North Dakota, Ohio, Oklahoma, South Carolina, South Dakota, Tennessee, Virginia, West Virginia.
(B) Those states having statutory regulations which are concerned with previously existing regulations related to fireworks:
Delaware, Idaho, Kansas, Maryland, Missouri, Montana, Pennsylvania, Vermont.
(C) Those states having statutory regulations related to explosives and related items:
Florida.
(D) Those states referencing 'Model Rocketry Safety Code':
Hawaii, Mississippi, New Mexico, Wisconsin.
(E) Those states referencing 'National Fire Protection Association Code of Model Rocketry' NFPA No. 41L-1968:
Alabama, Colorado, Kentucky, Maryland, Minnesota, Nebraska, Wyoming.
(F) States in which 'rockets' probably are forbidden by existing criminal statutes:
Utah.
(G) States regulating 'Rockets or Missiles' under State Aviation Commissions:
Indiana, Maine, Washington.
(H) States regulating 'Model Rockets' under Departments of Education:
Michigan.
(I) States regulating rockets under State Fire Marshal independent of fireworks regulations:
Oregon.
(J) States having explicit rocket regulations:
California, Connecticut, Massachusetts, New Jersey.
(–) No information:
Arkansas, Louisiana, North Carolina, Rhode Island, Texas.

From a scientific point of view, the absence of a statute relating to rocket safety indicates that the specifics relating to rockets, their manufacture, and use have not been promulgated.

Category 'B' of Table II shows those states in which statutory regulations, which are concerned with Fireworks, are cited in the survey. Fireworks displays are often part of festivities at which large numbers of the public are assembled in celebration or anniversaries, such as Independence Day.

Category 'C' of Table II shows that the State of Florida cites statutory regulations related to explosives and related items. These are contained in "Rules of the State Treasurer and Insurance Commissioner as State Fire Marshal, Chapter 4A20 (Fire Protection, Hazardous Chemicals), also Chapter 4A-2 (Explosives), also Chapter 791 (Sale of Fireworks)".

Category 'D' of Table II shows those states referencing 'Model Rocketry Safety Code'.

Mr R. C. Roberts, Supervisor of Science, Coordinator, Title III, NDEA, State of Mississippi Department of Education, Division of Instruction states: "State Department of Education does not have 'prescribed' regulations for conducting rocket experiments in the schools of Mississippi. – The State Department of Education has given some caution and guidance in this matter through a publication of the State Department of Education, The 'Science and Math Newsletter', Volume 6, April 1965. This 'Model Rocketry Safety Code' is one of their own."

Category 'E' of Table II shows those states referencing the 'National Fire Protection Association Code of Model Rocketry' NFPA No. 41L-1968.

The State of Minnesota, Department of Public Safety, Fire Marshal Division, states: "... we currently use the National Fire Protection Association's Pamphlet 'Model Rocketry' as our guide line. – We are not aware of any specific State Statute which gives any State agency the authority to strictly enforce Pamphlet No. 41L or any other regulation on model rocketry."

Category 'F' shows states in which 'rockets' probably are forbidden by existing criminal statutes.

Mr R. G. Ingersoll, Director of Utah Safety Council writes: "Your letter addressed to the Honorable Calvin L. Rampton, Governor of the State of Utah, regarding private rocket experiments by youth has been referred to this department for appropriate answer. The State of Utah, Criminal Statutes, adequately provide for penalties if any person constructs an apparatus calculated to endanger health, life, limb or property. The Utah Statutes covering explosives and rockets are herein quoted for your information:

76-18-9. *Infernal machine* defined. – An infernal machine is any box package, contrivance or apparatus containing or arranged with an explosive or acid or poisonous or inflammable substance, chemical or compound, or knife, loaded pistol or gun, or other dangerous or harmful weapon or thing, constructed, contrived or arranged so as to explode, ignite or throw forth its contents, or to strike with any of its parts, unexpectedly when moved, handled or opened, or after the lapse of time, or under conditions or in a manner calculated to endanger health, life, limb or property."

Category 'G' lists states regulating 'Rockets or Missiles' under State Aviation Commissions.

The State of Maine, Department of Aeronautics: "... is charged with the responsibility of enforcing the launching of rockets and missiles in the State of Maine as is set forth in Titel 6, Chapter 15 of the State of Maine Statutes relating to Aeronautics, Section 271."

The State of Indiana Aeronautics Commission under Rockets and Missiles Act, Acts of 1959 Chapter 221, requires the submission and approval of an "Application For Approval of Ramp, Rocket Site And Operational Range And Firing And Launching of Rockets." The form requires geographical coordinates, property data and designation of "supervisor in charge and responsible for operation".

Category 'H' lists states regulating 'Model Rockets' under Department of Education.

The State of Michigan, Department of Education administers the 'Model Rocket Law', Acts No. 333 to Public Acts 1965.

It should be noted that this law "... does not apply to the design, construction, production, maintenance, launch, flight, test, operation or use of, or any other activity in connection with a model rocket or model rocket engine, when carried on or engaged in by (a) the United States or this State, (b) a college, university or other institution of higher education or (c) an individual or an entity engaged in research, development or production of rockets, rocket engines or propellants, or components thereof, as a business."

Category 'I' – States regulating rockets under State Fire Marshal independent of fireworks regulations.

Governor Tom McCall states: "... this office is pleased to enclose herewith the regulations adopted by our Office of State Fire Marshal relating to the use of rockets in Oregon along with the standard permit application currently in use. There is no charge for rocket permits, and I am informed by the State Fire Marshal that the primary purpose of requiring permits is to assure that local and state officials will have basic information and control while at the same time holding restrictions to the minimum necessary."

Category 'J' – States having explicit rocket regulations.

The State of New Jersey is one of the states having explicit rocket regulations. Section 1 outlines Purpose and Scope, Section 2 provides Definitions, Section 3 General Provisions, Section 4 Permits, Section 5 Model Rocketry, Section 6 Amateur Rocketry, Section 7 Transportation and Storage.

In addition to the various regulations cited, Federal Aviation Regulations, Vol. VI Part 101 Subpart C, quoted below may apply:

SUBPART C–UNMANNED ROCKETS

§101.21 *Applicability*

This subpart applies to the operation of unmanned rockets. However, a person operating an unmanned rocket within a restricted area must comply only with subparagraph 101.23 (g) and with additional limitations imposed by the using or controlling agency, as appropriate.

§101.23 *Operating limitations*

No person may operate an unmanned rocket:
   (a) In a manner that creates a collision hazard with other aircraft;
   (b) In controlled airspace;
   (c) Within five miles of the boundary of any airport;
   (d) At any altitude where clouds or obscuring phenomena of more that five-tenths coverage prevails;
   (e) At any altitude where the horizontal visibility is less than five miles;
   (f) Into any cloud;
   (g) Within 1500 ft of any person or property that is not associated with the operations; or
   (h) At night.

§ 101.25 *Notice requirements*

No person may operate an unmanned rocket unless, within 24 to 48 h before beginning the operation, he gives the following information to the FAA ATC facility that is nearest to the place of intended operation:

(a) The names and addresses of the operators.

(b) The number of rockets to be operated.

(c) The size and weight of each rocket.

(d) The maximum altitude to which each rocket will be operated.

(e) The location of the operation.

(f) The date, time, and duration of the operation.

(g) Any other pertinent information requested by the ATC facility.

3. First European Conference Youth and Space

PARIS, FEBRUARY 23–27, 1972

The aim of this conference was to promote cooperation between various European youth organizations with rocketry on their program.

Association Nationale Des Clubs Aerospatiaux (ANCS) organized the conference, which was sponsored by CNES, UNESCO and the international youth science organisation CIC. Representatives from 12 European countries and observers from four non-European countries and several international organisations (UNESCO, ESRO, ELDO, IAF, UIT, FAI) were present.

The work of the conference was performed within two committees (Program Committee and Coordination Committee) and the results were approved by the Plenary Assembly as the recommendations of the conference.

THE PLENARY ASSEMBLY OF THE FIRST EUROPEAN CONFERENCE YOUTH AND SPACE:

Decided:

– to elect a Bureau for the period till the next conference. One member of the Bureau will be appointed observer by CIC.

– to hold the Conference annually.

– that the members of the Conference are the organizations represented and not the delegates.

– to invite all interested organizations to become members of the Conference.

– that the Bureau will be assisted by a secreteriat from ANCS.

Sets as a task for the Bureau:

– to study the structure of the Conference and give proposals to the members.

– to receive suggestions from the members and call for the next conference with a delay of three month.

– to establish a cooperation and negotiate with all national and international orga-

nizations which can contribute to the activities of the Conference (UNESCO, CIC, FAI, IAF etc.).
– to work out a questionaire in order to get the knowledge of the activities and experiences of the members.
– to be in contact with the members and other interested organizations.

Asks the represented organizations:

– to inform the Bureau of their activities.
– to make a list of different events concerning the exchange on a multilateral as well as bilateral level.
– to promote cooperation between responsible experts.
– to send a copy of each publication to the organisations named in the list mentioned in the next point.

Invites CIC:

– to publish an annual edition of its publication on the subject youth aerospace activities containing the following items:
    (a) A list of members and observers of the Conference.
    (b) The resolutions and recommendations of this Conference.
    (c) Articles proposed by the member organizations.
– to publish this edition in Russian too.
– to distribute this publications to all organizations mentioned in the list above.

Suggests that:

– the use of the FAI Model Rocket Sporting Code will be promoted among the members interested in model rocketry.
– a summer camp will be organized in 1973 for the information of the initiators.
– ANCS publishes in its magazine '3.2.1. ESPACE':
    (a) a list of magazines published by the members.
    (b) a list and summary of technical publications published by the members.
    The Bureau decides upon the language to be used.

Requests:

– the members and the Bureau to cooperate with the Safety in Youth Rocket Experiments (SYRE) Study Group of IAF concerning the establishment of international safety regulations for construction and launching of rockets and the use of these safety regulations.

Considers:

– a series of bilateral or multilateral projects to develop and launch experimental rockets during 1972–1975. Participating organizations are: ANCS (France), NERO (Netherlands), SAFR (Switzerland), DGLR (German Federal Republic), SPA (Poland), JSB (Belgium) and FEIC (Spain).

Decides that:

- the programs of cooperation will be regulated by contracts containing:
    - aims
    - persons responsible
    - distribution of the tasks
    - time schedule.
- the contracts will be registered by the Bureau which:
    (a) by means of coordination assists in the development of the programs,
    (b) regularly publishes to all European members information on project status, problems encountered and the solutions adopted.

Note:

The elected Bureau consists of

Mr Pierre Quetard (ANCS)	President
Mrs Galina Tchoubarova (USSR)	Vice President
Dr Roger Lo (DGLR)	Vice President
Mr Henry Lips (NERO)	Vice President
Mr Hervé Moulin (ANCS)	Secretary

PART IV

SECOND I.A.F. STUDENT CONFERENCE

LOW THRUST CONSTANT ACCELERATION TRAJECTORIES FOR A MERCURY ORBIT

WERNER-ANDREAS HEUSMANN

Deutsche Gesellschaft für Luft- und Raumfahrt eV (DGLR)

and

The AIAA, Münster/Westfalen, F.R.G.

Abstract. A probe launched into an orbit around the planet Mercury would be capable of giving new information about this planet. Using chemical propulsion systems, transfer orbits require rather high performance. Therefore the application of ion thrusters is necessary.

Based on parameters and the power supply of three German ion thrusters (RIT 10, ESKA 18 and HIT 5) the optimization of interplanetary trajectories in to a Mercury orbit is discussed. The different trajectories will be considered as trajectories in the ecliptic plane. The launch dates, the flight durations and the different mission profiles are presented.

Flight paths will start in Earth orbit. Parking orbits between 200 km and 1000 km are considered.

The interplanetary trajectories will end in an orbit around the target planet Mercury. The computations are made for trajectories under
- tangential acceleration,
- radial acceleration,
- constant acceleration.

All problems are analyzed, using an IBM-360/50 Digital computer and the final results are presented. The paper gives detailed information about the theoretical requirements and different equations of motion. Finally the basic computer program is discussed.

List of Symbols

M_0, M_i, M_I	total initial mass
M_P	propellant mass
M_W	power – plant
M_L	payload
M_{TE}	burnout or terminal mass
I_i	ion beam current
m	mass of the ion
e	elementary charge
η_m	factor
$\dot{M}_P$	mass flow
c	exhaust velocity
a	acceleration
b	acceleration
t	time
Δt	$t_2 - t_1$
F	thrust
g_0	gravitational acceleration of Earth
γ	gravitational constant
μ	$\gamma^* \, m_{\text{Earth}}$
κ	$\gamma^* \, m_{\text{Sun}}$
u	velocity of the circle orbit
I_{sp}	specific impulse
r, φ	coordinates
x, y	space coordinates

L. G. Napolitano et al. (eds.), Astronautical Research 1972, 335–347. *All Rights Reserved*
Copyright © 1973 by D. Reidel Publishing Company, Dordrecht-Holland

M or m	mass
$\dot{r}$	velocity (r – direction)
$\dot{\varphi}$	velocity (φ – direction)
$\ddot{r}$	acceleration (r – direction)
$\ddot{\varphi}$	acceleration (φ – direction)
b	acceleration vector
v	velocity vector
r	position vector
g	acceleration vector

Subscripts

0	at the departure planet
1	at the terminal planet
1	r – direction
2	φ – direction
r	radial – component
ϱ	tangential – component

Mercury is the smallest planet of our solar system and it moves nearest to the central force field Sun. Its numerical eccentricity has the greatest value $e=0.205\,615$, in comparison with the eccentricity of other planets. This is why the radial distance of Mercury exhibits large changes during one revolution. The geocentric distance from the Sun never exceeds $28°$. Moreover, the apparent angular diameter is about $10''$, making Earth-based Mercury observations very difficult. A Mercury probe offers good conditions for all observations.

Based on chemical propulsion systems, transfer orbits require high performance. Therefore the application of ion thrusters is necessary.

If we analyze the mass of such a probe we must consider the three different components

$$M_I = M_P + M_W + M_L, \tag{1}$$

where M_I is the initial mass of the satellite. The terminal mass will be determined by Equation (2)

$$M_{TE} = M_I - M_P, \tag{2}$$

hence for (3)

$$M_{TE} = M_W + M_L. \tag{3}$$

The thrust F will be determined by (4)

$$F = \frac{d}{dt}(Mc) = \dot{M}_P c = Ma. \tag{4}$$

We can substitute for $\dot{M}_P$ in Equation (1) and to get (5).

$$F = \dot{M}_P c = (M_P + M_W + M_L - \dot{M}_P t)\, a = (M_I - \dot{M}_P t)\, a. \tag{5}$$

Finally, we can write the mass flow in the form

$$\dot{M}_P = \frac{m_i}{e\eta_m} I_i, \tag{6}$$

m_i being the mass of the special atom. Moreover, the thrust F is now written as function of $\dot{M}_P$, c, η_m:

$$F = \eta_m \dot{M}_P c. \tag{7}$$

The term η_m characterizes the utilization of the propellant.

All computations have been made using performance data of the electrostatic engines RIT 10 and ESKA 18 and the Hall ion thruster HIT 4 which are listed in Table I.

TABLE I

Performance data	RIT 10	ESKA 18	HIT 4
Exhaust velocity	47.9 km s^{-1}	38 km s^{-1}	25 km s^{-1}
Propellant	Hg	Hg	argon
Ion beam current	100 mA	320 mA	5 A
Thrust	10 mN	25 mN	60 mN
Mass flow			2.5×10^{-6}

(See [1].)

Orbits in our solar system, especially motion in a central force field, are very complicated. Therefore it is difficult to determine the exact position of each body, its radius vector and the other elements which characterize the flight path. This is the reason why I made some simplifications in my calculations in order to determine the orbits
– the force field is spherically symmetrical,
– the central point of the Sun is looked upon as the centre of gravity,
– the influence of the seven other planets is neglected.
We can divide the flight path of a probe to an inner planet of our solar system into three phases:
– the departure from the planet Earth,
– the heliocentric transfer orbit,
– the arrival at the planet Mercury.
The radius r_1 of the departure orbit will be 200 km and 1000 km. The orbit is a circle. Therefore we can write for the orbit velocity

$$u = \sqrt{\frac{\mu}{r_1}}, \tag{8}$$

where u is the instantaneous orbit velocity of the departure orbit. The initial acceleration will be

$$a_i = \frac{F}{M_I}. \tag{9}$$

For the point of time t of escape we can say

$$t_{esc} = \frac{I_{sp}g_0}{b_0}\left(1 - e^{[\sqrt{\mu/r_1}]/I_{sp}g_0}\right). \tag{10}$$

The specific impulse is defined as

$$I_{sp} = \frac{c}{g_0}, \tag{11}$$

and therefore we can also write Equation (10) as

$$t_{esc} = \frac{c}{g_0}\left(1 - e^{[\sqrt{\mu/r_1}]/c}\right). \tag{12}$$

The propellant which will be need for the escape is determined by (13)

$$M_P = \frac{M_I a_i t_{esc}}{\eta_m c}. \tag{13}$$

We can say

$$M_{TE} = M_I - M_P \text{ (see Equation (2) and (3))}$$

and therefore we get

$$\Delta u = \eta_m c \ln\left(\frac{M_I}{M_{TE}}\right). \tag{14}$$

After having attained escape velocity, the transfer to the planet can begin.

We take as a basis the coordinates (x, y). These coordinates can be transformed in to the heliocentric polar coordinates (r, φ) by Equation (15)

$$\mathbf{r} = \begin{pmatrix} x \\ y \end{pmatrix} = \begin{pmatrix} r\cos\varphi \\ r\sin\varphi \end{pmatrix}. \tag{15}$$

The direction will be determined by e_1 and e_2:

$$\mathbf{r} = r\cos e_1 + r\sin e_2. \tag{16}$$

We compute the velocity by first order differentiation with respect to the time:

$$\frac{d\mathbf{r}}{dt} = \dot{r} =$$
$$= (\dot{r}\cos\varphi - r\dot{\varphi}\sin\varphi)\,e_1 + \tag{17}$$
$$+ (\dot{r}^2\sin\varphi + r\dot{\varphi}\cos\varphi)\,e_2,$$

and the second differentiation gives the acceleration:

$$\frac{d^2\mathbf{r}}{dt^2} = \ddot{r} =$$

$$= \{(\ddot{r} - r\dot{\varphi}^2)\cos\varphi - (r\ddot{\varphi} + 2\dot{r}\dot{\varphi})\sin\varphi\}\, e_1 + \qquad (18)$$
$$+ \{(\ddot{r} - r\dot{\varphi}^2)\sin\varphi + (r\ddot{\varphi} + 2\dot{r}\dot{\varphi})\cos\varphi\}\, e_2.$$

The different components are

$$\ddot{r} - r\dot{\varphi}^2 \qquad (19)$$

the radial acceleration and

$$r\ddot{\varphi} + 2\dot{r}\dot{\varphi} \qquad (20)$$

the tangential acceleration. Therefore we can write the acceleration in the form

$$b = \begin{pmatrix} b_1 \\ b_2 \end{pmatrix} = \begin{pmatrix} \ddot{r} - r\dot{\varphi}^2 \\ r\ddot{\varphi} + 2\dot{r}\dot{\varphi} \end{pmatrix}. \qquad (21)$$

In order to analyze the acceleration we must distinguish between three different kinds of acceleration:
 − the acceleration of the engines,
 − the acceleration of other planets,
 − the acceleration of the Sun.
This leads to Equation (22)

$$b = \begin{pmatrix} a_{\text{thrust}} \text{ and } a_{\text{perturbation}} \\ b_{\text{thrust}} \text{ and } b_{\text{perturbation}} \end{pmatrix}. \qquad (22)$$

The thrust will be analyzed by (23)

$$\mathbf{F} = \mathbf{F}_r + \mathbf{F}_\varphi. \qquad (23)$$

We can see the different directions of the thrust in Figure 1.

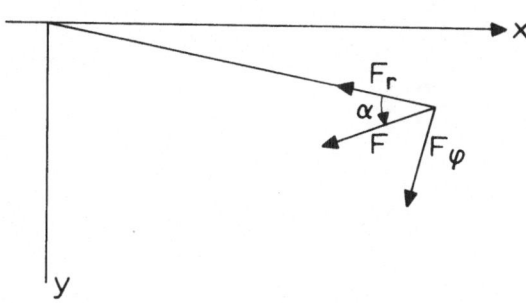

The angle between the r-direction component and the true acceleration direction is called α. Now we can write the following equations

$$\sin\alpha = \frac{F_\varphi}{F}$$

$$\cos\alpha = \frac{F_r}{F}$$

and this leads to

$$F_r = F\cos\alpha \qquad\qquad (24.1)$$

$$F_\varphi = F\sin\alpha \qquad\qquad (24.2)$$

The acceleration is determined by the general ratio

$$b = \frac{F}{M}.$$

Therefore we can transform Equation (21) by substitution of Equation (23) in Equation (25).

$$b = \begin{pmatrix} \dfrac{F_r}{M} \\[2mm] \dfrac{F_\varphi}{M} \end{pmatrix} = \begin{pmatrix} \dfrac{F}{M}\cos\alpha \\[2mm] \dfrac{F}{M}\sin\alpha \end{pmatrix}. \qquad\qquad (25)$$

We now consider the perturbation acceleration. We have already distinguished between the perturbation of the Sun, which will influence the radial component only, and the perturbation by the planets.

The perturbation caused by the Sun is

$$b = b_{\text{Sun}} = \frac{-K}{r^2}. \qquad\qquad (26)$$

Equation (25) leads to equation (27):

$$b = \begin{pmatrix} \ddot{r} - r\dot{\varphi}^2 \\[1mm] r\ddot{\varphi} + 2\dot{r}\dot{\varphi} \end{pmatrix} =$$

$$= \begin{pmatrix} \dfrac{F}{M}\cos\alpha - \dfrac{K}{r^2} \\[2mm] \dfrac{F}{M}\sin\alpha \end{pmatrix} \qquad\qquad (27)$$

and this leads to (28):

$$\begin{pmatrix} \ddot{r} \\[1mm] \ddot{\varphi} \end{pmatrix} = \begin{pmatrix} \dfrac{F}{M}\cos\alpha - \dfrac{K}{r^2} + r\dot{\varphi}^2 \\[2mm] \dfrac{F}{M}\sin\alpha - \dfrac{2\dot{r}\dot{\varphi}}{r}\cos\alpha \end{pmatrix}. \qquad\qquad (28)$$

But this result can also be derived from the Lagrange equation, which is written in the general form ([2, 3]):

$$\frac{d}{dt}\left(\frac{\partial E_k}{\partial \dot{x}_i}\right) - \frac{\partial E_k}{\partial x_i} - F_{x_i} = 0. \tag{29}$$

Using the coordinates r, φ we can write

$$x_1 = r \qquad \dot{x}_1 = \dot{r}$$
$$x_2 = \varphi \qquad \dot{x}_2 = \dot{\varphi}$$

With Equation (17) we obtain:

$$E_k = \tfrac{1}{2}(\dot{x}^2 + \dot{y}^2) = \tfrac{1}{2}(\dot{r}^2 + r^2\dot{\varphi}^2) \quad \text{and}$$
$$F_{x_i} = \frac{\partial U}{\partial x_i} + f_{x_i} \quad \text{with} \quad f = b = \frac{F}{M}.$$

For $i = 1$ we receive

$$\frac{\partial E_k}{\partial r} = r\dot{\varphi}^2$$

$$\frac{\partial E_k}{\partial \dot{r}} = \dot{r}$$

$$\frac{d}{dt}\left(\frac{\partial E_k}{\partial \dot{r}}\right) = \ddot{r}$$

$$F_r = \frac{\partial U}{\partial r} = -\frac{K}{r^2} + f_r$$

and this finally leads to Equation (28) first part and with $i = 2$ we obtain the second part of Equation (28)

$$\frac{\partial E_k}{\partial \varphi} = 0$$

$$\frac{1}{r}\frac{\partial E_k}{\partial \dot{\varphi}} = r\dot{\varphi}$$

$$\frac{1}{r}\frac{d}{dt}\left(\frac{\partial E_k}{\partial \dot{\varphi}}\right) = \dot{r}\dot{\varphi} + r\ddot{\varphi}$$

$$F = \frac{\partial U}{\partial \varphi} + f\varphi = f$$

It is very difficult to solve these two Equations (28) in a closed form on account of the unknown term $\dot{\varphi}$. This is why I tried to obtain the results by numerical integration.

We may write

$$\frac{F}{M} \cos\alpha \text{ as } \mathbf{b}_r \quad \text{and}$$

$$\frac{F}{M} \sin\alpha \quad \text{as} \quad \mathbf{b}_\varphi.$$

In the same way we can define

$$\mathbf{v}_{r0}, \mathbf{v}_{\varphi 0}, \mathbf{v}_{r1}, \mathbf{v}_{\varphi 1} \quad \text{and} \quad \mathbf{r}_0, \mathbf{r}_1,$$

where $\mathbf{r}_0$ is the vector of the departure point and in further numerical integration the vector of the first integration step $\mathbf{s}_0$ is the vector of the path which the satellite covers during the time increment Δt.

At $t=0$ we can write the following equations:

$$\mathbf{r}_0 = \begin{pmatrix} \mathbf{r}_1 \\ \mathbf{r}_2 \end{pmatrix} = \begin{pmatrix} \mathbf{r}_1 \\ 0 \end{pmatrix} \tag{30}$$

and for the velocity

$$\mathbf{v}_0 = \begin{pmatrix} \mathbf{v}_1 \\ \mathbf{v}_2 \end{pmatrix} = \begin{pmatrix} \mathbf{v}_r \\ \mathbf{v}_\varphi \end{pmatrix}. \tag{31}$$

Now the acceleration begins

$$\mathbf{b} = \begin{pmatrix} \mathbf{b}_r \\ \mathbf{b}_\varphi \end{pmatrix}. \tag{32}$$

After the increment Δt we must write:

$$\mathbf{s}_1 = \begin{pmatrix} \mathbf{s}_{11} \\ \mathbf{s}_{12} \end{pmatrix}. \tag{33}$$

We can transform Equation (33) to (34):

$$s_1 = \begin{pmatrix} \displaystyle\int_{t=1}^{t=2} \mathbf{v}_r \, dt \\[2ex] \displaystyle\int_{t=1}^{t=2} \mathbf{v}_\varphi \, dt \end{pmatrix} = \tag{34}$$

$$= \begin{pmatrix} \mathbf{v}_r \Delta t \\ \mathbf{v}_\varphi \Delta t \end{pmatrix}$$

The radius vector $\mathbf{r}_1$ after the time increment Δt will be determined by Equation (35):

$$\mathbf{r}_1 = \mathbf{r}_0 + \mathbf{s}_1$$
$$\mathbf{r}_1 = \begin{pmatrix} \mathbf{r}_{01} + \mathbf{s}_{11} \\ \mathbf{r}_{02} + \mathbf{s}_{12} \end{pmatrix}, \tag{35}$$

and for $\Delta\varphi$, the angle increment, we obtain:

$$\cos(\Delta\varphi) = \frac{\mathbf{r}_0 \cdot \mathbf{r}_1}{|\mathbf{r}_0| \, |\mathbf{r}_1|} = \tag{36}$$
$$= \arccos \frac{\mathbf{r}_0 \cdot \mathbf{r}_1}{|\mathbf{r}_0| \, |\mathbf{r}_1|}.$$

For the perturbation acceleration (Sun) we can now say:

$$\mathbf{b}_r = \frac{k m_{\text{Sun}}}{|\mathbf{r}_1^2|} \tag{37}$$

and with Equation (22.1) we obtain the acceleration with respect to the radial direction:

$$\mathbf{g}_{r_1} = \mathbf{b}_r + (\mathbf{F}_r / M) \tag{38}$$

and with (22.2) $F_\varphi = F \sin\alpha$ we receive the total amount of acceleration:

$$\mathbf{b}_1 = \begin{pmatrix} \dfrac{F_r}{M} + \mathbf{b}_r \\ \dfrac{F}{M} + 0 \end{pmatrix} = \begin{pmatrix} \mathbf{g}_{r_1} \\ \dfrac{F}{M} \end{pmatrix}. \tag{39}$$

First order differentiation in respect to the time leads to v_1:

$$v_1 = \begin{pmatrix} \displaystyle\int_{t_1}^{t_2} g_{r_1}\, dt \\ \displaystyle\int_{t_1}^{t_2} b_\varphi\, dt \end{pmatrix} = \begin{pmatrix} v_{r_1} \\ v_{\varphi_2} \end{pmatrix}. \tag{40}$$

Now we have gained $\mathbf{r}_1$ and $\mathbf{v}_1$ which are the new starting values for the next integration step.

A PL/1 program was written and was computed on an IBM-360/50 Digital computer. The program was interrupted when the radius distance of Mercury was attained.

```
PLAN    :PROC OPTIONS(MAIN);


        PLAN    :PROC OPTIONS(MAIN);
                DCL(    A01,    A02,    P,      V0R,    V0T,
                DT,     S0R2,   S0R1,   S0T2,   S0T1,   Q,
                S0R,    S0T,    S01,    S02,    RR,     R0,
                A12,    A11,    AL,     DS,     C,      PP,
                R2,     U,      DP,     BR,     FP,     FR,
                GR,     VR1,    VT1,    T,      TT,     DREI,
                PS,     M,      EIN,
                AM,     MI,     K,      F,      MPP,    B
                ) STATIC BIN FLOAT (53);
                OPEN FILE(SYSPRINT) PAGESIZE(72);
                GET FILE(SYSIN) EDIT(R0,P,V0R,V0T)(4(E(20,13)));
                GET FILE(SYSIN) EDIT(MI,K,AM,C)(4(E(20,13)));
                GET FILE(SYSIN) EDIT(DT,DREI,EIN,B)(4(E(20,13)));
                DS=3.600000000E+02;
                RK=2.0E+05;
                S0R2=1.104500000E+04;
                S01=R0+RK;
                S02=P;
                S0R1=V0R;
                Q=MI;
                MPP=B/C;
                MPP=MPP*8;
                B=B*8;
                DO AL=0 TO 45 BY 15;
                T=0.0E+00;
                R0=S01;
                P=S02;
                V0R=S0R1;
                V0T=S0R2;
                MI=Q;
                PUT FILE(SYSPRINT) EDIT(C,B,RK) (
                COLUMN(1),F(5,0),
                COLUMN(6),F(7,3),
                COL(14),F(7,0));
WEITER  :                               ;
                M=MI-(MPP*T);
                F=B/M;
                S0R=V0R*DT;
                S0T=V0T DT;
                A01=R0;
                A02=0.0E+00;
                A11=R0+S0R;
                A12=S0T;
                RR=(A01*A11)+(A02*A12);
                R0=(A01*A01)+(A02*A02);
                R0=SQRT(R0);
                R2=(A11*A11)+(A12*A12);
                R2=SQRT(R2);
                U=RR/(R0*R2);
                DP=EIN-(U*U);
                DP=(SQRT(DP)/U);
                IF DP<0.0E+00 THEN DO;
                DP=-DP;
                END;
                ELSE IF DP>0.0E+00 THEN DO;
```

PLAN   :PROC OPTIONS(MAIN);

```
STMT   LEVEL  NB T
48     1      2          DP=DP;
49     1      2          END;
50     1      1          DP=ATAND(DP);
51     1      1          BR=-(K/(R2*R2));
52     1      1          FP=F*SIND(AL);
53     1      1          FR=F*COSD(AL);
54     1      1          FP=-FP;
55     1      1          FR=-FR;
56     1      1          GR=FR+BR;
57     1      1          VR1=GR*DT+V0R;
58     1      1          VT1=FP*DT+V0T;
59     1      1          P=P+DP;
60     1      1          T=T+DT;
61     1      1          TT=T/DREI;
62     1      1          V0R=VR1;
63     1      1          V0T=VT1;
64     1      1          R0=R2;
65     1      1   DRUCK  :                          ;
66     1      1          PUT FILE(SYSPRINT) EDIT(AL,R0,P,V0R,V0T,TT,M)
                         COLUMN(22),F(2,0),
                         COLUMN(25),F(17,2),
                         COLUMN(43),F(18,13),
                         COLUMN(63),F(20,13),
                         COLUMN(84),F(20,13),
                         COLUMN(105),F(6,0),
                         COLUMN(112),F(6,0),SKIP(1));
67     1      1          IF R0>AM THEN GO TO WETTER;
69     1      1          ELSE GO TO AUS;
70     1      1   AUS    :                     ;
71     1      1          END;
72     1                 END;
```

Engine RIT 10*

$r_k = 200$ km

α	r_1	φ	t (h)
0	5.782734×10^7	27.510	1390
15	5.783433×10^7	27.489	1390
30	5.785525×10^7	27.467	1390
45	5.788865×10^7	27.446	1390
60	5.601268×10^7	27.833	1400
75	5.606508×10^7	27.819	1400
90	5.612137×10^7	27.810	1400

$r_k = 1000$ km

α	r_1	φ	t (h)
0	5.781147×10^7	25.906	1390
15	5.781848×10^7	25.884	1390
30	5.783941×10^7	25.862	1390
45	5.787282×10^7	25.842	1390
60	5.599636×10^7	26.206	1400
75	5.604878×10^7	26.192	1400
90	5.610507×10^7	26.183	1400

Engine ESKA 18*

$r_k = 200$ km

α	r_1	φ	t
0	5.7509×10^7	27.546	1390
15	5.7527×10^7	27.492	1390
30	5.7579×10^7	27.435	1390
45	5.7663×10^7	27.386	1390
60	5.7772×10^7	27.341	1390
75	5.7899×10^7	27.307	1390
90	5.6120×10^7	27.685	1400

$r_k = 1000$ km

α	r_1	φ	t
0	5.7493×10^7	25.940	1390
15	5.7511×10^7	25.885	1390
30	5.7563×10^7	25.831	1390
45	5.7647×10^7	25.780	1390
60	5.7756×10^7	25.736	1390
75	5.7884×10^7	25.702	1390
90	5.6103×10^7	26.058	1400

Engine HIT 4*

$r_k = 200$ km

α	r_1	φ	t
0	5.6731×10^7	27.632	1390
15	5.68054×10^7	27.500	1390
30	5.69321×10^7	27.367	1390
45	5.71343×10^7	27.243	1390
60	5.73979×10^7	27.136	1390
75	5.77047×10^7	27.053	1390
90	5.61173×10^7	27.393	1400

$r_k = 1000$ km

α	r_1	φ	t
0	5.67470×10^7	26.020	1390
15	5.67895×10^7	25.888	1390
30	5.69162×10^7	25.756	1390
45	5.71185×10^7	25.633	1390
60	5.73822×10^7	25.528	1390
75	5.76891×10^7	25.447	1390
90	5.610130×10^7	25.766	1400

* For all engines: $M_I = 7000$ kg.

References

[1] *Eurospace Memoranda*, Survey of electric propulsion systems for space application in Europe, No. 7, July 1970.
[2] Ehricke, K. A., *Space Flight, Vol. I, Environment and Celestial Mechanics*, D. Van Nostrand Company, Inc., Copyright 1960.
[3] Ehricke, K. A., *Space Flight, Vol. II, Dynamics*, D. Van Nostrand Company, Inc., Copyright 1960.
[4] Meschkowski, H., *Mathematisches Begriffswörterbuch*, B∗I∗99, Bibliographisches Institut, Mannheim/Wien/Zürich, 1971.
[5] Seifert, H. S., *Space Technology*, John Wiley and Sons, Inc., Copyright 1959.
[6] Stumpff, K., *Himmelsmechanik*, Bd. I und II, VEB Deutscher Verlag der Wissenschaft, Berlin 1959.
[7] Sagirow, P., *Satellitendynamik*, B∗I∗719/719a*, Bibliographisches Institut, Mannheim/Zürich/Wien, Mannheim 1970.
[8] Gellert, W., Küstner, H., Hellwich, M. and Kästner, H., *Großes Handbuch der Mathematik*, Buch und Zeit Verlagsgesellschaft M.B.H. Köln 1967.
[9] Heusmann, W. A., *Projected Study of a Probe for a Polar Mercury Orbit*, IAF-Student Conference 1971. Belgien.
[10] Rössger, E. and Zehle, H., *Interplanetare Flugbahnen kleiner Schubbeschleunigungen mit konstantem Schub*, Institut für Flugführung und Luftverkehr, Technische Universität Berlin, Bericht Nr. 26, Oktober 1963.
[11] Stuhlinger, E., *Ion Propulsion for Space Flight*, McGraw-Hill Book Company, Copyright 1964.
[12] Rottmann, K., *Mathematische Formelsammlung*, B∗I∗13, Bibliographisches Institut, Mannheim, 1960.
[13] Bohrmann, A., *Bahnen künstlicher Satelliten*, B∗I∗40/40a, Bibliographisches Institut Mannheim 1966.
[14] *Weltraumforschung I*, B∗I∗107/107a, Bibliographisches Institut, Mannheim, 1966.
[15] *DGLR, Deutsche Gesellschaft für Luft- und Raumfahrt*, 'Mitteilung 71-22', Bericht über das DGLR – Symposium elektrische Antriebssysteme am 22./23. Juni 1971 in Braunschweig.
[16] Brucerius, H., *Himmelsmechanik*, B∗I∗143/143a, Bibliographisches Institut, Mannheim, 1966.
[17] Metzger, R., *Interplanetare Missionen mit elektrischen Antrieben*, FM 386-Ö DGRR-Nr. 67-031.

THE STUDENT USE OF SATELLITES FOR
THIRD WORLD DEVELOPMENT

MARK R. BONSALL

The City University, on behalf of The Royal Aeronautical Society Graduates and Students Section
London, England

In the past few years the technical problems associated with geostationary satellites broadcasting directly to television receivers on the ground have to a large extent been solved. One use of these will be in establishing an educational service in the developing countries with widely dispersed populations.

Quite often a large majority of the population of these countries are illiterate and hence learning at present must either be by being taught to read or by the practical demonstrations and verbal communication of teachers who are present at the school. There are however not enough teachers to go round, educational television can overcome these two limitations by allowing practical demonstrations to be carried out under ideal conditions which are then seen by many groups simultaneously in TV clubs or community centres. This form of teaching can also overcome the shortage of teachers of reading and language by allowing sound and symbols to be learnt simultaneously as if a teacher were present.

Whilst the technical side of the problem of launching these satellites and modifying ground receivers for transmissions from them nears solution the problems in the human part of the system are not yet resolved. There are many differing ideas on what type of programmes should be broadcast. One suggestion is that the first step should be the teaching of a major world language. Only seven languages are spoken by over half the worlds population so this seems to be one possibility. Another is that the system should be used to disseminate information on farming techniques, farm management and population control. One favoured by some countries is using it as a news distribution network with the aim of promoting the social and cultural integration and development of the country.

India is the first country intending to deploy such a system and the government there intends to use its satellites for a least the second two of these possible uses. The situation in India is very favourable to the deployment of a geostationary direct broadcast system as it is near the equator and has a highly dispersed population. 560 000 villages dispersed in 1.2 million square miles. A full discussion of this project is given in H. P. Mama's article in the December 1971 issue of Spaceflight [1].

There is one other use of this system and that is a combination of information distribution, education and training. This is the diffusion of new but simple technology rather than just disseminating agricultural and other techniques already in use in the developed world. It is in this area that the large student population of the developed world can help with the content of the broadcasts.

First of all I will give a little of the background to the situation before dealing with the student side of this. At present there is a great debate going on as to the extent to

L. G. Napolitano et al. (eds.), Astronautical Research 1972, 349–352. All Rights Reserved
Copyright © 1973 by D. Reidel Publishing Company, Dordrecht-Holland

which the under-developed countries of the world should industrialize using capital intensive projects or whether they should utilize labour intensive techniques of production [2]. The argument is that heavy industry will create further capital for investment and with high wages a market for the goods produced. This will lead to a rapid increase in G.N.P. However it would create large disparities in wages between those working in cities and those working on farms with a consequent migration to the cities and unemployment. Pollution would increase and capital would not be available for rural development. Another form of economic structure favours the development of rural areas as a greater number of people would benefit. This would reduce the overall costs but it may reduce the rate of growth of the G.N.P. One unknown in both these situations is the employment policy of either governments or private employers as these two forms develop. In industrialized economies with nationalized industries government pressure is often brought to bear to keep price down and employment many times above its optimum. Hence there is no profit to tax or reinvest [3]. Further expansion thus has to come from further foreign aid which can then simply be covering a country with factories eating up its resources and trained manpower but not becoming a self sustaining entity. In low-capital projects intended to be labourintensive or at least to improve individual workers productivity the increase in productivity can lead not to an expansion in the total agricultural effort but to a reduction in the local labour force which whilst it may produce more food per unit of land causes greater problems of unemployment even resettlement.

Whatever the outcome of this debate there will be a role for both forms of development as there are certain areas where only one or the other is possible. Hence there will be a market for newly developed or rediscovered labour-intensive technology. This is often called 'intermediate technology'. In the agricultural sector the effect of this technology plus high yield crops and proper farm management is called 'The Green Revolution'. It is the Third Worlds equivalent of the Agricultural Revolution which occured before industrialization in the developed world. There are two further reasons why the Green Revolution is important. Firstly the world needs the extra food and secondly some countries especially in Africa have what are virtually one crop economies hence they need to develop this their main natural resource.

Agriculture is not the only area that can benefit from this technology. Building techniques and materials from natural products are possible. Small scale production of clothing, shoes, crockery and simple implements are also suitable areas. In many cases industry in the developed world regards the third world as an extension of its home market rather than a different market hence certain more advanced products are not made in the form necessary. One example is tractors as mentioned in reference [4] here no small efficient tractor tailored to the market has yet appeared. It is in the development of Intermediate Technology and its dissemination that the large student population of the developed world can help. This technology is characterized by a number of features. It must relatively speaking have a low initial cost and have little or no running costs. It must be simple, robust and above all easily used by previously untrained personnel.

Examples are small steam engines for saw mills or local electricity, making building materials such as bricks from natural products, solar cookers, simple irrigation pumps, prefabricated irrigation channels, and simple tools and farming implements for raising individual productivity. This has been a very successful application so far in rice growing and other crops.

In Britian there is a non-profit making organisation called the Intermediate Technology Development Group which provides an information service to developing nations and the governments of developed countries helping developing nations on the forms of intermediate technology available. They have published a Guide called 'Tools for Progress' [5] which lists many tools available for raising productivity in the rural areas of developing countries. The two problems with I. T. are its initial development for though it is based on existing technology it has to be adapted for the particular circumstances. Second there is the problem of its dissemination. Heavy industry can train its personnel as it gathers them in cities where they can be easily grouped together for the purpose. In disseminating I. T. the satellite system will have to be used.

At present many universities have small programmes for developing I.T. however these tend to be amongst research workers. I believe that students both Undergraduates and Postgraduates are suited to this form of work. As I.T. focuses on a small comprehensible component of the total system which is then improved or the need for a component is analysed and a solution found. It can be carried out in a time scale, say one or two years which is suitable for student involvement. This will give students experience in working in teams. It also gives experience of a real world problem rather than a contrived academic exercise of the type students are usually expected to do.

Having developed a particular piece of I.T. it is possible that the students by themselves could make the necessary TV programmes in their universities under the guidance of a body such as UNESCO. If the initial educational satellites teach people one of the major world languages this will be an easier task than at present with an estimated seven thousand languages in the world. There will be some problems of communication psychology, as the rate of flow of information from TV which is accepted by people in the developed countries may be too high for those with less initial education in under-developed countries. Also certain assumptions about the particular concepts comprehensible by the audience may be wrong if a study is not made of their customs and social organisation. All programmes will in any event have to be highly visual and practical in content. Obviously students from the region to which the transmission will go who are studying in universities in developed countries will have to be in these teams, together with students from the universities who will be able to understand the type of communication problems possible and will be able to advise on the presentation of the programmes.

The body of information on the subject of linguistics especially role of language learning in the developing of the frame in which ideas are organized in the mind will be of enormous importance here.

Before this proposal can be evaluated fully a much more detailed study will have to take place. I suggest that one way would be an international group of students set up by the IAF student conference to study this. They would research the possibilities in their own countries whilst each reading up about a specific aspect of the problem of underdevelopment. The idea would be to co-opt other students and professionals to the group and eventually to draw up a report. Cooperation with UNESCO officials and other U.N. departments would be necessary. This group could also look at other possible areas of student involvement in the use of satellites for world development, not just for the third world. It could then be in a position to inform any interested organisation as to the role which students could play in this area of astronautics applications.

References

[1] Mama, H. P., 'Indias Domestic Communications Satellite' *Spaceflight* **13**, No. 12. British Interplanetary Society. London 1971.
[2] Mountjoy, A. B., *Industrialization and Under-developed Countries*. Hutchinsons, 3rd Edition, London 1971.
[3] Fry, Maxwell J., 'Quest', *The Journal of The City University London*, No. 19, Christmas 1971.
[4] Yudelman, M., Buler, G., and Banerji, R., 'Technological Change in Agriculture and Employment in Developing Countries', Development Centre Studies, Employment Series No. 4. OECD Paris 1971.
[5] Yudelman, M., Buler, G., and Banerji, R., 'Tools for Progress' and 'Report on the Conference on the further Development in the U.K. of Appropriate Technologies for, and their Communication to Developing Countries'. Published by Intermediate Technology Development Ltd., London 1968.

STRUCTURE AND DEVELOPMENT OF
THE SLOVEN PROGRAM

MILAN NIKOLIĆ and DJORDJE BLAGOJEVIĆ

Academical Rocket Astronautical Club, Beograd, Yugoslavia

1. Introduction

ARAK (Academical Rocket Astronautical Club) is the student organization of the University in Belgrade. Its primary activity is theoretical and experimental work in the astronautical sciences, and especially rocketry. ARAK was established in March, 1963 so, it has existed nearly one decade.

Till now, ARAK has been working on three primary programs of research rockets – Strela (Arrow), Sonda (Probe) and Sloven (Slav). Every new program has been and will be based on previous work and thus facilitate the conquering of new sciences.

The Strela Program started in 1963, and it was completed in the year 1965, however rocket R-500 is retained in use to date. In its early versions it was a one-stage rocket, firstly with zinc-sulfur and later with double-base propellant. Later, several two-stage versions have been developed, using double-base and composite propellants as sustainers. Strela R-500 is now in use as standard rocket for testing electronic equipment components under rocket flight conditions.

The Sonda Program is a normal prolongation of the Strela. It has similar geometry and stage arrangement as Strela R-5. This program started in mid-1965. The first static tests of the first and second stage motors have been accomplished in LISIČJI JARAK, ARAK's test station near Belgrade. In the period 1966–1970 six Sonda rockets have been succesfully launched. Some of those rockets have been modified for special equipment (telemetry transmitters, electronic timers and measurement equipment).

About twenty students and engineers worked on these programs, with some help from professional Institutions. Working on those programs ARAK's team has obtained necessary knowledge and experience for the work on new programs.

The Sloven Program started in late-1968 with very ambitious objective – to develop the first Yugoslav sounding rocket using only the members of ARAK, without any assistance from outside. However, we plan intensive cooperation in future, with certain national and foreign organizations.

2. Sloven Program Conception

The Sloven Program is the primary activity of ARAK. New trends in theoretical and experimental research, new construction methods and new technology are the main characteristics of this program. For this reason, a great part of effort is expended in technology development, laboratory equipment and prospective systems research

L. G. Napolitano et al. (eds.), Astronautical Research 1972, 353–358. All Rights Reserved
Copyright © 1973 by D. Reidel Publishing Company, Dordrecht-Holland

and development. The program is very extensive (it contains design and development of seven types of rockets, technology base and ground equipment), so it is divided into four phases.

TABLE I

Basic characteristic of Strela and Sonda rockets

Designation	R-1	R-2	R-3	R-4	R-5	R-500	R-6	R-7
Diameter (m)	0.057	0.057	0.057	0.057	0.057	0.070	0.127	0.127
Length (m)	1.225	1.254	1.445	1.900	2.005	2.078	2.679	3.320
Initial mass (kg)	7.5	5.5	6.7	11.0	12.0	9.0	47.0	71.0
Payload (kg)	1.0	1.0	0.5	1.5	1.5	2.5	7.0	7.0
Max. altitude (m)	2000	4700	11600	5200	12300	2000	7800	15900
Number of flights	6	2	1	2	2	5	6	2
Booster motor	BM-1	BM-2	BM-2	BM-2	BM-2	BM-2	BM-6	BM-6
Propellant	Zn–S	JPN	JPN	JPN	JPN	JPN	JPN	JPN
Thrust (N)	4700	2500	2500	2500	2500	2500	22000	22000
Burning time (s)	0.28	0.7	0.7	0.7	0.7	0.7	1.1	1.1
Sustainer motor	–	–	M-3	BM-2	M-5	–	–	M-7
Propellant			GALCIT	JPN	GALCIT			GALCIT
Thrust (N)			200	2500	240			1100
Burning time (s)			7.0	0.7	14.0			15.0

The first phase of the program contained two projects: Sloven SAR-01 sounding rocket and its mobile launcher. The SAR-01 has been a classical sounding rocket configuration study (solid-booster, liquid-sustainer), based on American Aerobee rockets. It was fully designed in late-1968, but was not built on account of its high production cost and low performance.

The second phase, which started in April, 1970, after cancellation of the first phase, composed technology, ground equipment, and SAR-02 rocket, and represents the base for the whole program. The SAR-02 has a very different conception from the SAR-01. The basic construction material is filament wound fiberglass, with much better mechanical characteristics than metallic materials. The solid booster is eliminated as a result of high production cost of available solid propellants. The regeneratively cooled thrust chamber is replaced with a new, ablatively cooled one, with coaxial injector. Also, a new inertial stabilization system is adopted.

The third phase includes the development of SAR-03, SAR-02B, and SAR-03B rockets. These are uprated modifications of the SAR-02. The SAR-03 is a version with doubled engine burning time. The SAR-02B and the SAR-03B are SAR-02 and SAR-03 with hot gas pressurization system. Two types of gas generators are the objects of present R&D: a solid propellant one and ammonia vaporization one. The SAR-03B has also a hot gas secondary injection thrust vector control. This will be a step-by-step process which will complete the fourth phase and culmination of the program.

The fourth phase of the Sloven Program includes the development of a satellite launch vehicle, the first one in Yugoslavia. In this phase, development of two rockets

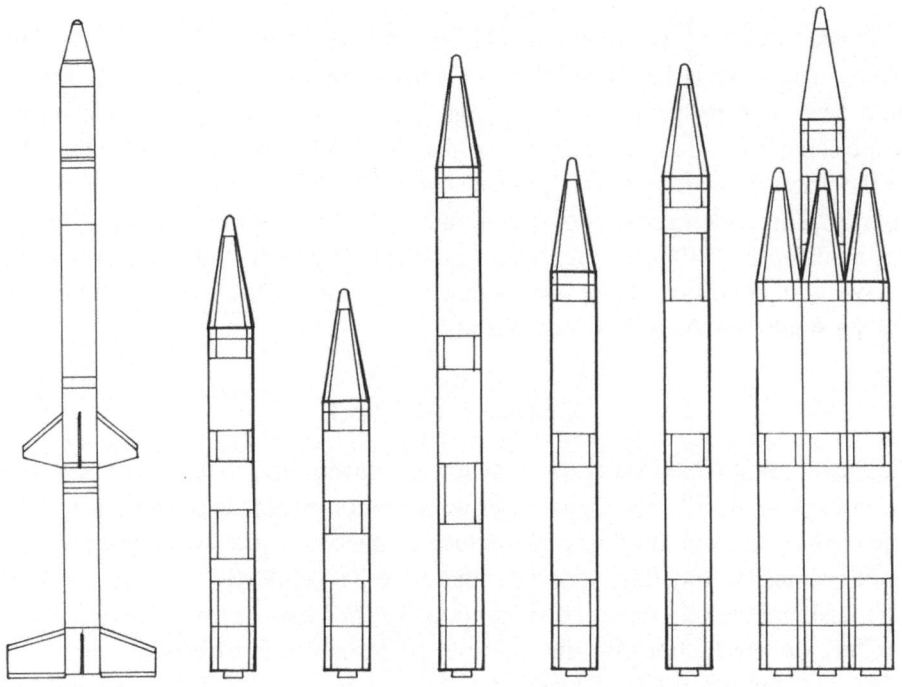

Fig. 1. Configurations of Sloven rockets.

TABLE II

Basic characteristics of Sloven rockets

Designation	SAR-01	SAR-02	SAR-02B	SAR-03	SAR-03B	SAR-04	SAR-05
Diameter (m)	0.380	0.500	0.500	0.500	0.500	0.500	1.520
Length (m)	7.050	4.780	4.000	6.460	5.700	6.420	6.920
Start mass (kg)	604	419	403	761·	703·	765	3380
Payload (kg)	40	40	40	40	40	40	40
Max. altitude (km)	150	240	310	490	850	1380	500[a]
Max. velocity (m s^{-1})	1510	1990	2290	2770	3580	4700	7870
Ist stage	CRPS-1	TRPS-2	TRPS-2B	TRPS-3	TRPS-3B	TRPS-3C	TRPS-3C[b]
Propellants	SOLID	RFNA JP-4	RFNA JP-4	RFNA JP-4	N_2O_4 UDMH	N_2O_4 UDMH	N_2O_4 UDMH
Thrust (N)	40000	10000	10000	10000	10000	11000	44000
Burning time (s)	8	60	60	120	120	110	110
IInd stage	TRPS-1	–	–	–	–	CRPS-2	TRPS-3D
Propellants	RFNA Aniline					SOLID	N_2O_4 UDMH
Thrust (N)	10000					5000	14000
Burning time (s)	40					50	120
IIIrd stage	–	–	–	–	–	–	CRPS-2
Propellants							SOLID
Thrust (N)							5000
Burning time (s)							50

[a] Earth orbit apogee.
[b] Cluster of four.

is planned: SAR-04 (two-stage test vehicle) and SAR-05 (three-stage satellite launch vehicle). The SAR-04 has a modified SAR-03B propulsion system as the first stage and a new, solid propellant, second stage. Testing stage separation, spin-stabilized second stage, inertial guidance system and other equipment designed for the SAR-05 are the primary mission of this vehicle. The SAR-05 is a culmination of the whole program. This rocket consists of four modified SAR-3B's strap-ons as the first stage, plus modified SAR-04 (with modified liquid rocket engine with high area ratio for near-vacuum conditions) core second and third stages. This vehicle will be capable to put a small satellite in low Earth orbit.

3. Sloven SAR-2 Rocket

The Sloven SAR-02 rocket is the base for all Sloven-type rockets, and will be presented in some detail. This is a modern-conception rocket with some innovations, which reduce cost and improve performance. In this conception we hope for maximum use of new technology (filament winding, and other reinforced plastics and ablative materials), and novel construction methods to increase the development cost, but to reduce the production cost and improve performance of rockets.

This is a liquid propellant rocket with an active inertial stabilization system and consists of three systems: Propulsion system, inertial stabilization system, and instrumental module.

3.1. Propulsion system

The propulsion system of the SAR-02 (TRPS-2) is a liquid-propellant cold-pressurized-gas-feed system. It consists of liquid rocket engine, pressurization subsystem, propellant subsystem and thrust-vector control subsystem. All systems will be described in some detail.

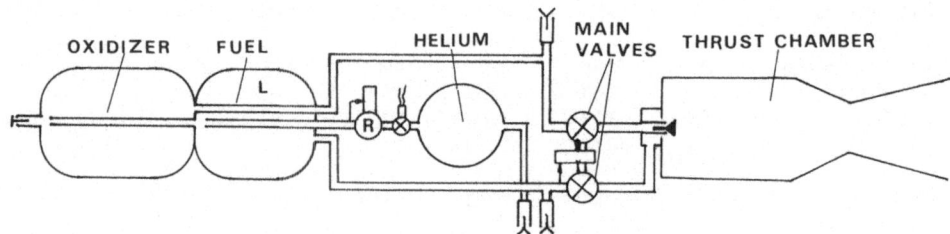

Fig. 2.   SAR-02 propulsion system diagram.

Liquid rocket engine TRM-2 is the first ablatively cooled liquid rocket engine in Yugoslavia. Propellants are Red-Fuming Nitric Acid and kerosene JP-4. This is not a novel combination, but at the moment, the only one available. As this combination is very sensitive to high frequency instabilities, we have designed a coaxial injector (very good for high frequency stability) for this engine. The thrust chamber is ablatively cooled. The materials used are: astrasile (phenolic-impregnated refrazile) for inner, ablative, liner, filament wound fiberglass for outer shell, and high strength aluminum alloy for end joints. Propellant feed equipment contains propellant main valves, propellant feed lines with flexibile joints and injector. All this details are made of $AlMg_5$ aluminium alloy.

Pressurization subsystem consists of pressurized helium tank, gas start valve, reducing valve, and gas filler valve. Pressurized gas tank is made of filament wound fiberglas, and all other details are made of high strength aluminium alloy ($AlZnMgCu_2$). The gas start valve is pyro-operated, and the reducing valve is a direct spring-loaded type.

Propellant subsystem consists of fuel tank, oxidizer tank, propellant filler valves, tank safety valve and pressurization and propellant feed lines. Propellant tanks are made of filament wound fiberglass. Internal lines are made of polyethylene and other details are made of $AlMg_5$ alloy.

3.2. INERTIAL STABILIZATION SYSTEM

Inertial stabilization system is a development of the SAR-05 satellite launch vehicle guidance system. It presents the simplest form of such system; stabilization is achieved in pitch and yaw planes.

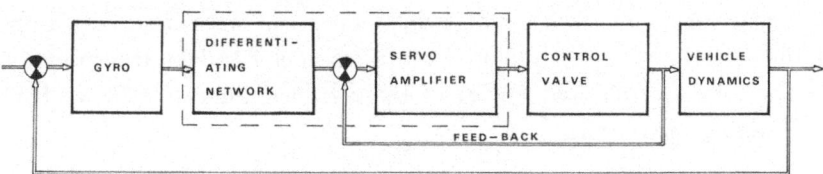

Fig. 4. Inertial stabilization system block diagram.

Reference subsystem consists of two 'strapped down' two-degree-of-freedom pyro-operated gyros. Wire potentiometer pick-offs generate error signals and send it to the electronic subsystem.

Electronic subsystem contains differentiating network and servo amplifier. It receives error signals from reference subsystem and sends amplified signals to the secondary injection thrust vector control subsystem.

3.3. INSTRUMENTAL MODULE

Instrumental module is an independent part of rocket and will be used, with some modifications, in all Sloven rockets excluding SAR-05 and some missions of SAR-04.

Nose cone structure is metallic structure with an ablative front cap. main structural element is a central tube, with radial consoles which carry internal equipment. This structure is designed to accomodate various equipment arrangements.

Re-entry and landing subsystem is basic subsystem of the module. It contains parachutes (braking and main) with ejection equipment and in some flights special sea-landing equipment.

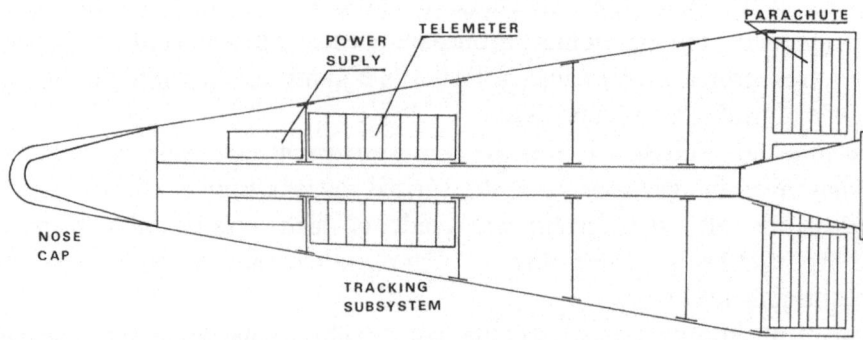

Fig. 4.   Instrumental module.

Electric power supply consists of four 12 V Ag–Zn bateries. This is a low weight – high performance power supply for all systems and measurement equipment.

Doppler tracking subsystem consists of four receivers, four doublers and four transmitters. It receives signals from four points from the ground and transmits four corresponding signals to the ground tracking station, where these signals enable computation of distance, velocity and acceleration.

Telemetry subsystem consists of two independent networks: 32-channel PCM transmitter for low-frequency data and 16-channel FM/FM transmitter for high frequency data. Carrier frequencies of these transmitters are 470 and 438 MHz with 3 W and 15 W output.

UNIT FOR MODULATING A LIQUID FLOW RATE BY DEFORMABLE WALLS

MICHEL MARCHETTI and PIERRE PLANQUES

Office National d'Études et de Recherches Aérospatiales (ONERA),
92320 Chatillon, France

Abstract. It is necessary to determine the dynamic characteristics of a hydraulic duct in order to predict the low frequency vibrations which may affect a space rocket using liquid propellant motors (POGO effect).

The transfer function of a duct can be determined experimentally by a sine variation of the flow, obtained either with a piston or with deformable walls.

A theoretical study of the latter solution, in incompressible fluid, and one- or two-dimensional plane flow, is presented for a practical case, and was checked by experiment.

1. Introduction

Within the framework of general studies of the POGO effect, ONERA undertook the design of units ensuring the sine modulation of a liquid flow rate. This paper presents a theoretical study of such a unit, using deformable walls. Two models, one one-dimensional and the other two-dimensional, describe the functioning of the unit in a realistic case of utilization.

2. Notations and Assumptions

2.1. Notations

The geometric parameters are defined Figure 1:

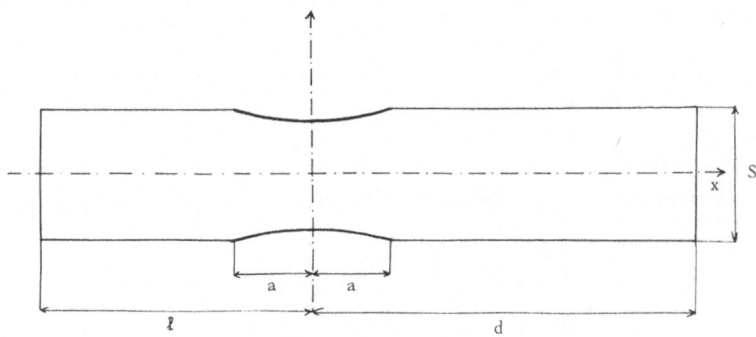

Fig. 1. One-dimensional diagram.

u_{-l}: velocity of the liquid in section of abscissa $-l$
u_d: velocity of the liquid in section of abscissa d
S: duct cross section, outside the deformable wall
$s(x, t)$: cross section of the deformable wall

L. G. Napolitano et al. (eds.), Astronautical Research 1972, 359–365. All Rights Reserved

ϱ: fluid density

φ: velocity potential.

2.2. ASSUMPTIONS

On the fluid:
  (i) incompressible and inviscid,
  (ii) gravity effect negligible.

On the deformation:
  (i) small relative variation of s,
  (ii) affine deformation, of a direction perpendicular to $0x$:

$$s(x, t) = S(1 + \varepsilon f(x) \, \alpha(t)).$$

2.3. UPSTREAM AND DOWNSTREAM BOUNDARY CONDITIONS

Pressures in $-l$ and d are only functions of velocity:

$$p_{-l} = \Phi(u_{-l})$$
$$p_d = \Psi(u_d).$$

3. One-Dimensional Model

3.1. BASIC EQUATIONS

Mass and momentum conservation equations on $(-a, +a)$ are written:

$$\frac{\partial(us)}{\partial x} + \frac{\partial s}{\partial_t} = 0 \tag{1}$$

$$u \frac{\partial u}{\partial x} + \frac{\partial u}{\partial t} + \frac{1}{\varrho} \frac{\partial p}{\partial x} = 0. \tag{2}$$

For the equations of the outlet velocity, integration in x of Equation (1) on the $(x, +a)$ segment leads to this expression for u:

$$u = \frac{S}{s} u_{+a} - \frac{1}{S} \int_x^a \frac{\partial s}{\partial t} \, d\xi. \tag{3}$$

Introducing u, as given by (3), in the partial derivative Equation (2), one obtains after calculation and utilization of the generalized Bernouilli formula:

$$A(t) \, u_a' + B(t) \, u_a + C(t) = \frac{p_{-l} - p_d}{\varrho} \tag{4}$$

in which $A(t)$, $B(t)$ and $C(t)$ are given by

$$A(t) = S \int_{-a}^{+a} \frac{dx}{s} + (l + d - 2a)$$

$$B(t) = 2S \int_{-a}^{+a} \frac{1}{s^2} \left[\frac{1}{s} \frac{\partial s}{\partial x} \int_{-a}^{x} \frac{\partial s}{\partial t} d\xi - \frac{\partial s}{\partial t} \right] dx$$

$$C(t) = \int_{-a}^{+a} \left\{ \frac{2}{s^2} \frac{\partial s}{\partial t} \int_{a}^{x} \frac{\partial s}{\partial t} \xi - \frac{1}{s} \int_{a}^{x} \frac{\partial^2 s}{\partial t^2} d\xi - \frac{1}{s^3} \frac{\partial s}{\partial x} \left[\int_{a}^{x} \frac{\partial s}{\partial t} d\xi \right]^2 \right\} dx +$$

$$+ \frac{l-a}{S} \int_{-a}^{+a} \frac{\partial^2 s}{\partial t^2} dx.$$

Assumptions on the boundary conditions in $-l$ and d on the one hand, Equation (3) written for $x = -a$ on the other, show that velocity at modulator outlet is given by the differential equation

$$A(t) u_a' + B(t) u_a + C(t) = G(u_a, t) \tag{5}$$

with:

$$G(u_a, t) = \frac{\Phi \left(u_a + \frac{1}{s} \int_{-a}^{+a} \frac{\partial s}{\partial t} dx \right) - \Psi(u_a)}{\varrho}. \tag{6}$$

3.2. LINEARIZATION OF THE COEFFICIENTS OF EQUATION (5)

The differential Equation (5) can be practically solved only in the case of small perturbations

$$s(x, t) = S[1 + \varepsilon f(x) \alpha(t)].$$

Neglecting terms in ε of an order higher than 2:

$$A(t) = l + d - \varepsilon \alpha(t) \int_{-a}^{+a} f(x) dx \tag{7}$$

$$B(t) = -2\varepsilon \alpha'(t) \int_{-a}^{+a} f(x) dx \tag{8}$$

$$C(t) = \varepsilon \alpha''(t) \left[(l-a) \int_{-a}^{+a} f(x) dx - \int_{-a}^{+a} dx \int_{a}^{x} f(\xi) d\xi \right]. \tag{9}$$

If we assume a symmetry for the modulation unit:

$$C(t) = \varepsilon \alpha''(t) \int_{-a}^{+a} f(x)\, dx. \tag{10}$$

4. Two-Dimensional Model (Figure 2)

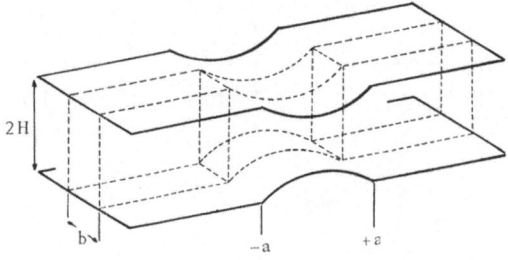

Fig. 2. Two-dimensional diagram.

Let us now consider a duct formed by two parallel planes, distant by $2H$ outside $(-a, +a)$, with a deformed zone $(-a, +a)$, of thickness $2z$ given by

$$z = H\left[1 + \varepsilon f(x)\, \alpha(t)\right].$$

For a longitudinal section of width b, the variation of the volume limited by sections $-l$ and d, between instants t and $t + dt$, is equal to the difference between the input and output fluid volumes; after calculations:

$$u_d - u_{-l} = -\varepsilon \alpha'(t) \int_{-a}^{+a} f(x)\, dx. \tag{11}$$

The potential difference $[\varphi]^d_{-l}$ can be calculated as the circulation of the velocity vector on a curve (T) entirely included within the fluid, and joining the equipotentials of abscissae $-l$ and d.

By taking a curve (T) a parallel to axis $0x$, of ordinate y, which never intersects the profile,

$$0 \leqslant y \leqslant H(1 - \varepsilon) = h$$

and by assuming the unit to be symmetrical and carrying out calculations to the first order in ε, the potential difference is written

$$[\varphi]^d_{-l} = u_d d + u_{-l} l - \varepsilon \alpha(t) \frac{u_d + u_{-l}}{2} \int_{-a}^{+a} f(x)\, dx. \tag{12}$$

As regards the differential equation of the outlet velocity, application of the generalized Bernouilli formula, taking account of Equation (11) and (12), leads to the same equation as for the one-dimensional model.

5. Resolution of the Differential Equation

Let w be the mean flow velocity. By writing

$$\lambda = \int_{-a}^{+a} f(x)\,\mathrm{d}x \qquad K_{-l} = \frac{\mathrm{d}\Phi}{\mathrm{d}u}(w) \qquad K_{-l} = \frac{\mathrm{d}\Psi}{\mathrm{d}u}(w)$$

the perturbation equation becomes

$$\frac{l+d}{\lambda}\,\tilde{u}'_d + \frac{K_d - K_{-l}}{\varrho\lambda}\,\tilde{u}_d + \alpha''(t) - \frac{K_{-l}}{\varrho}\,\alpha'(t) - 2w\alpha'(t) = 0. \tag{15}$$

In the case of a sine variation with time of the section: $\alpha(t) = \cos\omega t$, the velocity fluctuation at modulator outlet is:

$$\varepsilon\tilde{u}_a = \varepsilon U_a \cos(\omega t + \varphi)$$

with:

$$\varepsilon U_a = \varepsilon\lambda\omega \sqrt{\frac{l^2\omega^2 + \left(2w + \dfrac{K_{-l}}{\varrho}\right)^2}{(l+d)^2\omega^2 + \left(\dfrac{K_d - K_{-l}}{\varrho}\right)^2}}$$

$$\mathrm{tg}\,\varphi = \frac{\left(2w + \dfrac{K_{-l}}{\varrho}\right)\left(\dfrac{K_d - K_{-l}}{\varrho w}\right) - l(l+d)\,\omega}{\dfrac{K_{-l}}{\varrho} + (l+d)\left(2w + \dfrac{K_{-l}}{\varrho}\right)}.$$

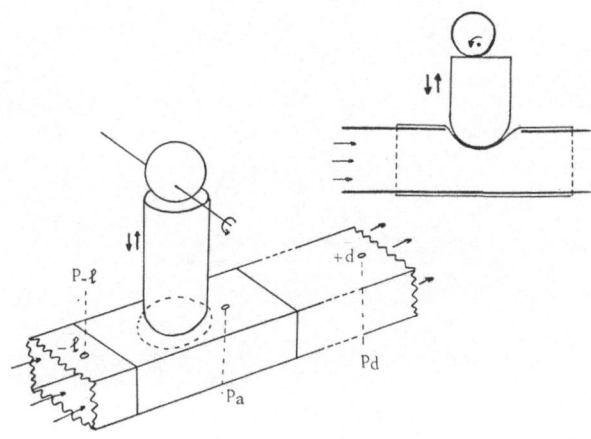

Fig. 3. Actual design.

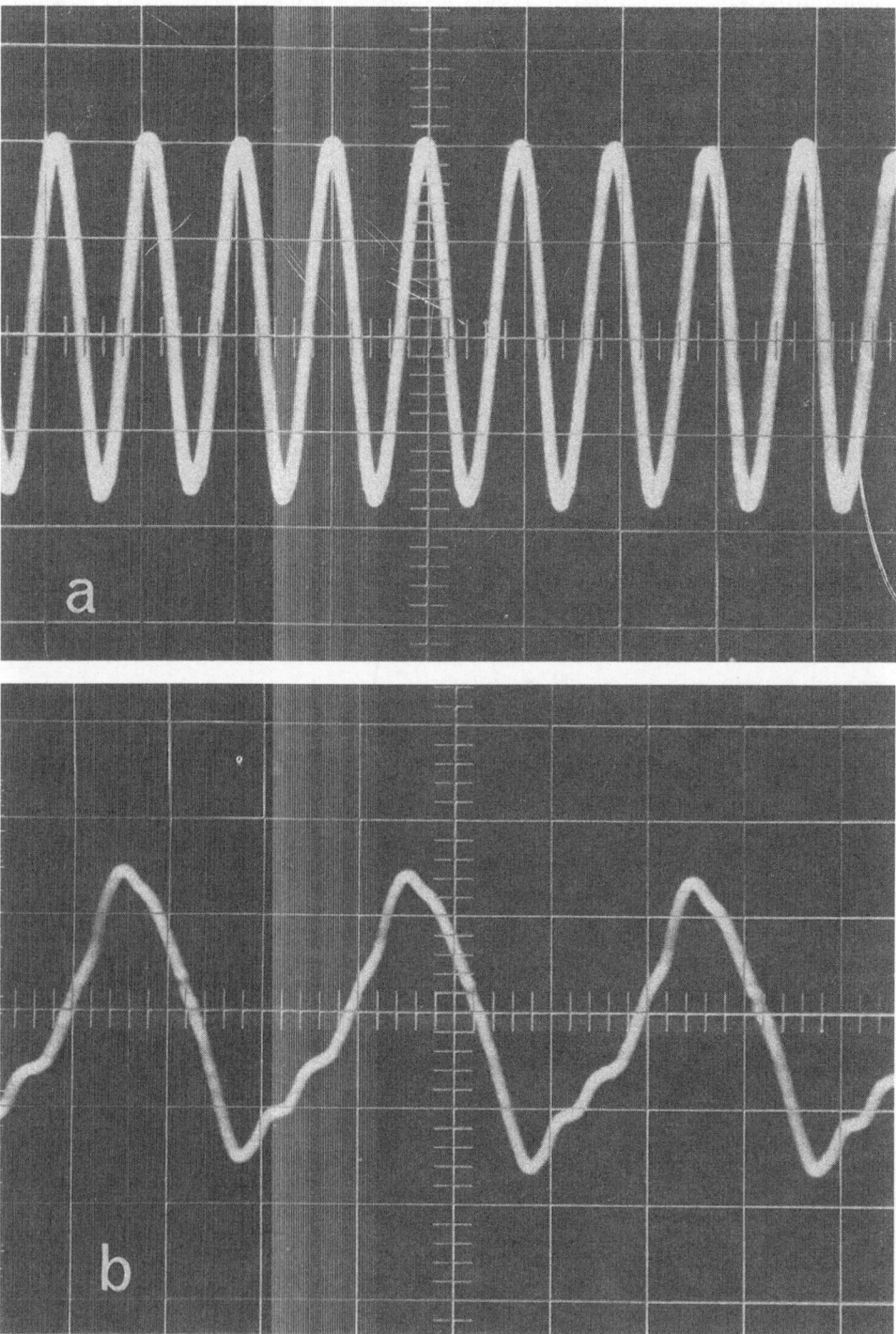

Fig. 4.   Recording of pressure fluctuations generated by the modulator. (a) $f = 10$ Hz, amplitude 1 mm;
(b) $f = 30$ Hz, amplitude 6 mm.

6. Conclusion

The two approaches just described show that, within the admitted assumptions, the modulation unit by deformable wall ensures a sine variation of the flow rate.

An experimental verification, done on a laboratory modulator (Figure 3) permitted to validate the theoretical calculations (Figures 4–5).

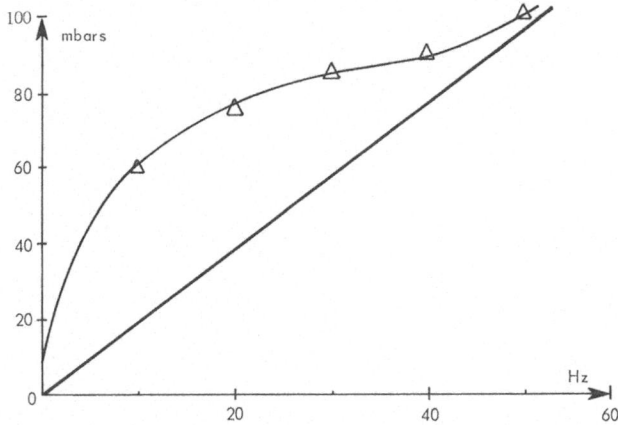

Fig. 5. Evolution of the pressure fluctuation amplitude recorded as a function of modulation frequency.

APPENDIX

List of Papers Presented at the 23rd I.A.F. Congress

FOURTH I.A.F. INVITED LECTURE

H. Guyford Stever: 'The Impact of Space on World Development'

FORUM SESSION: Space for World Development

Organizing Chairmen: R. W. Porter and F. De Mendonca

Panel Speakers: R. M. Hoffer, P. Moreira da Silva, O. Dominguez, R. H. Cannon, Th. P. Tromp, C. A. Berry, C. Terzani, J. Fry, W. von Braun, L. G. Napolitano, A. Hocker, Marvin Robinson.

MATERIALS PROCESSING IN ZERO GRAVITY

Organizing Chairman: W. H. Steurer

H. F. Wuenscher: 'Materials Processing in Zero Gravity – An Overview'
L. R. McCreight and L. Steg: 'Space Processing – Projections to 2000 A.D.'
J. L. Reger: 'Low Gravity Processing of Immiscible Materials'
R. A. Happe: 'Experiments Leading to the Production of New Glasses in Space'
I. N. Frantsevitch, V. S. Dvernyakov, V. V. Pasichny, N. Y. Shiganov, and Y. I. Korunov: 'Studies of the Possibilities to Use Solar Energy for Welding and Soldering in the Outer Space'
J. H. Bredt: 'New Space Processing Experiments for the Skylab Missions'

SAFETY IN SOUTH ROCKET EXPERIMENTS (SYRE)

Organizing Chairmen: G. S. James and A. I. Skoog

D. J. Malewicki, L. M. Cox, and D. C. Schwenn: 'A Pilot Program to Evaluate the Use of Model Rocketry in High School Aerospace Curricula'
E. D. Brown: 'The Estes Model Rocket Engine, Basic Methods of Manufacture and Quality Assurance'
H. Moulin: 'Results of the First European Youth and Space Conference Sponsored by the Association Nationale des Clubs Aérospatiaux'
A. I. Skoog: 'Preliminary Survey of European Rocket Safety Regulations'
E. Quaterman: 'Preliminary Survey of Rocket Safety Regulations in the Fifty States of the United States of America'
H. R. Lips: 'Zinc and Sulphur as a Rocket Propellant'

L. G. Napolitano et al. (eds), Astronautical Research 1972, 367–374 *All Rights Reserved*
Copyright © 1973 by D Reidel Publishing Company, Dordrecht-Holland

DESIGN ASPECTS OF SCIENTIFIC SPACECRAFT I

Organizing Chairmen: E. Stuhlinger and K. Y. Khodarev

R. J. Parks: 'Mariner 9 and the Exploration of Mars'
J. Martin: 'Project Viking'
J. Ortner, R. Pacault, and G. M. Israel: 'The European Venus Orbiter Study'
R. Hornung and H. Ory: 'Structural Problems of the Helios Solar Probe'
T. M. Spencer, W. H. Follett, L. T. Ostwold, and J. O. Simpson: 'A Decade of Improvements to Orbiting Solar Observatories'
O. C. Winzen: 'Twenty-Five Years of Plastic Research Balloons'
F. Unz: 'Helios Space Probe Development'

DESIGN ASPECTS OF SCIENTIFIC SPACECRAFT II

Organizing Chairmen: E. Stuhlinger and K. Y. Khodarev

R. W. Johnson and J. Naugle: 'Science in the Shuttle Era'
C. Dailey: 'Precision X-Ray Telescope'
P. M. Madon: 'Symphonie'
G. Arnik: 'The On-Board Computer of the Astronomical Netherlands Satellite'
R. J. Naumann: 'Design Principles for Contamination Abatement in Scientific Satellites' (included in extenso in the Proceedings, q.v.)
W. Wienss *et al.*: 'A Teleoperator System for Space Application'
A. Kutzer: 'Experience Gained in the German Scientific Spacecraft Program'

SECOND I.A.F. INTERNATIONAL STUDENT CONFERENCE

Conference Secretary: J. Slechten

M. Nikolić and D. Blagojević, Jr.: 'Structure and Development of the Sloven Program'
W. A. Heusmann: 'Low Thrust, Constant and Variable Acceleration Trajectories for a Polar Mercury Orbit'
R. Marazzi: 'Survey on Two Calculation Methods in The Transonic Regime'
Groupe amateur de réalisations et d'Études de fusées, Paris: 'AXIR II, fusée sonde expérimentale'
S. B. Sutton: 'A Study of Free Convection from a Vertical Plate with Sinusoisal Temperature Distribution'
V. C. Ting: 'An Engineering Estimation of Thermal Stresses in Dental Teeth'
M. Bonsall: 'The Student Use of Satellites for Third World Development'
Moulin: 'La Première Confèrence Européenne: Le Jeunes et l'Espace'
B. Debout: 'Théorie linéaire d'un propulseur à propergol solid à éjection modulée'
M. Marchetti and P. Planques: 'Étude d'un dispositif de modulation de débit dans une canalisation d'ergol'
Thebault: 'Méthode expérimentale de la granulométrie des particules d'alumine dans le jet d'un propulseur à propergol solide'
M. Breuning: 'Investigations at the Research and Training Centre for Rocketry'

A. J. Lèchner: 'The Advance of Special Solid Propellant Geometries for Hybrid Rocket Engines'

V. Horvat, M. Jelakovic, and B. Graf: 'Das Project CROATIA'

S. Vetrella: 'Interdisciplinary Research Program for the Earth Resources Satellite Applications in Southern Italy'

FLUID MECHANICS ASPECTS OF SPACE FLIGHT

Organizing Chairmen: S. Berndt and G. G. Chernyi

W. Schneider: 'Radiation Gasdynamics of Planetary Entry: Concepts and Recent Advances'

W. H. Hui: 'Axisymmetric and Two-Dimensional Flow with Attached Shock Waves'

G. R. Inger: 'Strong Ablation Effects on Planetary Entry Vehicles'

Y. Wu and B. Dutt: 'Kinetic Equation of a Dusty Gas Model and Its Application to Rarefied Internal Flow'

S. N. B. Murthy: 'Effect of a Line Energy Source at the Boundary of a Supersonic Flow'

J. D. Whitfield: 'Status of Hotshot Wind Tunnels for Hypersonic Aerodynamic Studies'

R. H. Page: 'Turbulent Supersonic Boundary Layer Flow in the Neighbourhood of a 90° Corner'

A. L. Gonor, V. I. Lapygin, and N. A. Ostapenko: 'Some Results of Calculations of Super-Hypersonic Flows around Conical Wings'

A. I. Zubkov, D. M. Voitenko, Ju. A. Panov, and A. I. Glagolev: 'Separated Flow Near Bodies at Supersonic and Hypersonic Speeds'

G. Koppenwallner: 'Force and Heat Transfer Measurements on Lifting Bodies in Hypersonic Low Density Flow'

R. A. Merz, R. H. Page, and C. E. G. Przirembel: 'Effect on Near Wake of Base Mounted Cylinders'

H. T. Uebelhack and J. Becker: 'Flow Model for the Determination of the Heat Transfer on the Base of Vehicles with Clustered H_2–O_2 Rocket Engines'

SPACE TRANSPORTATION PROBLEMS – EARTH TO ORBIT

Organizing Chairmen: G. K. C. Pardoe and G. M. Ulanov

R. F. Stengel: 'Optimal Guidance for the Space Shuttle Transition'

R. D. English: 'The Scout Launch Vehicle System'

H. O. Ruppe: 'Analytical Investigation of the Dual Propellant Mode'

L. W. Warzecha and S. Feldman: 'Modular Check-out and Test System – A New concept for Space Transportation and Payloads'

V. V. Pavlov: 'Properties of Ergotic Systems as Base of Designing Controlled Space Systems'

C. C. Kelber: 'Earth Orbit to Planetary Orbit by Nuclear-Electric Spacecraft'

SPACE TRANSPORTATION PROBLEMS – ORBIT TO ORBIT

Organizing Chairmen: M. Hunter, II and D. E. Koelle

M. M. Herardian: 'The Agena Orbit Transfer Stage'

G. J. M. Smith: 'Orbital Operations of the Space Tug'

D. J. Shapland and W. Müller: 'Study of Shuttle Based Systems for High Energy Planetary Missions'

K. Heiss: 'Shuttle Economics'

R. Reichert: 'Transportation of Radioactive Waste Materials into the Sun'

W. Tempelmen: 'Rendez-vous Targeting for Space Missions'

APPLICATIONS SATELLITES

Organizing Chairmen: L. Jaffe and K. Y. Kondratyev

Meteorology and Space Applications

H. Kaminski: 'Detection of Waste Water Effluences and of Their Surface Spread in the English Channel, the North Sea, and the Baltic Sea, through Determination of the Surface Temperature of the Sea by Means of Infrared Air Pictures Taken by Satellites'

A. Adelman: 'Development in Application of Remote Sensing to Hydrology'

J. J. Hall: 'Social, Economic and Political Factors Associated with Earth Resources Observation and Information Analyses'

P. A. Castruccio: 'The Contribution of Aerospace Technology to Alleviating Bay and Estuary Pollution'

F. Shahrokhi: 'A New Unified Non-Parametric Approach to Pattern Classification for Earth Resources'

W. Davis: 'Application of ERTS Data to problems of the U.S. Department of Commerce'

J. P. Kuettner: 'The Role of Satellites in the International Tropical Project Gate'

V. I. Sevastyanov and K. Y. Kondratyev: 'Observation of Glowing Particles from Manned Space Vehicles'

F. O. VonBun: 'Earth and Ocean Physics Application Program (EOPAP)'

Space Applications Technology

B. P. Miller: 'High Resolution Multispectral Television Camera System for ERTS A and B'

C. Fouche: 'Mésure de la fonction de transfert optique en ambiance spatiale d'un objectif embarquable'

G. Brachet: 'Evaluation et résultats du système opérationnelle de localisation EOLE'

D. Thieriet and A. Blanchard: 'Un système d'aide à la géodésie: le projet GEOLE'

A. Pinglier: 'Utilisation d'un système à satellites pour la navigation et le contrôle maritime'

R. A. Stamplf: 'Polar Orbiting Operational Weather Satellites'

L. Salter: 'Nacelle ballon pour photographie à très haute altitude'

E. P. Novoseltsyev: 'Influence of Aerosols in Space Meteorology'

I. S. Haas, B. T. Bachofer, and G. P. Fishman: 'Key Technological Challenges of the Earth Resources Technology Satellite Program (included in extenso in the Proceedings, q.v.)

Discussion Session: Latest Results from the Earth Resources Program

Session Chairman: W. Nordberg

W. Nordberg: 'The Earth Resources Technology Satellite-1 – Overview'

L. W. Morley: 'Application of ERTS-1 Observations to Resource Surveys in Canada'

Gene Thorley: 'ERTS-1 Applications to California Resource Inventory'

David Landgrebe: 'Automatic Classification of Soils and Vegetation with ERTS-1'

W. Davis: 'Hydrographic and Atmospheric Surveys with ERTS-1'

E. Risley: 'Geological Findings, Land Use Surveys, and Geographic Mapping with ERTS-1'

Fernando de Mendoca: 'Preliminary Results of ERTS-1 Observations of Amazonia'

J. Robert Porter, Jr.: 'Preliminary Analysis of ERTS-1 Data for Geologic, Forestry, and Land Use Applications'

L. G. Napolitano: 'Hydrological Applications of ERTS-1 Observations'

BIOASTRONAUTICS – LIFE SCIENCES

Organizing Chairmen: C. A. Berry and B. B. Egorov

C. A. Berry: 'Co-Chairman's Remarks: Fluid and Electrolyte Problems in Space Flight'

C. S. Leach, W. Carter Alexander: 'Fluid/Electrolyte Balance in Spaceflight'

S. Hulley: 'Bone Mineral Loss During Prolonged Bed Rest'

Yu. G. Nefyodov, V. P. Bichkov, V. G. Visotsky, A. N. Kochetkova, K. V. Smirnov, and A. S. Ushakov: 'Some Metabolic Indices as Connected with the Diet in Men Taking Part in a Year-Long Experiment'

K. H. Hyatt: 'Induction of Hemodynamic Deterioration by the Hypogravic State: An Evaluation of Mechanisms and Prevention'

K. K. Yarulin, T. N. Krupina, T. D. Vasilyeva, and D. A. Alekseev: 'A Comparative Study of Regional Hemodynamics During Tilt Test and Exposure to Lower Body Negative Pressure'

B. B. Egorov, N. A. Belaya, B. S. Katkovaki, and V. I. Pinogenov: 'Recovery of Human Performance Following a Prolonged Spaceflight'

V. I. Vashkov, N. V. Ramkova, Y. N. Nikiforova, G. V. Shcheglova, and L. N. Rogatina: 'The Complex of Measures Preventing the Introduction and Accumulation of Micro-organisms in Spacecraft'

BIOASTRONAUTICS – LIFE SUPPORT

Organizing Chairmen: B. A. Adamovich and A. O. Pearson

T. Weydeven: 'Regeneration of Oxygen from Carbon Dioxide and Water'

B. A. Adamovich and V. N. Sokolov: 'A System of Programs for Life Support System Optimization by the Minimum Equivalent Mass'

J. N. Pecoraro, H. E. Podall, and J. M. Spurloch: 'Progress in the Development of Spacecraft Water Recovery Subsystem'

S. V. Chizhov, Z. P. Pak, N. N. Sitnikova, and Y. C. Koloskova: 'Problems of Potable Water Decontamination and Conservation in Spacecraft Water Management Systems'

N. C. Willis and J. M. Neel: 'The Space Station Prototype Program, the Development of a Regenerative Life Support System for Extended Duration Missions'

J. R. Howell and C. J. Huang: 'System Design of a Near-Self-Supporting Space Colony'

Y. G. Nefedov, V. P. Savina, and N. L. Sokolov: 'Exhaled Air as a Source of Carbon Monoxide Contamination of Spacecraft Environment'

BIOASTRONAUTICS – SPACE MEDICAL TECHNOLOGY APPLICATIONS
Organizing Chairman: D. Flickinger

R. H. Green: 'Application of Planetary Quarantine Methodology and Spacecraft Sterilization Technology to Improved Health Care Delivery'

L. F. Wailly, J. W. Watters, and P. B. Carter: 'New Cancer Therapy Treatment Techniques Using Space Dosimetric Concepts'

J. W. Watters, L. F. Wailly, and P. B. Carter: 'Utilization of Space Developed Thermolunimescent Dosimetry Systems in Medicine'

B. A. Adamovich, V. N. Golovin, and V. G. Chuchkin: 'Bioengineering Principles of Building Space Greenhouses and Their Use for Food Production on an Industrial Scale'

B. B. Egorov, A. N. Nazin, and O. P. Buadze: 'The Experimental Development of a Method for Chronic Implantation of Plastic Catheters into Various Compartments of the Cardiovascular System'

ASTRODYNAMICS – NATURAL MOTION
Organizing Chairman: G. N. Douboshine

P. Guillaume: 'L'application de la théorie du Matching de Breakwell-Perko pour un satellite de la Lune'

V. K. Kaisin: 'Some Problems of the Theory of Latitudinal Space Vehicle Motion in a Noncentral Gravitational Field'

B. Garfinkel: 'Global Solution of the Ideal Resonance Problems'

P. E. Elyasberg, B. V. Kugaenko, V. M. Sinitsyn, and M. I. Voiskovsky: 'The Estimation of Accuracy of Short-Term Atmosphere Density Prediction'

S. P. Altman: 'Velocity Space Maps and Transforms of Tracking Observations for Orbital Trajectory State Analysis'

V. Szebehely: 'Explicit Series Solutions for the Frequencies of Motion Around the Lagrangian Points in the Restricted Problem of Three Bodies'

Yu. V. Batrakov and E. N. Makarova: 'An Encke-Type Method for Computing the Trajectories in Space'

P. G. D. Barkhan, V. J. Modi, and A. C. Soudack: 'The Concept of Reference Loci Applied to Four Body Dynamics'

P. Bielkowicz: 'Use of Synchronous Satellite for Ecological Survey'

B. D. Tapley and B. E. Schuts: 'Estimation of Unmodeled Forces on a Lunar Satellite'

ASTRODYNAMICS – OPTIMIZATION

Organizing Chairman: P. Contensou

J. W. Griffin, Jr. and N. X. Vinh: 'Optimal Three Dimensional Manoeuvering of a Rocket Powered Hypervelocity Vehicle'

D. McMillen and R. E. Lohfeld: 'Optimization of Altitude and Inclination Change Schedule during Low Thrust Ascent to Geosynchronous Orbit'

G. M. Anderson and R. C. Tubbs: 'Some Necessary Conditions for Impulsive Thrust-Coast-Thrust Minimum-Time, Fixed-Fuel Transfers between Coplanar Orbits'

J. F. Nichols: 'The Path Independence of Orbit Inclination Momentum'

D. Biausse and J. P. Marec: 'Optimization de l'impulsion d'apogée lors de la mise en place d'un satellite géostationnaire'

C. Marchal: 'Les tests du second ordre dans les problèmes d'optimisation'

A. Alfonz: 'Examination of Rocket Control Systems by Means of Analog Computer'

ASTRODYNAMICS – MOTION AROUND CENTER OF MASS

Organizing Chairman: R. E. Roberson

G. Campion, D. Johnson, and P. Y. Willems: 'Stability Analysis of a Class of Deformable Gyrostats Based on Sturm's Theorem'

K. Tsuchiya and H. Saito: 'Dynamics of Spin-Stabilized Satellites Having Flexible Appendages'

V. M. Modi and S. K. Shrivastava: 'Aerosol: A Semi-Passive Hybrid Attitude Control System'

P. C. Hughes: 'Dynamics of Flexible Satellites with Active Attitude Control'

J. Wittenburn: 'Relative Equilibrium Positions and Their Stability for a General Gyrostat-Satellite in a Circular Orbit'

MATERIALS AND STRUCTURES

Organizing Chairman: H. Ashley

J. H. Argyris, J. St. Doltsinis, J. F. Gloudeman, K. Straub, and K. J. William: 'Aspects of the Finite Element Methods as Applied to the Design of Air- and Spacecraft'

R. Valid, R. Ohayon, and H. Berger: 'The Computation of Elastic Tanks Partially Filled with Liquids for the Prevision of the POGO Effect'

K. O. Brauer: 'Utilization of Advanced Composite Materials for Spacecraft and Space Shuttle Applications'

D. S. Brown: 'The Preparation of Large Optical Surfaces'

H. U. Schuerch: 'Structures for Large-Sized, Orbiting Radio Observatories'

R. A. Anderson and E. T. Kruszewski: 'Application of Space Structures and Materials Technology to the Civil Sector'

PROPULSION

Organizing Chairmen: M. Barrère and R. Monti

P. Santini and R. Barboni: 'Stability of Rockets with Fuel Sloshing'

B. Petropoulos: 'Projets de construction des propulseurs ioniques à champs magnétique obtenu à l'aide de bobines supra conductives et étude spectroscopique des caractéristiques du plasma à hydrogène utilisé'

G. Palumbo: 'Study of a Kaufman Type Engine Discharge'

B. T. Zinn, W. A. Bell, and B. R. Daniel: 'Experimental and Theoretical Determination of the Admittance of a Family of Nozzles Subjected to Axial Instabilities'

N. I. Kidin, V. B. Librovitch, and G. M. Makhviladze: 'Transition Non-Steady Burning of Solid Propellant in a Channel Throttled by a Convergent-Divergent Nozzle'

R. Hess: 'On the Possibility to Stabilize Atomic Hydrogen'

P. Dordain and R. Lhuillier: 'Modulation de la poussée des propergols solides par injection de liquide'

W. U. Roessler and E. M. Landsbaum: 'Stagnation Pressure Losses at Abrupt Contractions with Transverse Flow'

S. Rubin: 'A General Study of the POGO'

H. J. Sternfeld: 'Experimental Performance of Coaxial Injectors in Thrust-Variable LO_2/GH_2 Rocket Engines'

A. Seidel: 'The Development of GH_2/GO_2 Pulse Mode Rocket Engines in the Thrust Range of 6660–9340 N'

W. Tabakoff, W. Hosny, and A. Hamed: 'Performance and Flow Properties Change Through a Rocket Turbine by Presence of Solid Particles'

D. D. Evans, R. D. Cannova, and M. J. Cork: 'Development and In-Flight Performance of the Mariner-9 Spacecraft Propulsion System'

EDUCATION AND EARTH RESOURCES

Organizing Chairman: L. G. Napolitano

Part I

Round Table on Education Opportunities in Earth Resources: 'Innovation and/or New Structures Required in Universities to Cope with Earth Resources – Training Earth Resources Specialists and/or Users'

Panelists: A. H. Abdel-Ghani, H. G. S. Murthy, T. M. Tabanera, R. Kling, P. Santini, O. A. Arnas.

Part II

M. Haavelsrud: 'Assessment of a Satellite Information Network for Education'

M. Saarlas and G. Chang: 'An Approach to Spacecraft Design and Performance and Its Use in Education'

K. S. Kleinknecht and J. E. Powers, Jr.: 'Skylab Student Project'